Next-Generation Actuators Leading Breakthroughs

Toshiro Higuchi • Koichi Suzumori
Satoshi Tadokoro
Editors

Next-Generation Actuators Leading Breakthroughs

Springer

Editors
Toshiro Higuchi, Prof.
University of Tokyo
Graduate School of Engineering
7-3-1 Hongo, Bunkyo-ku
Tokyo 113-8656
Japan
higuchi@aml.t.u-tokyo.ac.jp

Koichi Suzumori, Prof.
Okayama University
Graduate School of Natural Science
and Technology
3-1-1 Tsushima-naka, Kita-ku
Okayama 700-8530
Japan
suzumori@sys.okayama-u.ac.jp

Satoshi Tadokoro, Prof.
Tohoku University
Graduate School of Information Sciences
6-6-01 Aramaki Aza Aoba, Aoba-ku
Sendai 980-8579
Japan
tadokoro@rm.is.tohoku.ac.jp

ISBN 978-1-4471-5970-4
ISBN 978-1-84882-991-6 (eBook)
DOI 10.1007/978-1-84882-991-6
Springer London Dordrecht Heidelberg New York

British Library Cataloguing in Publication Data
A catalogue record for this book is available from the British Library

Softcover re-print of the Hardcover 1st edition 2010

Cover design: eStudioCalamar, Figueres/Berlin

Printed on acid-free paper

Springer is part of Springer Science+Business Media (www.springer.com)

Preface

The conventional motors driven by the electromagnetic force have been playing and will play the most important roles in motion control of automated machines like robots. However, for the advanced machines and instruments with higher performance, the conventional motors seem to be difficult to satisfy the coming sophisticated demands. And new outstanding actuators are expected to cause technological innovations in such broad fields of their applications as industry, basic science, medicine, welfare, and global environment. So, the development of innovative actuators is recognized as one of the most important key technology for next generation. In accordance with the requirement, the five year national research project "Next-Generation Actuators Leading Breakthroughs" was organized and granted by Ministry of Education, Culture, Sports, Science and Technology of Japan. This 5 year national research project started in September 2004 and finished in March 2009.

Since development of innovative actuators requires a comprehensive and interdisciplinary approach, the project was conducted by many researchers from various fields such as mechatronics, robotics, control engineering, MEMS, new material, processing, production technology, and bio technology.

The project was operated cooperatively by the following five research groups: (1) High precision and nano actuators, (2) Micro actuators, (3) Smart actuators, (4) Power actuators, and (5) Actuators for special environments. Each group consists of about 8 subjects. So, over 100 researchers participate the project. In order to seek the applications of the new actuators, research and development of some subjects have been propelled in cooperation with industry.

The objective of this book is to introduce the activity of the project with the latest results. It contains various kinds of new actuators like electrostatic actuators, micro actuators, surface acoustic wave motors, ultrasonic motors, bio-actuated micro pump, intelligent actuators, pneumatic soft actuators, polymer ion gel actuators, piezoelectric actuators, functional fluid actuators, thermal actuators, multi DOF actuators and magnetostrictive actuators. Readers of the volume are expected to obtain up-to-date knowledge of these new actuators.

As the leader of the project, I appreciate the authors for their contributions. I would like to express special thanks to Dr. Takefumi Kanda for his dedicated achievement as an associate editor.

University of Tokyo
January, 2010

Toshiro Higuchi

Contents

Chapter 1
Next-Generation Actuators

Toshiro Higuchi[1]

Abstract The conventional motors driven by the electromagnetic force have been playing and will still play the most important roles in motion control of automated machines like robots. However, for the advanced machines and instruments with higher performance, the conventional motors seem to be difficult to satisfy the coming sophisticated demands. So, the development of innovative actuators is recognized as one of the most important subjects of the key technology for next generation. Present state of the technology related to innovative actuators is introduced with some examples. Key points about how to develop a new actuator and to put it to practical use are also described.

1.1 Introduction

Accompanied by the progress of computers, numerical control of machines has been developed since 1950. In this half century the technology related to automated machines like NC machine tools and robots has progressed steadily to the new technology called by mechatronics that combines mechanical engineering and electrical engineering. In 21st century, machines have to work with much more intelligence and performance in various environments. For example, a humanoid robot will be developed in several decades. To realize such an advanced robot toward the next generation, innovations are necessary in actuators that drive the robot and control its motion. In industry and scientific instruments, precise and high speed positioning is one of the most important technologies. Especially in the production of semiconductors and flat panel displays, dust-free transporting systems for wafers and thin glass plates are needed to avoid generation of dusts. In peripheral machines for computers like disc memories and mobile devices, small and light actuators are necessary to satisfy the demand of reduction of thickness and weight of the products. The conventional motors driven by the electromagnetic force will still play the most important roles. However, in some cases these

[1] Toshiro Higuchi

Department of Precision Engineering, University of Tokyo

conventional actuators seem to be difficult to satisfy the new and advanced demands. Therefore, seeking for new actuators has been activated. In this article, a perspective of the research and development of new actuators is described.

1.2 Micro and Nano Actuators

Accompanying to the growth of MEMS (micro electro mechanical systems), many kinds of micro actuators have be developed. As prominent rules of physics in the small-scale world are different from those of normal size, invention of new principle of actuators is expected for micro and nano actuators. For example, electrostatic force and deformation related to heat can be applied to the drive mechanism of micro actuators. Micro actuator has such an advantage that it is allowed to employ precious materials. We should also pay attention to the research activity related to nano structure like carbon nano-tubes. Research of mechanism of muscle and driving mechanism of very small creatures give us a clue for a new nano actuator. Manipulation of small objects is also important technology necessary for assembling micro machines and for processing cells.

1.3 Actuators Using Piezo Elements

The success of STM(scanning tunneling microscope) and AFM proves us that we can manipulate probes and specimen with an atomic scale. In order to get precise positioning with a resolution of from several nanometers to sub microns, piezoelectric elements are the most useful actuators. The probe of STM should be positioned with the resolution of nm even angstrom. Now only piezoelectric elements can satisfy the requirements commercially. As for the frequency response, a piezo element itself can deform much faster than usual electric linear motors like a voice coil motor. So, the piezoelectric element seems to be the most convenient actuator when we need to position a rather small object with high accuracy. On the other hand, the maximum deformation of a piezo element itself is limited to very small, like 10 micrometers for a 10 mm long piezo element.

Since the displacement of a piezo itself is limited, several methods have been developed to realize longer or boundless movement by combining some mechanism with piezo elements. The typical mechanisms are “inchworm”, “impact drive mechanism” and ultrasonic motors. As piezo elements can be used also to generate vibration of high frequency, driving mechanism of ultrasonic motors has been developed for continuous rotation and long distance movements. Since ultrasonic motors have the property of low speed and high torque motors, they are suitable for direct drive applications.

By using surface acoustic wave, a new type of ultrasonic linear motor was developed. It can achieve acceleration of a mover at 1000 G and positioning with resolution of several nanometers. This new thin linear motor has a possibility to be used for head access mechanism of a future disk memory.

As the material of piezo element, PZT(lead zirconate titanate) is widely and commonly used to obtain a large piezoelectric constant. Since PZT contains lead that should be eliminated from all consumer goods, development of lead-free piezo elements with good property for actuators becomes a serious and urgent problem.

1.4 Artificial Muscle and Muscle Like Actuators

The most interesting subject in the next decade related to robotics is realization of humanoid robots. In order to complete the ultimate robots that can think and act exactly like human, we have to develop many new technologies. One of the most important elements for motion control of humanoid robots is such actuators that can work like muscles. Muscles of animals have some splendid properties that have not been obtained by the ever-developed actuators. Comparing to the usual electrical motors, human muscle has the following advantages.

1. High efficiency of energy conversion ----- Coolness
2. High power density and lightweight ----- Lightness
3. Elasticity and softness of its structure ----- Softness
4. Self-diagnosis and self-repairing -----Toughness

Since the mechanism of the human muscle has been investigated profoundly in the molecular scale, we should develop such an actuator of which mechanism is similar to the muscle where numberless micro energy-conversion units are accumulated. Electrostatic actuators, mechano-chemical actuators, ultrasonic motors, shape-memory alloy actuators, and pneumatic actuators have the possibility to advance to the muscle-like actuators.

1.5 Actuators in Unconventional Environments

According to the advancements of science and technology, the use of unconventional environments such as super-clean, ultra high vacuum, high temperature and cryogenic environment must be increasing. For the material handling and processing in the unconventional environments, conventional electric motors do not always act well. We have to develop new actuators that are suitable for these severe environments.

As a permanent magnet loses its potential of magnet at Curie temperature, conventional electromagnetic actuators are difficult for use in high temperature. We have to develop new kinds of actuators for high temperature by applying new materials or by devising new mechanism.

Unconventional environments do not always restrict the conventional methods but sometimes give us means that can work well only in the severe environments. For instance, in cryogenic environments, we can utilize super-conductive material for coils of motors without heat loss and for magnetic bearings and levitations.

The demand for the actuators that do not contaminate a super-clean environment and vacuum is increasing in such fields as semiconductor processing, biotechnology and pure processing of new materials. Since fine particles and contaminants are generated mainly form bearings, ball screw, and guides of slider, actuators without mechanical contacts should be developed for super clean use. Lubricants should be also eliminated from the actuators. Combination of direct drive motors and contact free bearings like magnetic bearings is a typical solution. In the case where magnetic field is avoided, electrostatic motors combined with electrostatic levitation seem to be promising. Especially in vacuum, we can use higher electrostatic filed than in normal air. Our group succeeded in electrostatic levitation of aluminum plate of 20 mm thickness in vacuum.

1.6 Development and Research of New Actuators

The desire of the researchers related to innovative actuators is to develop a new actuator which works in real products for consumers. Here I would like to present notions and views about how to develop a new actuator.

1.6.1 Invention of a New Actuator

In order to develop a new actuator, we have to invent or find a new principle of drive mechanism in which conversion to mechanical energy from other energy like electrical energy and chemical energy is accompanied. In addition to the knowledge of electromagnetism, mechanical engineering, and control engineering, comprehensive understanding of chemistry, biology, and material is necessary for creating a new method of actuation. Especially the performance of actuators is dominated by the property of materials. We have to watch progress and innovation of material. For instance, a big evolution must occur in electric motors when room-temperature super conductive material emerges. Reevaluation of abandoned idea proposed in old time may yield a new valuable actuator by applying resent technology.

1.6.2 Importance of Production Engineering

Production technology to provide the new-developed actuators with an appropriate cost is also important to be accepted in market. For example printing technology has been tried to produce electrodes of electrostatic actuators.

1.6.3 Control and Drive System

Since sensors and power drivers are indispensable to control the actuators, we have to develop these elements to achieve good performance. In order to provide small driving system, not only an actuator but also its driving circuit and power amplifier should be made small.

1.6.4 Standardization of Evaluation

To make an innovative actuator to be used commonly, standardization of tests of performance is necessary. We have to develop measuring methods to evaluate the actuators. For instance, how to measure a tiny torque of an actuator is an important issue among the researchers of micro actuators.

1.7 Conclusion

Actuators will play very important roles in the advanced mechatronics in this new century. Since research and development of actuators has relationship with various fields, cooperated research projects organized by universities and industry should be propelled for innovations of actuators.

International collaboration of research and development for innovative actuators is also necessary to cope with rapid and extensive demands.

Chapter 2
Surface Acoustic Wave Motor Modeling and Motion Control

Minoru Kuribayashi KUROSAWA[1]

Abstract For miniaturization of ultrasonic transducers, a surface acoustic wave device has an advantage in rigid mounting and high-power-density operation. A surface acoustic wave (SAW) motor has been investigated, and its superior performances have been demonstrated. From investigations based on experiments, it was found that slider surface texture affects motor performances such as speed and thrust. Theoretically, however, the effect of the physical property of a slider-textured surface on motor performance had not been investigated sufficiently. A physical modeling of the SAW motor has been attempted, one slider projection was modeled including the compliance of the slider and stator materials, and also the stick and slip at the boundary. Using the slider projection modeling, operations of the SAW motor were simulated, and then, the results were compared with the experimental results. For servo control system application, a feed back controller compensating a nonlinear dead zone of the motor is reported. The feed back controller is simple and very effective. For an advanced motion control, a precise modeling of the SAW motor has been studied.

2.1 Introduction

For miniaturization of ultrasonic transducers, a surface acoustic wave device has an advantage in rigid mounting and high-power-density operation. A surface acoustic wave (SAW) motor has been investigated [1-4], and its superior performances have been demonstrated: a speed of 1.5 m/s [5], a thrust of 12 N [4], and a stepping motion of 0.5 nm [6, 7] in the case of using a 60x14x1 mm3 plate piezoelectric transducer. In addition, to increase the application range, a higher operation frequency of 100 MHz [8] and a two-dimensional design [9] were investi-

[1] Minoru Kuribayashi KUROSAWA

Interdisciplinary Graduate School of Science and Engineering, Tokyo Institute of Technology gated. In addition, to reduce power consumption, energy circulation driving has been proposed [10] and investigated [11].

From investigations based on experiments, it was found that slider surface texture affects motor performances such as speed and thrust [3-5]. Theoretically, however, the effect of the physical property of a slider-textured surface on motor performance had not been investigated sufficiently [12].

Physical modeling of the SAW motor has been attempted, one slider projection was modeled including the compliance of the slider and stator materials [13], and also the stick and slip at the boundary. Using the slider projection modeling, operations of the SAW motor were simulated, and then, the results were compared with the experimental results.

For servo control system application, not only for the efficiency improvement from electric power to mechanical output, an energy circulation driving is important [14]. Traveling wave excitation and the motor operation has been reported already. In the following section, feed back controller compensating a nonlinear dead zone of the motor is reported [15, 16]. The feed back controller is simple and very effective. For an advanced motion control, a precise modeling of the SAW motor has been studied.

2.2 Principle

If we use a Rayleigh wave for a surface acoustic wave actuator, a basic device construction for the motor is shown in Fig.2.1. The actuator has only a thin plate transducer and a thin friction material. The plate transducer is a surface acoustic wave device to generate a Rayleigh wave using two interdigital transducers (IDTs) to excite a bidirectional traveling wave. One IDT at a time is driven to excite the wave, and then, to drive the slider in one direction. The active IDT is changed for the alternative linear motion of the slider. For friction force, which generates the linear motion of the actuator, pressing force acting on the slider is given, so that the slider is pressing against the stator. This pressing force is called the preload.

To explain the principle of the actuator driving mechanism, a drawing in which the wave motion is enhanced is convenient to show the image. Actual wave motion is too small to present in the drawing; therefore, the displacement of wave

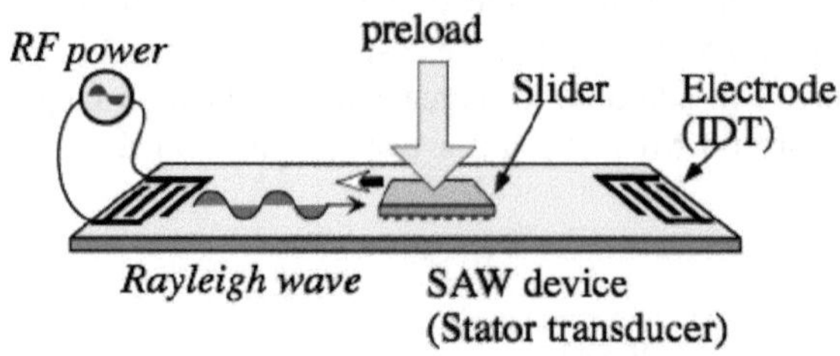

Fig. 2.1 Schematic of surface acoustic wave motor

motion is enlarged about 10,000-fold in Fig.2.2. Regarding the actual dimensions of wavelength and vibration displacements, the stator surface is almost flat; in the case of 9.6 MHz driving frequency, for example, the wavelength and vibration displacements are about 400 μm and 20 nm, respectively. Physically, therefore, the contact between the stator and the slider contains elastic deformation, stick, and slip. Under actual driving conditions, the driving frequency and the preload are maintained to be constant, and then, driving voltage is changed to control slider speed by changing wave amplitude, namely, the vibration velocity at the crests.

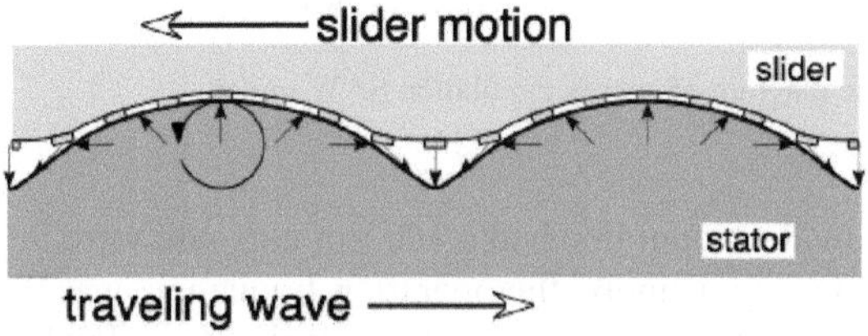

Fig. 2.2 Schematic drawing of contact between traveling wave in stator and projections on slider surface

The wave motion has two direction components; direction normal to the boundary and tangential direction. Owing to the wave motion in the stator and slider stiffness, the pressing force between the slider and the stator has distribution. Hence, around the crest of the wave, the normal force becomes large. On the contrary, at the valley of the wave, the slider surface is off. The large normal force around the wave crest is important for the frictional force in the actuator principle.

At the crest, the particles in the stator have the maximum vibration velocity in the tangential direction, which is opposite to the wave traveling direction. The large frictional force and the large tangential vibration velocity generate the linear motion of the actuator. The limit of the actuator speed is the vibration velocity of the wave crest, which depends on vibration amplitude. The limit of the actuator output force is the friction force; at the maximum, the product of the preload and the frictional coefficient between the stator and the slider.

For stable friction driving conditions, many projections are fabricated on the surface of the slider. The diameter of the projections was 20 μm or smaller to obtain a large thrust and a high speed close to vibration velocity in the case of 9.6 MHz driving frequency. It has been found from experiments that the larger the projection diameters, such as 30 or 50 μm, the lower the performance in terms of speed and thrust [4]. For the experiment, several types of silicon slider were fabricated with the dry etching process. The slider dimensions were 5x5 mm^2; 2 μm high projections were fabricated in a 4x4 mm^2 area.

For experiments [4], the material of the SAW device was 128 degree y-rotated x-propagation $LiNbO_3$. The dimensions of the device were 60 mm long, 14 mm wide, and 1 mm thick. The electrodes were twenty pairs of IDTs that were 100-

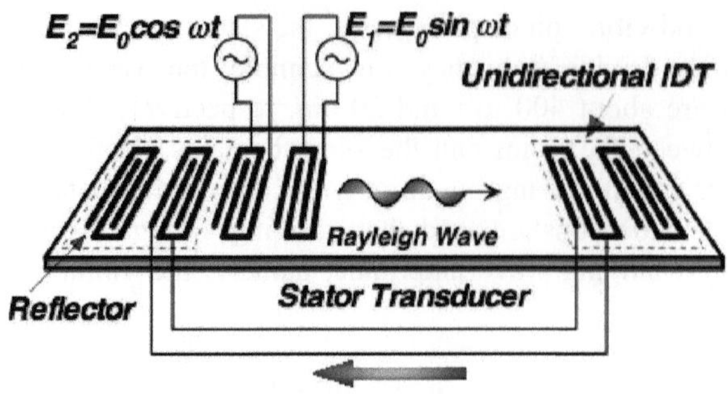

Fig. 2.3 Schematic diagram of energy circulation SAW motor

μm-wide chromium/aluminum line had a 100 μm gap, and were 9 mm in aperture. The resonance frequency, namely, the operation frequency, was 9.61 MHz. When the driving voltage was 125 $V_{0\text{-}p}$, the vibration amplitude and velocity were 21 nm in the direction normal to the surface and 1.1 m/s in tangential direction, respectively; the input power was 70 W.

A basic constructions of a surface acoustic wave motor is to reduce the electrical power from RF power sources two phase driving method have been proposed as shown in Fig.2.3. The principle is almost same as traveling wave excitation in circular ring/plate ultrasonic motors except for the energy conversion at the both end parts of the plate; sin and cos generators are used as drawn in Fig.2.3.

2.3 Modeling of Contact

A physical modeling of frictional drive of a SAW motor has been carried out [17-21] on the basis of contact mechanics [22]. For the first step of the modeling, a slider elastic body, a rigid projection, and a stator elastic body were expressed using four springs, one rigid body connected to the elastic slider part, and frictional boundary surfaces, as shown in Fig.2.4. In the modeling drawing shown in Fig.2.4, a_v, a_h, P, and Q are the vertical wave amplitude, the horizontal wave amplitude, the vertical direction force acting on the projection, and the thrust force acting on the projection surface, respectively. The equivalent spring constant of the slider and the stator are indicated by k_{pn}, k_{pt}, k_{sn}, and k_{st} in the normal and tangential directions, respectively.

The two springs in the sliders k_{pn} and k_{pt} can be written in the forms

$$k_{pn} = 4\, G_{Si}\, a[\ln(3\text{-}4\, \nu_{Si})/(1\text{-}2\, \nu_{Si})] \tag{2.1}$$

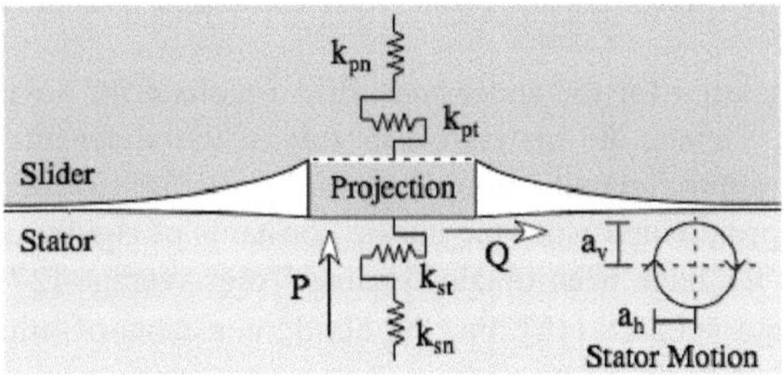

Fig. 2.4 Physical contact modeling of projection on slider and stator surface

$$k_{pt} = G_{Si}\ a/[1/8+(1-2\nu_{Si})/(8\pi\kappa)]. \tag{2.2}$$

In these equations, G_{Si}, ν_{Si}, and a are the slider material shear modulus, Poisson's ratio, and projection radius. In eq. (2.2), $\kappa = \ln(3-4\nu_{Si})/\pi$. The other two springs in the stators k_{sn} and k_{st} can be written in the forms

$$k_{sn} = 4\ G_{LN}\ a/(1-\nu_{LN}) \tag{2.3}$$

$$k_{st} = 8\ G_{LN}\ a/(2-\nu_{LN}). \tag{2.4}$$

In these equations, G_{LN} and ν_{LN} are the stator material shear modulus and the Poisson's ratio.

Using the physical model of one projection, we carried out a numerical simulation in time domain including preload, friction coefficient, vibration amplitude, and so on [23]. From the simulation, the thrust between the stator and the slider were obtained, then, the mean speed and thrust of the slider were estimated [24]. It is understood that the friction drive has two parts: sticking with elastic deformation in the nanometer range and slipping at the boundaries.

2.3.1 Estimation on Physical Model

Simulations of a SAW motor operation were carried out using the projection contact model and results were compared with experimental results at 9.6 MHz motor operation. The driving voltage was 125 $V_{0\text{-}p}$, so that the vibration amplitude and velocity were 21 nm in the direction normal to the surface and 1.1 m/s in tangential direction. The slider projection number was changed from 1089 to 10000 in the case of a 20-μm-diameter slider. Then, performance differences depending on the slider projection diameters ranging from 20 to 50 μm with same contact surface area of about 3 mm^2 in 4 by 4 mm^2 were compared with the experimental results [4, 25, 26].

The silicon material for the slider and lithium niobate for the stator are anisotropic materials. Hence, the material constants of the slider and the stator were simplified to be approximated isotropic elastic constants [13] for use in eqs. (2.1) to (2.4). The approximated isotropic elastic constants of rigidity modulus G_{Si} and Poisson's ratio ν_{Si} have been obtained using Voigt average [27], namely, G_{Si} = 68.0 GPa and $\nu_{Si} = 0.218$ [18]. For the elastic constants of lithium niobate, we then approximated to the isotropy from the propagation velocity of the Rayleigh wave and the longitudinal wave [28]; substituting those velocities to the equations of velocities of an isotropic material extracted the approximated isotropic elastic constants [28], which were $G_{LN} = 93.7$ GPa and $\nu_{LN} = 0.05$ [19].

It is difficult to maintain a uniform contact condition for each of the projections distributed in a 4x4 mm^2 area, owing to the small vibration amplitude and elastic deformation in the nanometer range. Hence, the effective factor of projection contact was investigated using 20 μm diameter, 80 projections in the line by 80 projections in the column slider. In the simulation, the number of projections was reduced by the effective factor in the line and column. For example, in the case of the effective factor of 0.6, the actual slider of 80x80 projections was reduced to 48x48 projections. In addition, in the simulation, the projection distribution in the wave traveling direction was ignored. Namely, the attenuation of the wave by the friction drive was neglected to simplify the calculation.

Using five different slides with projection numbers of 10000, 6400, 4356, 2500, and 1089, thrust and speed that were estimated by the simulation were compared with experimental results [29]; the projection diameter was 20 μm. The thrust at null speed, namely, maximum thrust, depends on the preload and the number of

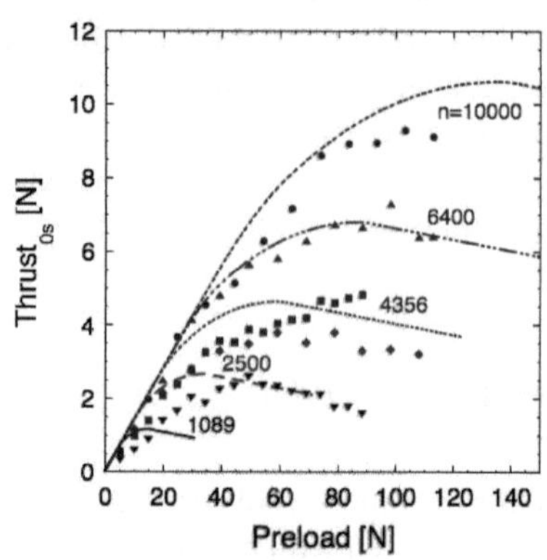

Fig. 2.5 Output force characteristics of 20-μm-diameter sliders with change in numbers of projections

the projections, as indicated in Fig.2.5. There are five curves obtained for simulation and plots for experimental results, as indicated in Fig.2.5 as follows: "n=10000", "6400", and so on. From the simulation, it is understood that the stiffness at the slider contact surface is important for a large thrust. This is because the stiffness is in proportion to the number of projections. The stiff surface is suitable for high-speed operation.

A difference in slider projection diameter as a function of the output force of the SAW motor indicated interesting result, as shown in Fig.2.6 [29]. The simulation results are indicated with a dotted line, a dashed line, and a solid line for 20, 30, and 50 μm projection diameter sliders, whereas the experimental results are indicated with circles, triangles, and rectangulars, respectively. Three sliders had the same contact area of 3 mm^2 but different projection diameters; the 20-μm-diameter projection slider had 10000 projections. The small diameter projections have lower stiffness than the large projections. However, the total stiffness becomes higher if the total projection contact surface area is the same as the large projection slider. It is clearly indicated in eqs. (2.1) to (2.4) that the stiffness is proportional to the radius of projections, not to the square of projection radius. Thus, if the total projection areas are the same, the smaller projection slider has a higher surface stiffness. The high surface stiffness of slider induces a large thrust. The higher surface stiffness due to small-diameter projections provides superior performance in terms of speed.

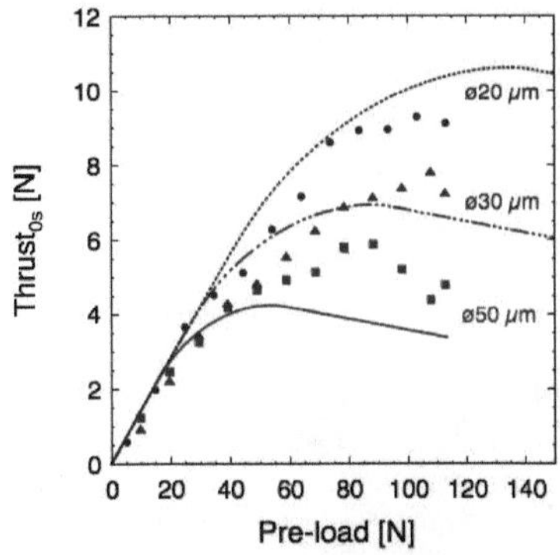

Fig. 2.6 Differences in output force characteristics between 20-μm-, 30-μm- and 50-μm-diameter projection sliders

2.4 Feedback Control

2.4.1 Experimental Setup

Material of the stator is 128 degree rotated Y-cut $LiNbO_3$. The dimensions of the device were about 80 mm long, 13 mm wide and 1 mm thick. The IDTs on the stator were designed for the energy circulation drive SAW motor [15, 16]. From the dimensions of the electrode, the resonance frequency of the IDT was about 13.34 MHz.

The slider material was Si. The contacting surface was dry etched to be fabricated a lot of projections for friction drive. The contact area was 4 by 4 mm^2. Thin film material were used as friction material on Si slider. The materials were diamond like carbon. Thickness of the film was 0.1 μm. The slider used for the experiment had 5 μm diameter projections with intervals of 14 μm on their friction drive surface. The height of the projections was 0.5 μm.

An experimental setup is shown in Fig.2.7. On the stator device, the silicon slider is placed with applying pre-load by a plate spring. The pre-load was designed to be 8.7 N. The motion of the slider is guided with a ball bearing linear slider placed beneath the stator SAW device.

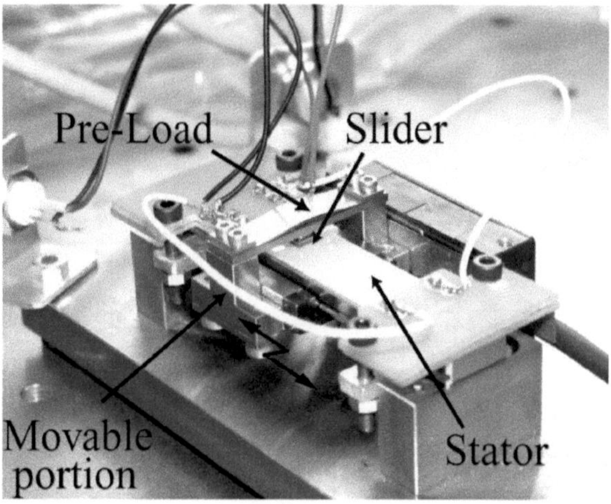

Fig. 2.7 Experimental setup

2.4.2 Response of Motor

For characterize the motor speed response, transient responses of the motor were measured by changing the driving voltage, namely, driving power. The speed of a SAW motor depends on the driving power.

Due to the unsymmetrical construction of the electrodes arrangement, the propagating power without a slider was already imbalanced. In addition, in case of forward propagation of the wave, the driving wave is directly radiated from the driving IDTs. However, in case of backward propagation, the wave generating power is once transmitted to the circulation electrodes. Hence, the losses at the unidirectional IDTs cause the power loss for the backward drive.

A problem of the SAW motor as a plant of a feed back controller is a nonlinearity. It means that the response has dead zone, for example, from -10 V_{rms} to +10 V_{rms}. The dead zone should be considered to make a controller.

2.4.3 Controller and Response

The SAW motor was modeled by nonlinear part and a first order delay term as shown in Fig.2.8. This model is simplified for the first order approximation. Actually, acceleration at the starting points depended by the driving power, namely, driving voltage. Against the lower driving power, acceleration is smaller. Hence, the time constant indicated by 'T' in Fig.2.8 has dependence on the driving volt-

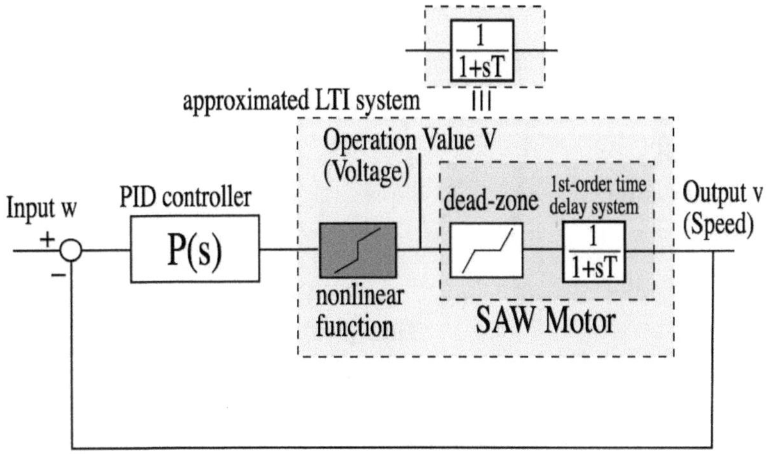

Fig. 2.8 Speed control system diagram

age. But this time, the time constant was simplified to be independent on the driving voltage.

The plant of the surface acoustic wave motor was linearized simply using nonlinear function as shown in Fig.2.8. Then, the linearized plant that has first order time delay was controlled with a simple PID feed back control system. The sinusoidal motor speed response indicated in Fig.2.9 was almost no switching distortion at around null speed.

2.4.4 Improved Modeling

The energy circulation drive SAW motor model has been improved and represented with a simple block diagram [30, 31], in case of the motor force-speed characteristic is linear. The block diagram includes the driving force function, the damping coefficient and the slider mass. By giving the driving voltage alone, the motor response in time domain, from the stating up to the stop, was obtained. The simulation model response was almost same as the actual motor response. This simple model has an advantage to build a servo control system for the SAW motor. In addition, for the versatile modeling, the block diagram will improved further.

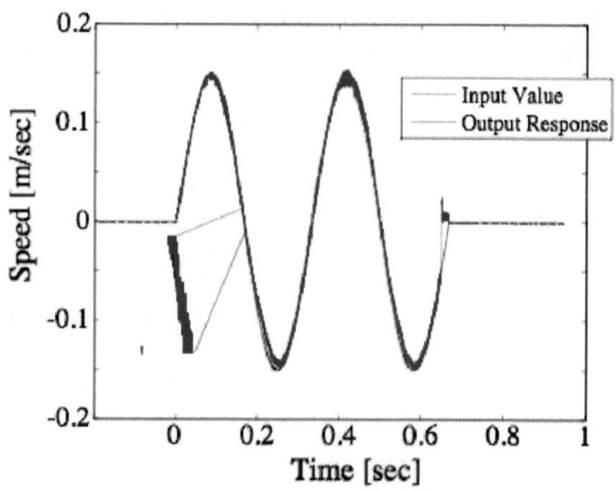

Fig. 2.9 Sinusoidal response of speed controlled SAW motor

2.5 Conclusions

It has been unclear for several years why a smaller-diameter projection slider has superior performances in terms of speed and thrust. From the physical modeling of the SAW motor based on contact mechanics, the motor operation has successfully explained about the difference in slider surface texture. The stiffness of the slider surface depends on projection diameter and the amount of projections; the smaller-diameter slider has a stiff surface if the contact areas are the same. This is because the stiffness of one projection is in proportion to the projection radius, not the square of the radius.

We demonstrated that the SAW motor response is linearized by simple nonlinear function to apply the simple PID controller for a feed back system. The motor response was improved in quick response and linearity. However, the other nonlinearity such as thrust and other factors were not considered in this work. An improved, in case of the motor force-speed characteristic is linear, simple block diagram has been proposed.

Acknowledgments This work was supported with the Ministry of Education, Science, Sports and Culture.

References

1. Kurosawa M, Takahashi M, Higuchi T (1996) Ultrasonic linear motor using surface acoustic wave. IEEE Trans. Ultrason. Ferroelectr. Freq. Control 43(5): 901-906
2. Kurosawa MK, Takahashi M, Higuchi T (1998) Elastic contact conditions to optimize friction drive of surface acoustic wave motor. IEEE Trans. Ultrason. Ferroelectr. Freq. Control 45(5): 1229-1237
3. Asai K, Kurosawa MK and Higuchi T (2000) Evaluation of the driving performance of a surface acoustic wave linear motor. Proc. IEEE Ultrasonics Symp.: 675-679
4. Kurosawa MK, Itoh H, Asai K, Takasaki M, Higuchi T (2001) Optimization of slider contact face geometry for surface acoustic wave motor. Proc. of IEEE MEMS: 252-255
5. Nakamura N, Kurosawa MK, Shigematsu T, Asai K (2003) Effects of ceramic thin film coating on friction surfaces for surface acoustic wave linear motor. Proc. IEEE Ultrasonics Symp.: 1766-1769
6. Shigematsu T, Kurosawa MK, Asai K (2003) Nanometer stepping drives of surface acoustic wave motor. IEEE Trans. Ultrason. Ferroelectr. Freq. Control 50(4): 376-385
7. Shigematsu T, Kurosawa MK, Asai K (2003) Sub-nanometer stepping drive of surface acoustic wave motor. Proc. IEEE-NANO: 299-302
8. Shigematsu T, Kurosawa MK (2006) IEEJ Trans. Sens. Micromach., 126-E: 166
9. Iseki T, Shigematsu T, Okumura M, Sugawara T, Kurosawa MK (2006) Two-dimensionally self-holding deflection mirror using surface acoustic wave motor. Opti. Rev. 13: 195-200
10. Asai K, Kurosawa MK (2003) Energy circulation methods for surface acoustic wave motor. IEICE Trans. Fundam., J86-A: 345-353 [in Japanese]
11. Kurosawa MK, Miyazaki Y, Shigematsu T (2007) Study of scattering by surface acoustic wave motor slider using finite element method simulation. J. Jpn. Appl. Phys. 46: 4915-4920
12. Asai K, Kurosawa MK (2003) Simulation model of surface acoustic wave motor considering tangential rigidity. IEICE Trans. Fundam., J86-A: 1442-1452 [in Japanese]

13. Shigematsu T, Kurosawa MK (2006) Friction drive modeling of SAW motor using classical theory of contact mechanics. Proc. Actuators: 444-447
14. Asai K, Kurosawa MK, Higuchi T (2004) Energy circulation methods for surface acoustic wave motor. Electronics and Communications in Japan, Part 3, 87(2): 345-353
15. Suzuki T, Kurosawa MK, Asai K (2005) Control of a surface acoustic wave motor using PID controller. Proceedings of LDIA: 326-329
16. Kurosawa MK, Suzuki T, Asai K (2007) Surface acoustic wave motor using feed back controller with dead zone linearization, J. Jpn. AEM, 15(2): 125-131 (in Japanese)
17. Shigematsu T, Kurosawa MK (2008) Friction Drive of an SAW motor Part I: Measurements, IEEE Trans. Ultrasonics, Ferroelectrics, and Frequency Control, (57)9: 2005-2015
18. Shigematsu T, Kurosawa MK (2008) Friction Drive of an SAW motor Part II: Analyses, IEEE Trans. Ultrasonics, Ferroelectrics, and Frequency Control, (57)9: 2016-2024
19. Shigematsu T, Kurosawa MK (2008) Friction Drive of an SAW motor Part III: Modeling, IEEE Trans. Ultrasonics, Ferroelectrics, and Frequency Control, (57)10: 2266-2276
20. Shigematsu T, Kurosawa MK (2008) Friction Drive of an SAW motor Part IV: Physics of Contact, IEEE Trans. Ultrasonics, Ferroelectrics, and Frequency Control, (57)10: 2277-2287
21. Shigematsu T, Kurosawa MK (2008) Friction Drive of an SAW motor Part V: Design Criteria, IEEE Trans. Ultrasonics, Ferroelectrics, and Frequency Control, (57)10: 2288-2297
22. Johnson KL (1985) Contact Mechanics. Cambridge University Press, Cambridge, U.K.
23. Asai K, Kurosawa MK (2002) Performance estimation of surface acoustic wave motor using simulation model of friction drive. IEICE Trans. Fundam., J85-A: 1428-1439 [in Japanese]
24. Mano T, Tsukimoto T, Miyake A (1992) IEEE Trans. Ultrason. Ferroelectr. Freq. Control 39: 668
25 Kurosawa MK, Itoh H, Asai K (2001) Influence of elastic deformation in surface acoustic wave motor friction drive. Proc. Transducers: 726-729
26 Kurosawa MK, Itoh H, Asai K (2003) Elastic friction drive of surface acoustic wave motor, Ultrasonics 41(4): 271-275
27. Hull R (ed) (1999) Properties of Crystalline Silicon. Inspec, London
28. Kushibiki J, Takanaga I, Arakawa M, Sannomiya T (1999) IEEE Trans. Ultrason. Ferroelectr. Freq. Control 46: 1315
29. Kurosawa MK, Shigematsu T (2008) Friction drive simulation of surface acoustic wave motor characteristics based on contact mechanics, Jpn. J. Appl. Phys., 47(5): 4287-4291
30. Okano M, Kurosawa MK (2007) Study on modeling of surface acoustic wave motor. Proc. of IEEE Int. Symp. on Industrial Electronics: 1508-1513
31. Okano M, Kurosawa MK (2008) Model based position control of surface acoustic wave motor. Proc. of Actuator: 172-175

Chapter 3
AZARASHI (Seal) Mechanism for Meso/Micro/Nano Manipulators

Katsushi FURUTANI[1]

Abstract This article deals with AZARASHI (Seal) mechanism. A device with three degrees of freedom (DOFs) consists of two controlled electromagnets connected by two piezoelectric actuators with an electromagnet that generates a constant friction. The friction at the controlled electromagnets is alternated. The whole device wriggles by deforming the piezoelectric actuators. A displacement control method of the piezoelectric actuator by driving with a series of current pulses is developed for the high resolution throughout the deformable range. In the feed-forward control, the hysteresis and nonlinearity in the drive by the current pulse were much smaller than those by a voltage linear amplifier. It was applied to the driving method of Extension device of a 1-DOF device with nanometer-accuracy. An application to a positioning stage of a manipulator is introduced.

3.1 Introduction

Multiple degrees of freedom (DOFs) devices that have wide movable range with a high resolution are required for a fine motion stage in manipulators and scanning probe microscopes (SPMs). A coarse motion device is generally combined with a fine motion device to build such a positioning device. Inchworm Mechanism [1][2] and Impact Drive Mechanism [3][4] have been proposed for a coarse motion mechanism. With the increase of DOFs of the devices, the number of controlled actuators is increase, the size of a whole device becomes large and a structure of the device becomes complex. The author has proposed AZARASHI (Seal) Mechanism and made a 3-DOF device [5]. Though the mechanism has smaller number of controlled devices, it can move with micrometer order steps in the *x*-, *y*- and θ-directions.

[1] Katsushi FURUTANI
Department of Advanced Science and Technology, Toyota Technological Institute

In this article, the principle and performance of AZARASHI Mechanism are introduced. Then, a driving method of a piezoelectric actuator with current pulses is introduced. Finally some applications of AZARASHI mechanism are described.

3.2 AZARASHI Mechanism

3.2.1 Principle of Movement

3.2.1.1 One-DOF Device

Fig.3.1 shows a moving principle of a 1-DOF device [6, 7]. The 1-DOF device put on a base consists of Friction devices A and B connected by Extension device.

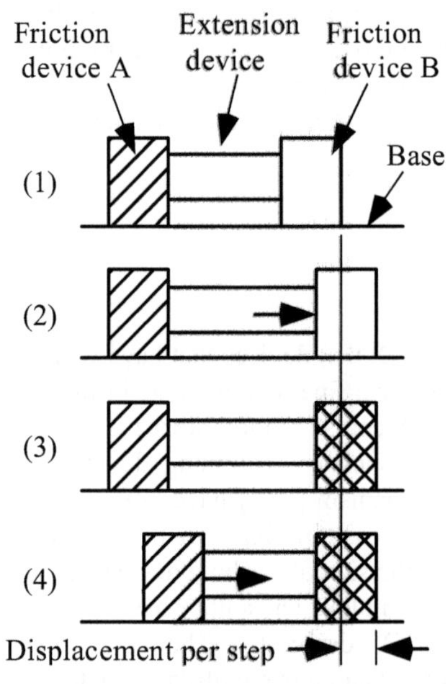

Fig. 3.1 Principle of movement of 1-DOF device

Brightness of Friction devices indicates strength of a frictional force. Friction device A applies a constant frictional force and Friction device B is controlled by an on-off action. The following relation must be satisfied:

$$F_{Off} < F_C < F_{On} \tag{3.1}$$

where F_C is a constant frictional force at Friction device A, and F_{On} and F_{Off} are frictional forces in the cases of clamping and releasing, respectively.

The device moves rightward by repeating the following cycle.

1. Extension device is contracted and Friction device B is turned off.
2. Friction device A is stationary and Friction device B wriggles rightward during Extension device is expanded..
3. Friction device B is turned on.
4. Friction device B is stationary and Friction device A is drawn toward Friction device B during Extension device is contracted.

The device can move leftward by exchanging expansion and contraction of Extension device in the cycle above. Because the device wriggles like a seal (AZARASHI in Japanese), it is named "AZARASHI Mechanism." The steplike movement can be interpolated by changing the length of Extension device continuously. Each is called Coarse mode and Fine mode, respectively.

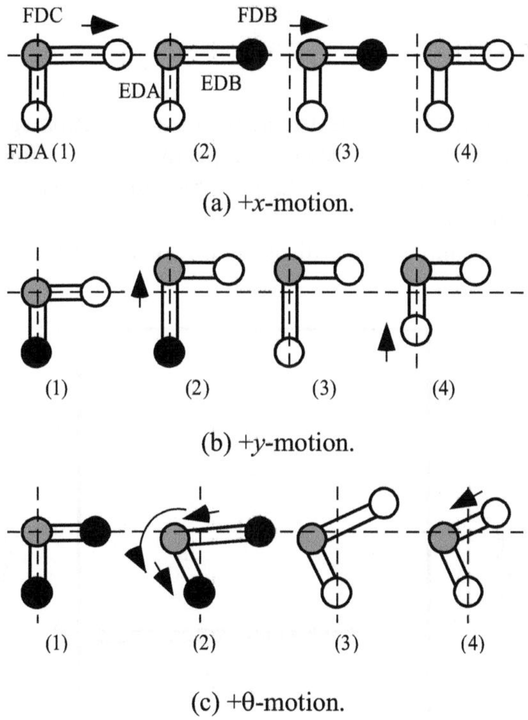

Fig. 3.2 Principle of movement of 3-DOF device

3.2.1.2 Three-DOF Device

Fig.3.2 shows a structure of 3-DOF device of AZARASHI Mechanism and illustrates the movement principle on a plane. It consists of Friction devices A (FDA) and B (FDB), Friction device C (FDC) connected FDA and FDB by Extension devices A (EDA) and B (EDB) with a right angle. FDA and FDB are controlled by an on-off action, and FDC, which works as a passive device, applies a constant frictional force. The device can move in the +*x*-direction as shown in Fig.3. 2 (a).

1. FDA and FDB are turned off. FDB is fed in the +*x*-direction and the whole device is stationary during EDB is expanded.
2. FDB is turned on.
3. The whole device wriggles in the +*x*-direction during EDB is contracted.
4. FDB is turned off. Then the whole device returns to the initial state (1).

Figs. 3.2 (b) and (c) illustrate the movement principle in the *y*- and θ-directions. These sequences are similar to that for the movement in the *x*-direction. Either EDA or EDB is used to rotate in the θ-direction.

3.2.2 Performance

In a 3-DOF device, stacked piezoelectric actuators and electromagnets were used for Extension devices and Friction devices, respectively. The whole device measures 63×63×23 mm and weighs 36 g. The movable range of AZARASHI mechanism is infinite in principle.

Fig.3.3 shows an example of the displacement driven in the *x*-direction. The

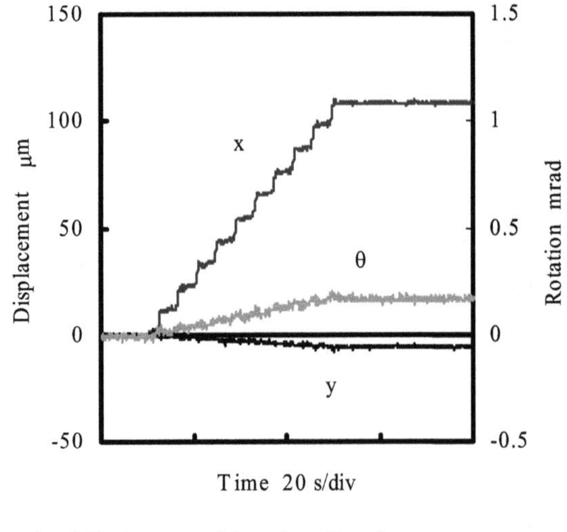

Fig. 3.3 Example of displacement driven in *x*-direction

driving conditions were 100 V as the applied voltage amplitude to EDB and 1 s as the transition time and all the waiting times. Although the device moves with 11-μm steps in the *x*-direction. The displacement is proportional to the voltage amplitude. The displacement in the orthogonal direction to the driving one is reduced by deforming both Extension devices simultaneously [6].

3.3 Driving Method of Piezoelectric Actuator with Current Drive

3.3.1 Driving Principle

Stacked piezoelectric actuators extend several tens of micrometers in maximum, and they have a potential of a nanometer-order or finer resolution. The applied voltage to the piezoelectric actuator is usually used to control the displacement of the piezoelectric actuator [7]. In a simple feed-forward control, the hysteresis of the displacement to the applied voltage to the piezoelectric actuator is often observed. The resolution of the DAC restricts the dynamic range of the driving signal. On the other hand, a charge control has been proposed for the feed-forward control, in which the hysteresis of the displacement is not observed [8]. The author also has proposed a charge control by detecting induced charge on the electrodes attached at the ends of a piezoelectric actuator [9]. It is expected that a driving

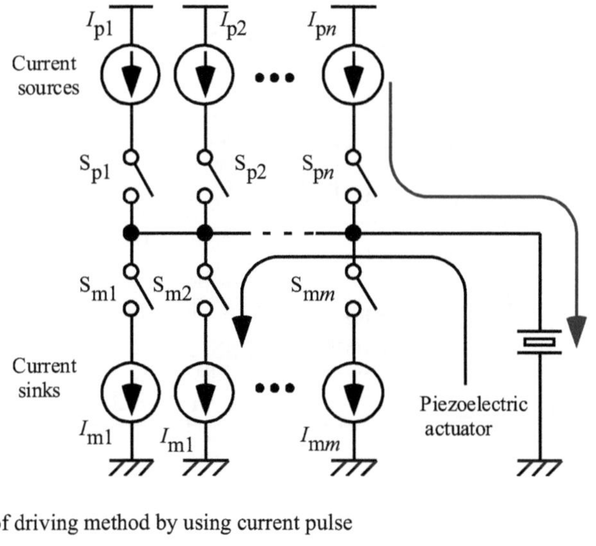

Fig. 3.4 Principle of driving method by using current pulse

method with a series of current pulses may allow better performance because the piezoelectric actuator is a capacitive load.

Fig.3.4 shows a driving circuit of the piezoelectric actuator by the current pulse. It consists of several current souses and sinks with different peaks connected by switches in parallel. The output current pulse from the driving circuit follows the driving signal provided by the controller to select one of the current sources or sinks. The deformation of the piezoelectric actuator can be controlled by supplying the charge. The corresponding current pulse with the error is provided. This method is a kind of the charge control methods because the charge amount is represented by the product of the pulse duration by the current value. Consequently, the displacement with little hysteresis in any range of the applied voltage is promised. The displacement can be predicted by counting the given pulses without the displacement feedback.

3.3.2 Experimental Setup

Fig.3.5 illustrates an experimental setup. The driving circuit consists of three sets of the current sources and sinks: coarse-, medium- and fine-step ones. A set of the current source and sink were built on a board and three sets were connected in parallel. The measured minimum pulse duration of the current sources and sinks was 16 μs. The displacement of a piezoelectric actuator with a capacitance of 1.4 μF is magnified by a flexure hinge mechanism. It deforms 70 μm at an applied voltage of 100 V with a hysteresis of 14%. The first and second natural frequencies were 770 and 1800 Hz, and the anti-natural frequency was 1200 Hz in the case of attached a small steel plate as a sensor target. The space around the piezoelectric actuator was filled with gel to passively suppress the residual vibration.

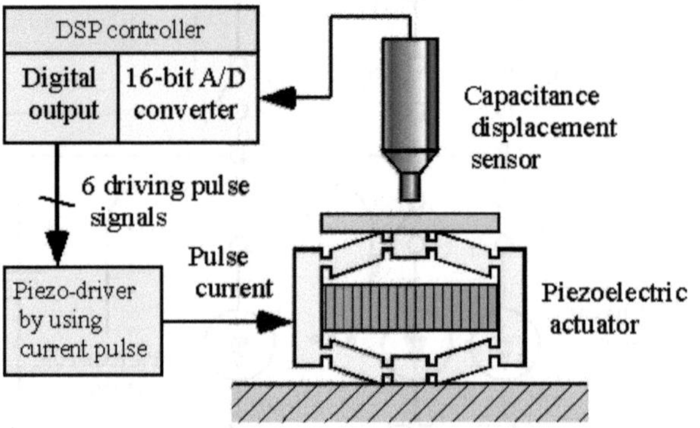

Fig. 3.5 Experimental setup

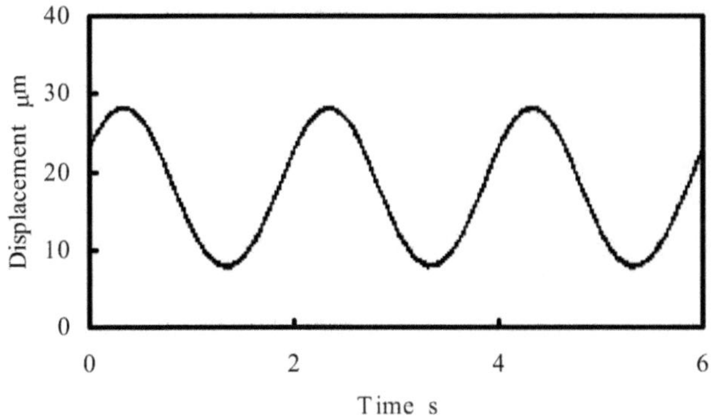

a Displacement

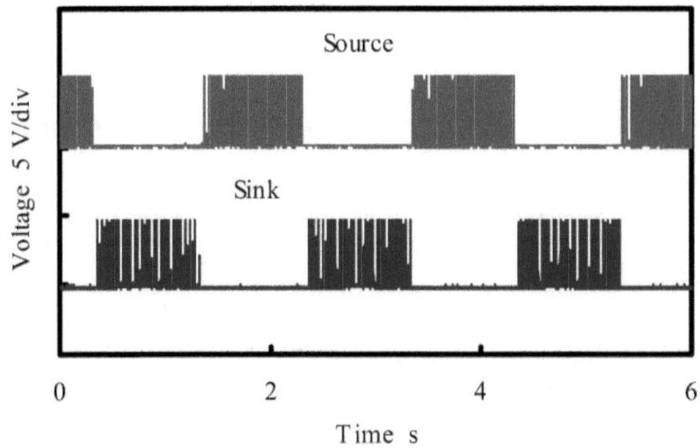

b Driving signals

Fig. 3.6 Displacement of piezoelectric actuator by feed-forward control with current drive circuit

3.3.3 Performance

The pulse durations were adjusted to equalize the displacements by the source and sink each other. The cycle time was set to 1 ms for the coarse-step circuit and 100 μs for the medium- and fine-step circuits. The step a pulse was 6 nm/pulse, which is equivalent to a 13-bit DAC.

The step heights ware calibrated at an amplitude of 20 μm both in the current pulse drive and voltage linear drive. Fig.3.6 (a) shows an example of the displacement driven by the feed-forward control through the fine-step circuit and Fig.3.6 (b) shows driving signals. The reference displacement was set to the sinusoidal wave with a frequency of 0.5 Hz and an amplitude of 10.0 μm. Because the piezoelectric actuator performs as an integrator of the current pulse, the driving signal was generated based on the Δ conversion for PDM. The driving pulse signals were separately provided to each source and sink. The density of the current pulses was sparse at the beginning and end of a series, and dense in the middle. The cycle time of the every current pulse was set to 100 μs. The displacement mainly contained a frequency element of 0.5 Hz though the driving signals contained higher ones.

3.4 Applications

3.4.1 Precision Positioning with Nanometer-Accuracy

3.4.1.1 Design of Device

Fig.3.7 shows a structure of AZARASHI device with 1 DOF. Stacked piezoelectric actuators were used for Extension device and Friction one. Pads of Friction

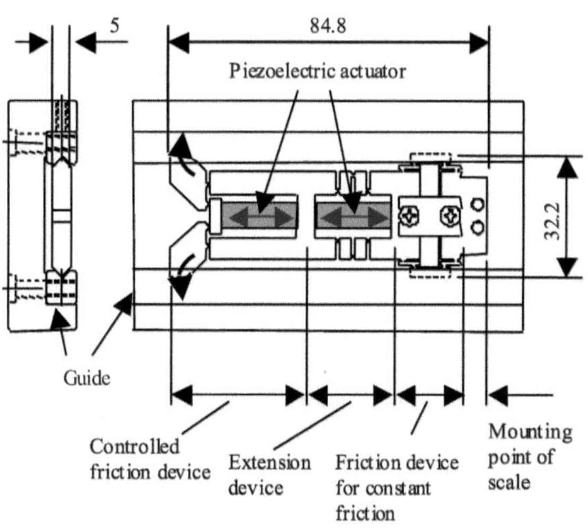

Fig. 3.7 Structure of 1-DOF device

devices were pressed against rails of a V-guide fitted outside the movable part. The displacement of the piezoelectric actuator was expanded twice with a lever mechanism in the controlled friction device. The thinnest part of the notch hinge measured 0.5 mm with a radius of 0.5 mm. The spaces for the piezoelectric actuators in Extension and Friction devices tightly fitted to apply the preload to the actuators.

The current pulse sources and sinks of 0.5 mA, 20 mA and 1 A were connected each other in parallel. The cycle time was set to 100 μs for all sources and sinks. Extension device was driven with the current pulse in both Coarse and Fine modes. Friction device was driven with a voltage linear amplifier by an on-off action. The device was controlled with a personal computer. The driving signal for Extension device was generated by the Δ conversion. The displacement was measured with a linear encoder with a resolution of 0.07 nm.

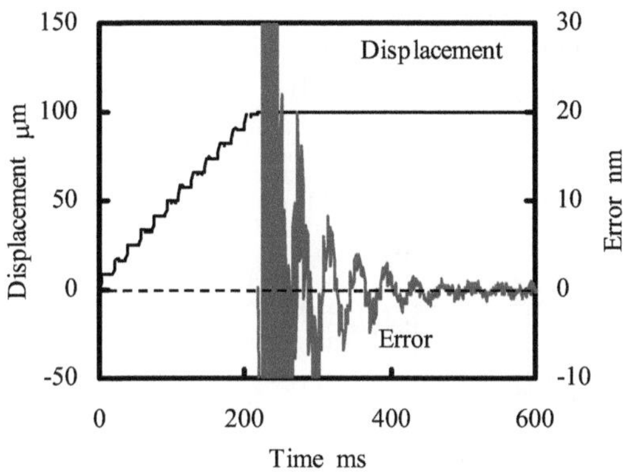

Fig. 3.8 Example of positioning (Reference: 100 μm)

3.4.1.2 Experiments

Fig.3.8 shows an example of positioning with a reference of 100 μm. When the error became smaller than the threshold displacement, the current was switched to smaller one. In Fine mode, Friction device was clamped and Extension device was extended and contracted. The residual vibration was observed though the frictional force at the uncontrolled friction device is insufficient. The residual vibration affected the dispersion of the settling time. By driving Extension device with a voltage linear amplifier, the driving cycle could be shortened to 6 ms. The maximum velocity was 1.8 mm/s in this case.

3.4.2 Micromanipulation

3.4.2.1 Experimental Setup

Fig.3.9 shows an experimental setup for the manipulation. It consists of a 3-DOF AZARASHI device for positioning a stage, and a one-DOF device mounting micro-tweezers that consist of a frame, 2 bimorph piezoelectric actuators and tips for holding micro-objects. When the tweezers holds an object during one of the actuators is oscillated, another detects the vibration through the object. The images of the objects and stage were monitored with a color CCD camera, then captured in a personal computer as 640×480-pixel image. The maximum and minimum resolutions were 1.0 and 5.9 μm/pixel, respectively.

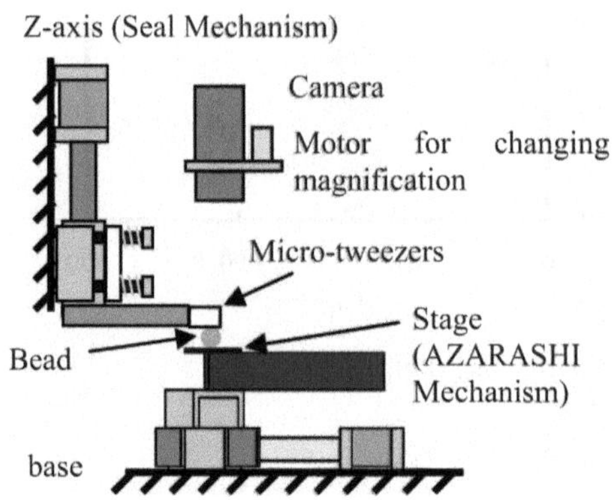

Fig. 3.9 Configuration of micromanipulator

3.4.2.2 Manipulation Applying LMS Method

Glass beads with a diameter of 60-70 μm were used as the objects to be manipulated in the following experiments. The local machining station (LMS) method [10] that improves the positioning accuracy by using local coordinate systems was applied to position beads. In a view from the camera, a bead is located on the stage with some bright points and the tips of the tweezers approaches. A stage and local coordinate systems are defined by using each pair of bright points. Fig.3.10 shows the manipulation sequence. By repeating the steps (5) to (7), precision positioning can be carried out near the local coordinate system. The manipulation area applying LMS method expands by 4 times larger than that without LMS method.

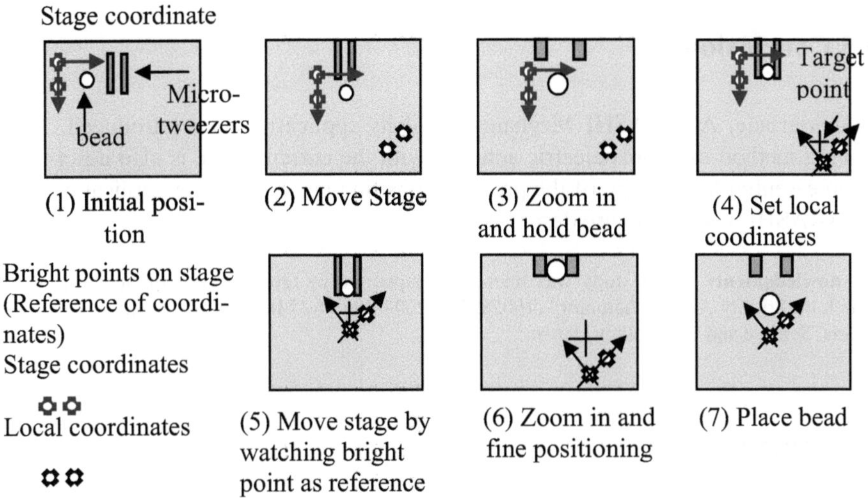

Fig. 3.10 Manipulation sequence applying LMS method

Table 3.1 Results of bead positioning

Method	Magnification	Standard deviation μm (Local coordinates)	
		x	*y*
Without LMS	2	21.3	16.8
LMS	2-12	9.5	8.4

Firstly, the resolution was fixed at 5.9 μm/pixel in order to evaluate the performance of the bead positioning without LMS method. The humidity was regulated from 15 to 20%. The positions of the bead, the bright points for setting the stage coordinates and the tweezer tips were detected by the calculation of their centroids after binarizing the captured image. Then the stage was positioned based on the stage coordinate system. After the bead was placed, the position of the bead was measured with the camera at the highest magnification. The manipulation was repeated until the bead was successfully manipulated 10 times. Table 3.1 shows the bead positioning precision. The errors caused the low measurement accuracy at the lowest magnification, and those in the *x*- and *y*-directions are almost the same each other. Next, the performance of LMS method was also evaluated. The same pair of the bright points as that used the above experiment was set to the origin of the local coordinate system for LMS method. The results applying LMS method are also shown in Table 3.1. The positioning accuracy was improved by LMS method.

3.5 Conclusions

In this article, AZARASHI Mechanism and its applications are introduced. The driving method of a piezoelectric actuator with the current pulse is also described as an essential technology of the precision positioning. Then some applications of AZARASHI mechanism are described.

Acknowledgments This study was financially supported by Grant-in-Aid for Scientific Research on Priority Areas, "Actuator" (16078214), 2004-2008 by Ministry of Education, Culture, Sports, Science and Technology, Japan.

References

1. May Jr WG (1975) Piezoelectric Electromechanical Translation Apparatus. US Pat 3902084
2. Sugihara K, Mori I, Tojo T et al (1989) Piezoelectrically Driven XYΘ Table for Submicron Lithography Systems. Rev Sci Instrum 60: 3024-3029
3. Higuchi T, Watanabe M, Kudoh K (1988) Precise positioner Utilizing Rapid Deformations of Piezoelectric Elements. J Jpn Soc Precis Eng 54: 2107-2112 [in Japanese]
4. Shim JY, Gweon DG (2001) Piezo-driven metrological multiaxis nanopositioner. Rev Sci Instrum 72: 4183-4187
5. Furutani K, Ohta N, Kawagoe K (2003) Coarse and Fine Positioning Performance of an L-shaped Seal Mechanism with Three Degrees of Freedom. Meas Sci Technol 15: 103-111
6. Furutani K, Kawagoe K (2006) Improvement of Positioning Performance of AZARASHI (Seal) Mechanism with Three Degrees of Freedom. Trans IEEJapan 126E: 131-136 [in Japanese]
7. Xu WL, Han L (1999) Piezoelectric actuator based active error compensation of precision machining. Meas Sci Technol 10: 106-111
8. Newcomb CV, Flinn I (1982) Improving the Linearity of Piezoelectric Ceramic Actuators. Electron Let 18: 442-444
9. Furutani K, Urushibata M, Mohri N (1998) Displacement Control of Piezoelectric Element by Feedback of Induced Charge. Nanotechnology 9: 93-98
10. Furutani K, Kenjo Y, Mohri N (1998) Electrical Discharge Machining by Local Machining Station Method - Improvement of Positioning Accuracy for Machining on Large Workpiece -. Proc 1998 Jpn-USA Sympo Flex Autom Ohtsu, Shiga, Japan, II: 833-840

Chapter 4
Disturbance Observer Design Based on Frequency Domain

-Application to Robot Manipulator Control Using Brain Wave Signal-

Masatake SHIRAISHI[1] and Akihiro ITO[2]

Abstract Generally, the specifications of the system robustness and disturbance or noise applied to the system are often provided in a frequency domain. In this paper, a reduced-dynamic-order observer based on a transfer function in frequency domain is proposed, which is composed of a diophantine equation, and only three steps are required for design. This method is practically used to extract the brain wave signals and various experimental results show the effectiveness of the proposed approach even in the presence of noises among the brain wave signals.

Key words: frequency domain observer, brain-wave signal, robot manipulator, robustness

4.1 Introduction

Based on a basic observer system due to a diophantine equation, we describe a frequency domain observer, which is equivalent to Gopinath method. This approach obtains the observer by using polynomial algebra to solve diophantine equations from only the input/output signals of that transfer function. Our method [1] can analyze its characteristics in the frequency domain. By employing the redundancy in the frequency domain, the designed observer also improves a particular feature in which the observer frequency characteristic is directly assigned with the robustness to the disturbance or noise.

[1] Masatake SHIRAISHI

Faculty of Engineering, Ibaraki University

[2] Akihiro ITO

Nihon Densan Sankyo Co. Ltd.

On the other hand, when sensory engineering is applied to control systems, the key point to consider is the method for mapping and evaluating the senses. In this study we address the use of sensory brain wave signals to operate a multi-joint robot manipulator by applying the frequency domain observer mentioned above. In measurement of brain-wave signals, various noises and disturbances are major causes for the deterioration of wave signals. The proposed observer was, therefore, used to extract precise and required brain wave signals only. In the experiments, we focused on α–brain wave signals under comfortable and uncomfortable auditory stimuli conditions to the subjects and analyzed and quantified them to produce control signals for use in driving a robot manipulator.

4.2 Disturbance Model

4.2.1 Two Kinds of Disturbances

The time domain representation of a linear time invariant n th-order system is given by

$$y(s) = \{r(s)/p(s)\}\ u(s) + d(s). \tag{4.1}$$

where u(s) is a scalar input, y(s) is an observable output, r(s) is a polynomial equation, p(s) is a monic-polynomial, and d(s) is an output disturbance. Since the brain wave signal is observable, p(s) and r(s) are relatively prime. The problem is to estimate the unknown output disturbance d(s) from u(s) and y(s).

Let us consider the following two kinds of output disturbances d(t) in a form of [a] and [b].

$$\text{[a]}\ d(t) = d_0 + d_1 t + \cdot\cdot + d_i t^i. \tag{4.2}$$

In a Laplace transform of Eq. (4.2),

$$d(s) = D_0/s + \cdot\cdot + D_i/s^{i+1}. \tag{4.3}$$

$$\text{[b]}\ d(t) = d_0 \sin(\omega t). \tag{4.4}$$

In a Laplace transform of Eq. (4.4),

$$d(s) = D_0/(s^2 + \omega^2). \tag{4.5}$$

D_0 is the corresponding coefficient after transformation.

4.2.2 Observer for Output Disturbance Estimation

By using the input u(s) and output y(s), the fundamental observer design for estimating disturbance d(s) can be done by the following steps.

1) To establish an arbitrary and stable m th-order monic- polynomial as

$$q(s) = s^m + \ldots\ldots + q_1 s + q_0. \tag{4.6}$$

This is the characteristic polynomial of the observer. Here, $m = n + i$ in the case of disturbance [a], and $m = n + 1$ in the case of disturbance [b]. n is the system order.

2) To define a m th-order polynomial $h(s)$ and $(m-1)$ th-order k(s) by comparing each coefficient in order to satisfy the following equation (4.7),

$$k(s)\, p(s) + h(s)\, r(s) = 0. \tag{4.7}$$

Here, $k(s) = k_{m-1} s^{m-1} + \ldots\ldots + k_1 s + k_0$.
$h(s) = h_m s^m + \ldots + h_1 s + h_0$.

In the case of disturbance [a], $h(s)$ satisfies the following relation.

$$h_j = q_j, \quad j = 0 \sim i. \tag{4.8}$$

In the case of disturbance [b],

$$q(s) - h(s) = H(s)(s^2 + \omega^2). \tag{4.9}$$

$H(s)$ is determined to satisfy the equation (4.9).

3) To determine an estimation observer of $d(s)$ as

$$w(s) = \{k(s)/q(s)\}\, u(s) + \{h(s)/q(s)\}\, y(s). \tag{4.10}$$

4.3 Application to a Robot Manipulator Operation by Using Brain Wave Signals

People are becoming increasingly concerned about the lack of human qualities in new technologies. More attention should thus be paid to characteristics such as "harmony" and "symbiosis" instead of concentrating solely on the development of technology for its own sake. A typical example of this supplementation in the design field is sensory engineering. When sensory engineering is applied to control systems, the key point to consider is the method for mapping and evaluating the senses. Specifically, this involves measuring human senses, processing and quantifying the results of these measurements, and mapping them in some way to facilitate their evaluation (including the creation of control signals). Methods for sensory measurement have been widely and actively studied both in Japan and elsewhere. References [2, 3, 4, 10] are typical examples. However, the vague and objective nature of human senses has so far made it impossible to systematize them in a scientific fashion. This makes it difficult to process and quantify them and to develop suitable methods for mapping and evaluating them (For example, references [5, 6]).

With this background into consideration, the application addresses the use of sensory brain wave signals to operate a multi–joint robot manipulator, with the ultimate goals of developing symbiotic control methods that are more human like and designing robotic appliances for people with disabilities based on these methods. Preliminary experiments showed that people subjected to comfortable auditory stimuli produced α–brain wave signals rather than other brain wave signals. We thus studied α–brain wave signals under comfortable and uncomfortable auditory stimuli conditions and analyzed and quantified them to produce control signals for use in driving a robot manipulator arm. The resulting manipulator motions were then evaluated by the 20 test subjects. Two evaluations were conducted: a subjective one based on questionnaire responses and an objective one based on pulse fluctuation detection. Both evaluations indicated that this approach to controlling a manipulator produces a comfortable and relaxed feeling. In the fields of medicine and welfare, attempts have already been made to treat illnesses with brain wave stimuli and to operate computers by using brain waves. However, there have so far been no reported studies in which brain wave signals resulting from sensory activity have been directly applied to the operation of robots, thus this will be the first challenge in the robotics area.

4.3.1 Characterization of Senses Using Brain Waves

A number of attempts to evaluate human senses either qualitatively or quantitatively have been reported. For example, it is suggested that healthy adults with their eyes closed generally exhibit α–brain wave fluctuations corresponding to comfortable and uncomfortable emotions [9]. Further, one of the authors conducted a study in which metal surface finishes were visually evaluated. Those judged by the test subjects to have "favorable surface properties" were found to exhibit $1/f$ fluctuation characteristics, and the characteristic features of the fluctuations were found in the test subjects' α–wave patterns [8]. Since $1/f$ fluctuations engender comfortable feelings in people [7], we focused on the α–wave frequency fluctuations in brain wave signals. The evaluation was made by applying these signals to the locus through which a robot manipulator arm was driven. As the basis for this evaluation, three cases were specifically assumed.

1. White–noise fluctuation: random manipulator motion that might be said to convey an unexpected or uneasy feeling to the viewer.
2. $1/f$ fluctuation: manipulator motion that is somewhat predictable but is accompanied by natural variations.
3. $1/f^2$ fluctuation: manipulator motion in which the variations are more strongly correlated, resulting in greater predictability and repetitiveness, which is usually tiresome to look at.

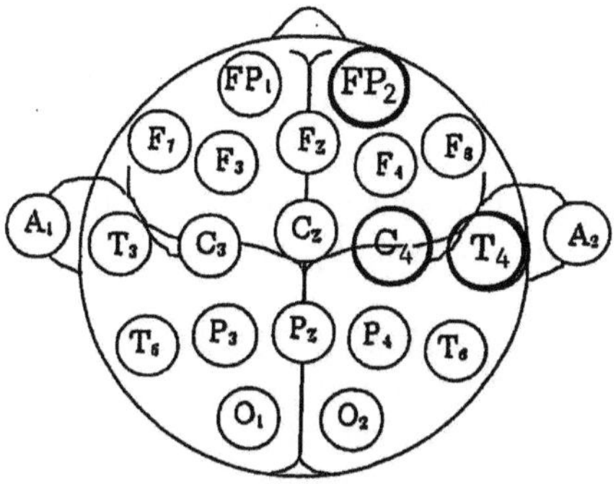

Fig. 4.1 Measurement points of brain wave signals

4.3.2 Detection of α-Wave Frequency Fluctuations

The brain waves of the test subjects were measured to extract α–waves in the frequency range of 8 to 13 Hz. while they were at rest with their eyes closed. Two types of auditory stimuli were used as the inputs for sensory detection: "comfortable" and "uncomfortable". Comfortable music was used for the former, and noise and tiresome music were used for the latter. Since the characteristics of fluctuations in modulation width and intensity appeared in both types of stimuli, the music selected for the comfortable stimulus had a $1/f$ fluctuation acoustic pressure (dB). The test subjects listened to it at their preferred volume for three minutes in a shield room. For the uncomfortable stimulus, a normal random number series was converted into a time series of acoustic pressures, and the subjects listened to it at the maximum tolerable volume (white-noise fluctuation). Also, as the uncomfortable stimulus, an arbitrary music with the same acoustic pressure was repeatedly inputted to the subject ($1/f^2$fluctuation). To clarify the changes in brain waves accompanying environmental changes, the comfortable auditory stimulus was applied first for five minutes. The uncomfortable stimulus was then immediately applied for five minutes. Although we distinguish "relaxed" (relaxation) from "comfortable" (good feeling) in this study, each is classified as "$1/f$ fluctuation". The measurements were restricted to the right side of the brain and were performed at three points following the international 10-20 method as shown in Fig.4.1: frontal lobe *FP2,* which is responsible for emotional feelings and thoughts, parietal lobe *C4,* which is responsible for high order faculties such as perception, recognition, and thought, and temporal lobe *T4,* which is thought to play a role in the recognition of music.

The α–wave frequencies were extracted from the brain wave signals measured at these three points and then analyzed in terms of frequency fluctuations. A representative example of measured behavior is shown in Fig.4.2. The original waveform of the α–wave frequency variations measured from temporal lobe *T4* is shown in (a). The power spectrum obtained by calculating a moving average from a 1024 point FFT shifted at 512 point intervals is shown in (b). There are pronounced differences between the characteristics corresponding to the comfortable and uncomfortable stimuli at frequencies below 1 Hz. In particular, with a comfortable auditory stimulus, the gradient of power is close to approximately –1 (i.e., $1/f$ fluctuation).

Similarly, the three points for each of the 20 subjects were evaluated to analyze the gradient of power. Using the least squares method, we calculated a linear approximation of the data between 0.05 and 1.0 Hz. The emphasis was placed on the range of gradients obtained with comfortable (*$1/f$ fluctuations*) and uncomfortable audio stimuli (*white noise and $1/f^2$ fluctuations*). Then it was found that the negative gradients with the comfortable stimuli ($1/f$ fluctuation) were found to range from 0.7 to 1.0 at *FP2*, 0.5 to 0.7 at *C4*, and 0.5 to 0.8 at *T4*. With the uncomfortable stimuli (White–noise fluctuation and $1/f^2$ fluctuation) the negative gradients at all three points ranged from 0.1 to 0.4, indicating that the characteristics of α–wave fluctuations generally corresponded to the auditory stimuli.

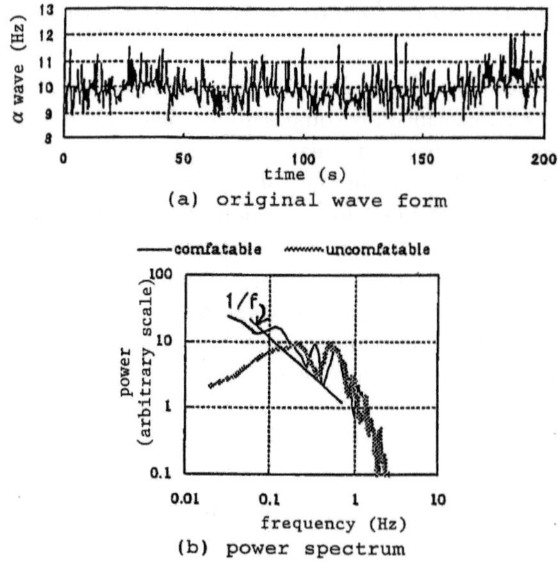

Fig. 4.2 Fluctuation behavior of α-wave signal

4.3.3 Example of Brain Wave Signal Through Disturbance Observer

Figure 4.3 shows an example of typical brain wave signal obtained by using a disturbance observer mentioned in Chapter 4 when a disturbance of Eq. (4.3) was applied. The horizontal axis implies a "time" and the vertical axis means an "arbitrary magnitude." This signal was estimated through frontal lobe FP2. As shown in the figure, noises applied in the measurement are eliminated through the observer.

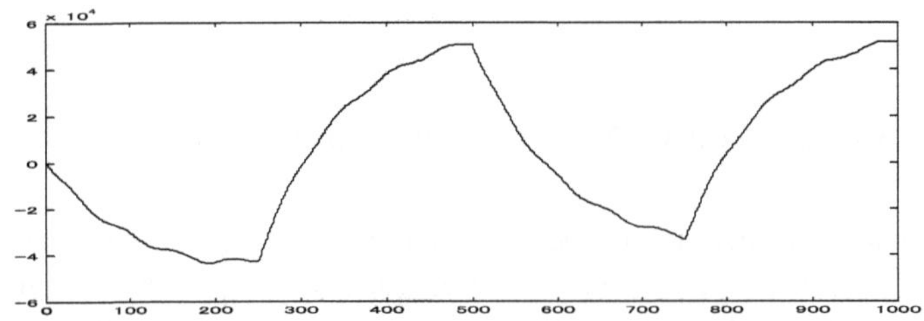

Fig. 4.3 Brain wave signal through disturbance observer

4.3.4 Quantification of Fluctuations for Manipulator Operation

A position/attitude control system with five degrees of freedom was used in the experiments. This mechanism is composed of a manipulator driven by DC motors with optical encoders. The manipulator was operated with three degrees of freedom (the θ_2, θ_3, and θ_4 joints in Fig.4.4(a)), confining its motion to up and down. The parameters of the manipulator operation control were inputted as angular positions to each joint in real time, meaning that the fluctuation characteristics had to be mapped to valid drive input ranges at each joint. Based on the results of preliminary experiments, the fluctuation waveforms were mapped by normal quantification to maximum and minimum values of ±25° as the size of the angular position input to the DC motor in each joint.

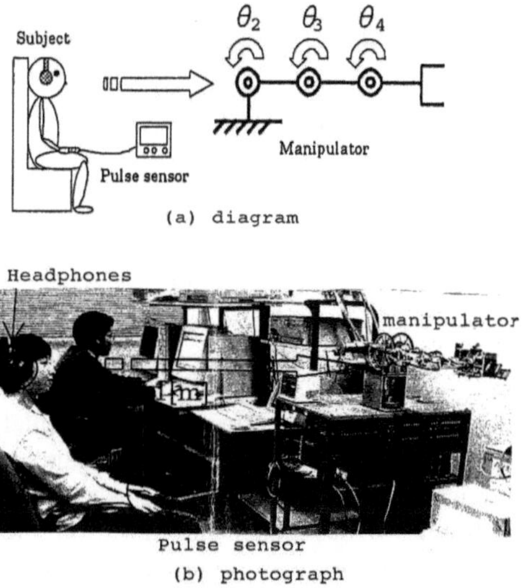

Fig. 4.4 Experimental setup

4.4 Evaluation Test of Manipulator Operation

We evaluated the operation of the manipulator by subjecting the same 20 subjects to auditory stimuli that created comfortable and uncomfortable feelings. Specifically, the α–wave fluctuation information obtained in the evaluation described above was used to control the motion of a manipulator, and the test subjects were asked to evaluate the resulting motion.

4.4.1 Test Method

Figure 4.4 shows the overall configuration of the experimental setup. The motor rotation angle of each joint was transformed into a pulse signal by an encoder and then inputted to a control computer through an interface. Inside the computer, the control laws were calculated for the voltage command values and inputted to the driver of each DC motor to drive the motors. To make it easier to see the natural motion of the manipulator created by the fluctuating inputs, only proportional control was used for the motors. The fluctuating waveforms from *FP2*, *C4,* and *T4* were applied to joint angles θ_2, θ_3, and θ_4, respectively, as the target input angle commands.

The test subjects comprised 16 males and 4 females (all twenties); they evaluated the manipulator motion by visual observation. The manipulator arm was initially aligned horizontally, as shown in the photo in Fig.4.4(b); the angular position commands were then inputted to each of the three joints. The distance between the subject and the manipulator was 1.5 m. A pulse sensor was fitted to the left index finger of the test subject, and the variations in the pulse were measured with a sampling interval of 1.0 s.

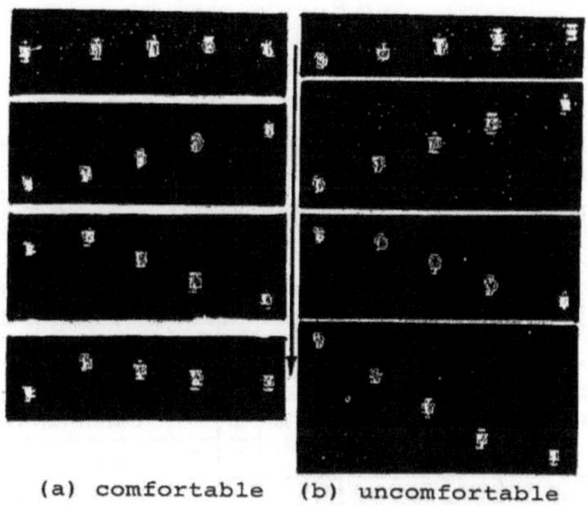

Fig. 4.5 Light photograph of manipulator motion

To make it easier to see the motions made by the manipulator, a miniature light bulb was mounted at the end of the arm, at each joint, and at the mid points of the joints (five in all). Figure 4.5 shows two sets of photographs of the manipulator arm with these lights switched on and the ambient light blocked out. Each figure of (a) and (b) shows the motions corresponding to the comfortable and uncomfortable stimuli, respectively. These photographs were extracted from images of the continuous sequential motions taken from overhead; the vertical amplitude of the motion was clearly larger in the uncomfortable case (In Fig.4.5, we only see the difference in the range and speed of the motion but there exists a significant difference in irregularities).

4.4.2 Subjective Evaluation and Results

Using the semantic differential method (SD) and the questionnaire shown in Fig. 4.6, we subjectively evaluated how the test subjects felt visually after the manipulator had finished moving. They were asked to make a selection from seven levels of intensity, which were then converted into points. To avoid subjective influences from the test subjects and to extract specific trends, a t-test was applied using the following formula. The values of X_i are the scores corresponding to the questionnaire responses, $\overline{X}$ is the average of these values, and U^2 is the sample variance. Since there are 20 responses, there are 19 degrees of freedom.

$$t = \frac{\bar{x}\sqrt{20}}{u}. \tag{4.11}$$

$$\bar{x} = \frac{x_1 + x_2 + \dots\dots\dots x_{20}}{n}. \tag{4.12}$$

$$u^2 = \frac{1}{19}\sum_{i=1}^{20}(x_i - \bar{x})^2. \tag{4.13}$$

tension	-3	-2	-1	0	1	2	3	relaxed
uninteresting	-3	-2	-1	0	1	2	3	interesting
dislike	-3	-2	-1	0	1	2	3	prefer
uncomfortable	-3	-2	-1	0	1	2	3	comfortable
sleepy	-3	-2	-1	0	1	2	3	awake
mechanical	-3	-2	-1	0	1	2	3	human like

evenly=0, a little=1, pretty much=2, quite=3

Fig. 4.6 Questionnaire for evaluation

Assuming the significance level to be 5%, the percentage point is found to be about 2.1 by referring to a t-distribution table. Thus, when the value of Equation (4.11) is greater than 2.1, it is judged that the impression of the corresponding evaluation item is strongly felt.

As shown in Fig.4.7, significant differences are seen in the responses for “relaxed” and “sleepy” for motion with a comfortable auditory stimulus. Conversely, in cases where an uncomfortable auditory stimulus is applied, a significant difference is found in the response for “mechanical”. No significant differences are seen for the other items. Therefore, the motion of the manipulator when operated using a comfortable auditory stimulus can be described as a gentle motion that is relaxing to look at. Because of the small number of female subjects, we found no obvious differences between the male and female evaluation results. In the experiments, however, all the female subjects answered "good feeling" for the comfortable auditory stimulus. This point should be further investigated by increasing the number of female subjects and by changing ages. In particular, there will be significant difference between those subjects.

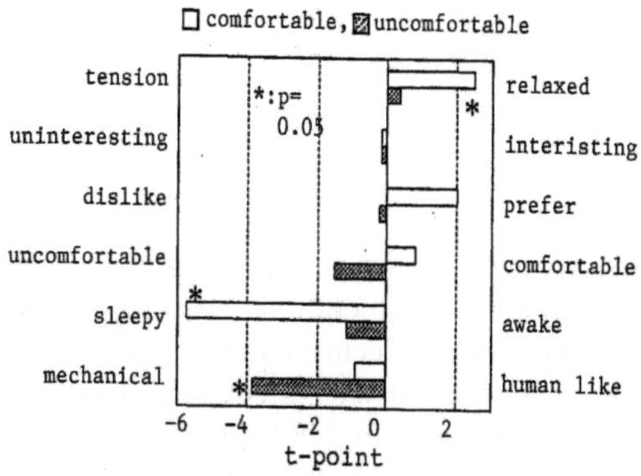

Fig. 4.7 Results of t-test

4.4.3 Objective Evaluation and Results

Changes in emotion are accompanied by physiological responses caused by the propagation of physiological control signals from the brain to the autonomous system and are thus accompanied by changes in the pulse rate. The measurements were performed in terms of pulse fluctuations detected using a pulse sensor. Each test subject had rested for five minutes with his or her eyes closed and with no applied stimuli. As an example, Fig.4.8 shows the results for a test subject in which pronounced pulse fluctuations occurred with both comfortable and uncomfortable auditory stimuli. The pulse rate varies more when the test subject listened to music with white–noise properties listened to music with white–noise properties.

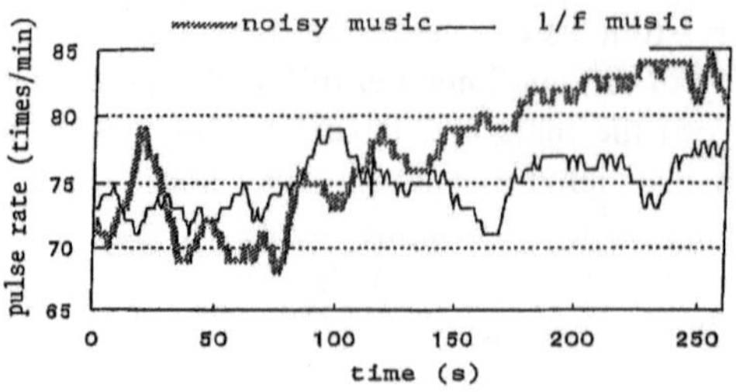

Fig. 4.8 Fluctuation in pulse rate

4.5 Conclusion

In the presence of disturbances, the proposed output disturbance observer was found to be effective in the measurement of α–wave signal. By applying auditory stimuli to test subjects, α–wave fluctuation time series were created that corresponded to the feelings of comfort and discomfort identified from the brain wave signals resulting from these stimuli. These time series were then used as target values for controlling the motion of each joint of a robot manipulator arm. Evaluation tests were performed in which the resulting motion was visually evaluated by test subjects. From the results of a questionnaire study, it was determined that the observation of comfortable motions engenders feelings of "relaxation" and "reduced wakefulness" with a significance level of 5%. Also, frequency analysis of the pulse variations showed that a stronger correlation of the manipulator motion is exhibited when a comfortable auditory stimulus is applied than when an uncomfortable one is applied. In this study, the motion of the manipulator arm was restricted to two dimensions.

This is the first stage of our challenge. In future studies, we plan to look at three dimensional motion and to use a broader range of test subjects in terms of age, sex, and other attributes. There may be the mean and standard deviation of the age of the subject (i.e., some age-related difference in the results) and some envisage possible differences between male and female subjects. We should focus on these problems.

References

1. Ito A, Shiraishi M (2000) Trend and Some Simple Design Examples of Robust Control in Mechatronics Systems. JSPE (in Japanese) 66 5:209-213

2. Chantelau K (1997) Segmentation of moving images by human visual system. Biological Cybernetics 77: 89-101
3.Fermuller C et al (1997) Visual space distortion. Biological Cybernetics 77:323-337
4. Genno H et al (1996) Using Facial Skin Temperature to Objectively Evaluate Sensations. Int J Industrial Ergonomics 19 2:161-165
5. Hanneton S et al (1997) Does the brain use sliding variables for the control of movements? Biological Cybernetics 77:381-393
6. Lusted H S, Knapp R B (1997) Controlling Computers with Neural Signals. Scientific American October: 64-71
7. Musha T (1985) Bio-information and 1/f Fluctuation. Applied Physics (in Japanese) 54 5: 429-435
8. Shiraishi M (1992) A Possibility of 1/f Evaluation on Surface Finishes Based on Fluctuation Behavior. Trans ASME J Eng for Industry 114 2:207-212
9. Yoshida S (1990) Fluctuation Measurement of Brain Wave and Evaluation of Relaxation. Japanese J Acoustics 46 1:914-919
10. Yoshida T (1995) Frequency Fluctuation of Brain Wave and Emotional Evaluation. J JSME 98 918: 403-406

Chapter 5
Development of High-Speed Actuator for Scanning Probe Microscopy

Yasuhiro Sugawara[1], Yan Jun Li[1], Yoshitaka Naitoh[1] and Masami Kageshima[1]

Abstract A novel closed-loop regulation of a piezoelectric actuator is presented to implement wide-band and hysteresis-free motion required for high speed operation of scanning probe microscopy. Velocity of the actuator's motion is detected via induced current, and converted to displacement by integration. By appropriately applying both velocity and displacement feedbacks, the fundamental resonance of the actuator was completely eliminated and other subresonances were significantly suppressed. A bandwidth of ca. 300 kHz for dynamic operation was achieved consequently. Actuator's intrinsic hysteresis was also significantly suppressed.

5.1 Introduction

A piezoelectric actuator is one of the most popular electro-mechanical device and widely used for applications where multi-dimensional positioning in subnanometer- to micrometer-scale is required. A scanning probe microscope (SPM) is one of its most demanding application as it requires subnanometer-scale stability, controllability and dynamic response. In SPM a quantity sensitive to separation between the probe and the sample like tunneling current or atomic force is detected. Commonly a feedback loop is activated to regulate the probe-sample separation so that these quantities are kept constant while the lateral two-dimensional scan is carried out. Therefore, the dynamic performance of the actuator that takes the role of this gap-regulation is one of major limiting factors of the performance, especially when high-speed operation is concerned. The fundamental property that limits dynamic response of a piezoelectric actuator is its mechanical resonance that is closely related to its dimension. Since the range of its spatial displacement

[1] Yasuhiro Sugawara, Yan Jun Li, Yoshitaka Naitoh and Masami Kageshima

Department of Applied Physics, Graduate School of Engineering, Osaka University is also determined by its dimension, an attempt to raise the resonance by reducing the dimension would collide with requirement of the dynamic range.

An alternative solution is to adopt a closed loop between the input signal and the resultant spatial motion so that the excess motion at resonance peak is suppressed. Recently, Kodera et al. employed this active damping technique in their high-speed biological atomic force microscopy (AFM) to suppress a resonance at ca. 100 kHz and attained an imaging speed of 21 flames/s. In their setup an LCR network was placed parallel to the z actuator to electrically simulate the resonance. Its response to the regulating voltage was regarded as representing the actuator's motion and was used to generate a feedback signal for active damping.

Although this mock piezo thus greatly contributed toward implementation of the high-speed AFM, it cannot solve every problems caused by the resonance. An actuator has a number of vibrational modes besides the fundamental one. Since the most demanding feedback motion required in an SPM is a response to a step-like signal that is a synthesis of a number of high-frequency components, in order to prevent damage to the tip or the sample by inaccurate response to a step signal, it is essential to extend the feedback bandwidth as wide as possible. It means multiple numbers of mock piezos must be added to the system in order to cancel all the resonances within the required feedback bandwidth, which would result in unfavorable complexity of the system.

Another problem that the above mock piezo approach cannot solve is hysteresis of the actuator, which is intrinsically caused by irreversible response of ferroelectric domains to the external electric field. In applications in which accuracy of the displacement is essential, influence of the hysteresis is often eliminated by monitoring the displacement with other precision measurement devices and feeding it back to the input voltage [2]. This setup would, however, lead to substantial complexity in the apparatus and have a negative effect in the dynamic performance of the SPM. Another solution to this problem is use of mathematical procedure to anticipate the response of the actuator [3]. Although this approach proved useful, time lag due to calculation may delay the response. In addition, it cannot circumvent time-lapse change in the property of the actuator.

It should be noted that all the problems stated above are originating from rather indirect monitoring of the spatial motion of the actuator. A more direct and simple method is required to attain wide-band and hysteresis-free regulation of the actuator. In the present article a method to directly detect motion of the actuator with accuracy and a wide bandwidth using the current induced by strain in the actuator is presented. Procedure and results of resonance suppression and hysteresis cancellation based on this approach are also shown.

5.2 Displacement Detection of Piezoelectric Actuator Using Induced Current

To understand the mechanism of the present approach, it would be helpful to trace back to a fundamental mechanism of a piezoelectric actuator. When a voltage is applied between the two electrodes of a piezoelectric actuator, a strain proportional to the electric field is generated. This is the origin of the macroscopic displacement of the actuator. This strain causes deviation of charge distribution in the actuator and induces charge on its electrodes. Therefore, the amount of charge is proportional to the displacement of the actuator [4]. It is expected that accurate detection of actuator's motion can be implemented using the induced charge or its derivative, i.e., the current.

In the present approach, as described later, both the displacement signal and its derivative, i.e., the velocity are needed. There are three possible configurations for obtaining the both signals as shown in Fig.5.1. One is to measure the induced current proportional to the velocity via a current-voltage converter [5, 6] and to integrate it (Fig.5.1(a)), and another is to measure the induced charge proportional to displacement via a charge amplifier and to differentiate it (Fig.5.1(b)). The latter is, however, not appropriate considering that differentiation spoils S/N ratio. The other scheme is to detect the driving current as a voltage drop across a small resistor inserted before the actuator while measuring the displacement through the induced charge (Fig.5.1(c)). In the present research, mainly for simplicity reason, the scheme of Fig.5.1(a) was adopted. However, the scheme of Fig.5.1(c) can also be the candidate and would be worth trying. Since the inverting pin of the I-V converter or the charge amplifier placed subsequent to the actuator is virtually grounded in any of the above three cases, the current or charge induced on actuator's cold electrode should be same as that if the electrode is directly grounded.

In the measurement examples shown hereafter, a multilayered piezoelectric actuator (NEC Tokin Co., AE0203D04) was used. Here, as proposed by Ando et al.[7], two same actuators were glued in symmetry onto both sides of a thick aluminum substrate and were driven simultaneously so that effect of impulsive force caused by fast motion of the actuator cancels each other.

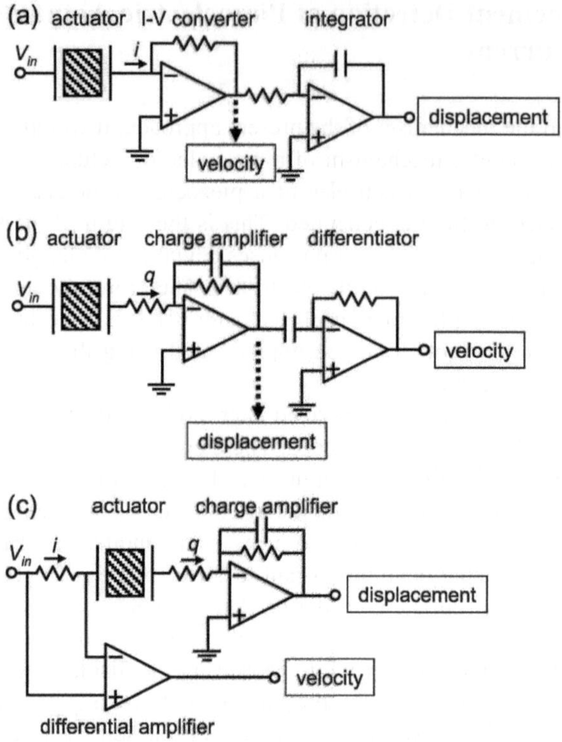

Fig. 5.1 Three possible schemes to detect actuator's displacement and velocity simultaneously. q and i represent charge and current, respectively. Velocity is detected through induced current and is integrated to displacement (a), displacement is detected through induced charge and is differentiated to velocity (b), and both the velocity and displacement are measured independently using a differential amplifier and a charge amplifier, respectively (c). In the present article, scheme (a) was adopted, although (c) can be another potential candidate

5.3 Actuator Regulation by Combined Velocity and Displacement Feedbacks

The active control for quality factor of a resonating system is a well-known technique, and is recently utilized in controlling the effective quality factor of an AFM cantilever [8-11]. Here its simplest principle is described briefly. An oscillating system is modeled as a harmonic oscillator with an effective mass m and a spring constant k. When it is driven with external sinusoidal force $Fe^{i\omega t}$ in a dissipating medium, its equation of motion is

$$m\ddot{z} + \gamma\dot{z} + kz = Fe^{i\omega t} \tag{5.1}$$

where γ is the viscous drag coefficient. If another external force proportional to the velocity $\dot{z}$ is applied to the system as

$$m\ddot{z} + \gamma\dot{z} + kz = Fe^{i\omega t} + A\dot{z} \tag{5.2}$$

the effective drag coefficient γ_{eff} is expressed as $\gamma_{eff} = \gamma - A$. Thus the effective quality factor of the system $Q_{eff} = m\omega / \gamma_{eff}$ is apparently enhanced or suppressed, depending on sign and magnitude of A. If the oscillation is sinusoidal, the velocity signal is simply proportional to the deflection with a phase shift of π/2. Thus a feedback loop with an appropriate gain and a phase shifter is enough to actively control the resonance. However, if the signal of interest is nonsinusoidal, i.e., superposition of Fourier components with various frequencies, the above solution fails as shift amount of an analog phase shifter is mostly frequency-dependent. In the present study, since the velocity signal itself is readily obtained from the induced current, active damping is possible even for nonsinusoidal oscillation of the actuator, which is an advantage of the present approach.

Another feedback needs to be introduced in order to cancel the hysteresis of the actuator and the above-mentioned phase lag caused by active damping. Because both of these are problems in relatively low frequency regime, this feedback should be based on actuator's displacement rather than its velocity. Thus, displacement signal is fed back to V_{in}.

It proved that the best performance for the present goal with well-suppressed resonance and hysteresis can be achieved by combining the velocity and the displacement feedbacks. Since here fast response to small signal is especially concerned, a prototype system was built based on ±15 V OP-amps. Figure 5.2 shows a diagram of the system with both velocity and displacement feedbacks. The inverter inserted between the summing points of the two feedback loops is for correcting polarity inversion caused by the integrator. The optimum feedback amounts for the both loops are determined with the variable resistor placed in each loop. Due to a 1 MΩ resistor placed parallel to the feedback capacitor to shunt circuit's unfavorable DC voltage offset or drift, the integrator has a low cutoff of 1.6 Hz above which it acts as a proper integrator. For linearization of slow motions below this frequency, the system should be combined with another detection device sensitive to static displacement like a strain gauge etc.

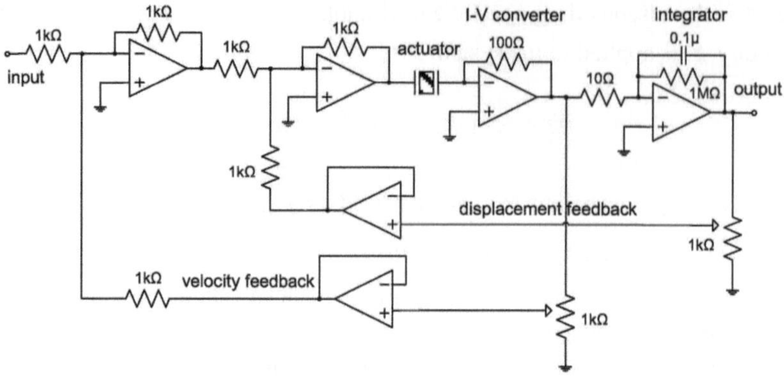

Fig. 5.2 Schemtic diagram of the prototype circuit to evaluate performance of the present regulation technique. The integrator has a shunt resistor that sets its lower limit of integration to 1.6 Hz. An inverter is inserted between the summation nodes for two feedbacks loops to compensate polarity inversion by the integrator. In the present setup the summation is made with ratio of 1:1 and gain of 1 for the both nodes

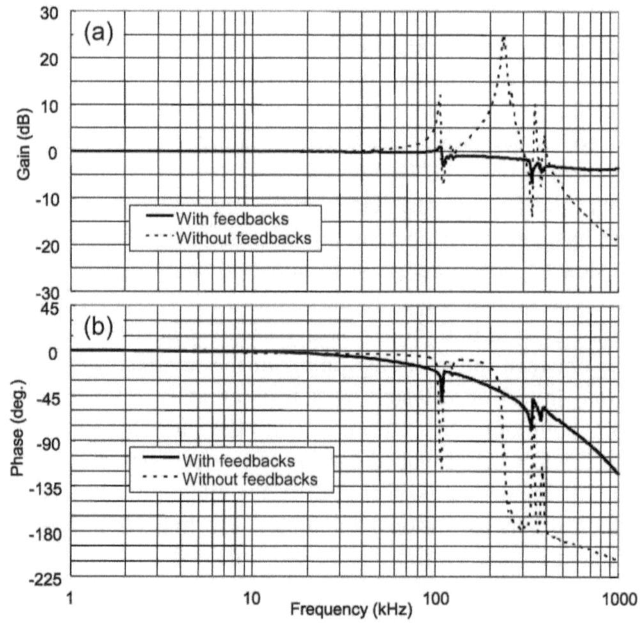

Fig. 5.3 Effect of combined velocity and displacement feedbacks on the resonance characteristics of the transducer. Gain normalized with DC displacement (a) and phase (b) are shown. Broken line indicates uncompensated data whereas solid line the one compensated with the combined feedbacks in both (a) and (b)

Figure 5.3 shows gain and phase of V_{out} with the both feedback loops activated with respect to V_{in}. Here, the feedback amount for each loop has been determined so that the gain exhibits the best flatness over the bandwidth of 1 MHz. The fundamental resonance has been completely suppressed and error in gain has been mostly less than ± 5 dB. The -45° bandwidth is ca. 250 kHz ignoring the slight dip at 105 kHz. It has been confirmed that it can be tuned to ca. 300 kHz at a cost of gain flatness by altering the valance of the two feedback loops.

In order to evaluate the effect of feedbacks to settling time of the system, its response to burst oscillation input was measured. The input frequency was chosen to the 105 kHz resonance that is one of the major residual resonance modes after the fundamental one was suppressed by the feedback. Figure 5.4 shows comparison of the responses to input with amplitude of 100 mV with and without the feedback loops. From Fig.5.4 (a) the time constant of the response signal τ proved to be 88 μs without the feedbacks, implying that the quality factor of the system $Q = \pi f \tau$ at this resonance frequency f is 29. With the feedbacks adjusted to the flattest gain condition of Fig.5.3, the response was drastically improved as shown in Fig.5.4(b). It is obvious that now τ is smaller than a quarter of the period 2.4 μs, which implies Q of at most 0.75. The steady-state amplitude of the response signal has been reduced to less than 10 % of the initial value by the feedbacks. This is partly due to suppressed resonance of 105 kHz and partly to decrease in net voltage applied to the hot electrode of the actuator caused by the displacement feedback. The latter will be readily improved by optimizing the gain distribution at the summation node of the displacement feedback.

Figure 5.3 shows gain and phase of V_{out} with the both feedback loops activated with respect to V_{in}. Here, the feedback amount for each loop has been determined so that the gain exhibits the best flatness over the bandwidth of 1 MHz. The fun-

damental resonance has been completely suppressed and error in gain has been mostly less than ± 5 dB. The -45° bandwidth is ca. 250 kHz ignoring the slight dip at 105 kHz. It has been confirmed that it can be tuned to ca. 300 kHz at a cost of gain flatness by altering the valance of the two feedback loops.

In order to evaluate the effect of feedbacks to settling time of the system, its response to burst oscillation input was measured. The input frequency was chosen to the 105 kHz resonance that is one of the major residual resonance modes after the fundamental one was suppressed by the feedback. Figure 5.4 shows comparison of the responses to input with amplitude of 100 mV with and without the feedback loops. From Fig.5.4 (a) the time constant of the response signal τ proved to be 88 μs without the feedbacks, implying that the quality factor of the system $Q = \pi f \tau$ at this resonance frequency f is 29. With the feedbacks adjusted to the flattest gain condition of Fig.5.3, the response was drastically improved as shown in Fig.5.4(b). It is obvious that now τ is smaller than a quarter of the period 2.4 μs, which implies Q of at most 0.75. The steady-state amplitude of the response signal has been reduced to less than 10 % of the initial value by the feedbacks. This is partly due to suppressed resonance of 105 kHz and partly to decrease in net voltage applied to the hot electrode of the actuator caused by the displacement feedback. The latter will be readily improved by optimizing the gain distribution at the summation node of the displacement feedback.

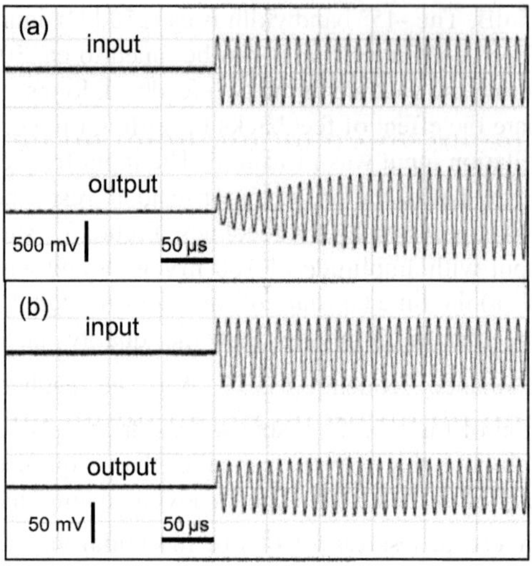

Fig. 5.4 Oscillographs of the system's response to burst oscillation input with an amplitude of 100 mV and a frequency of 105 kHz. Uncompensated system (a) and one compensated with the combined feedbacks (b) are compared

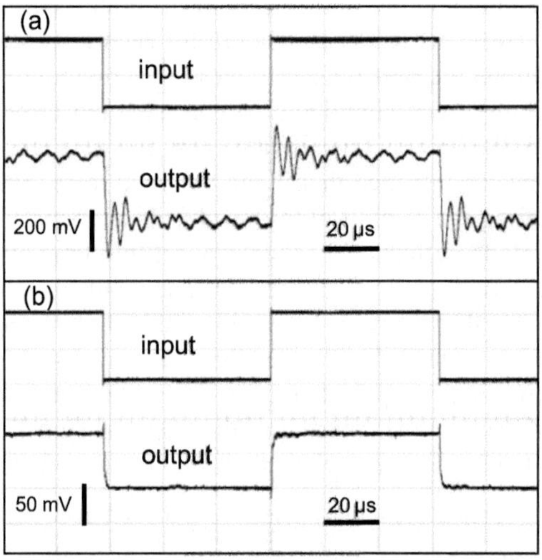

Fig. 5.5 Oscillographs of the system's response to a stepped wave signal with an amplitude of 100 mV and a frequency of 8 kHz. Uncompensated system (a) and one compensated with the combined feedbacks (b) are compared. The vertical scales are for output signals only

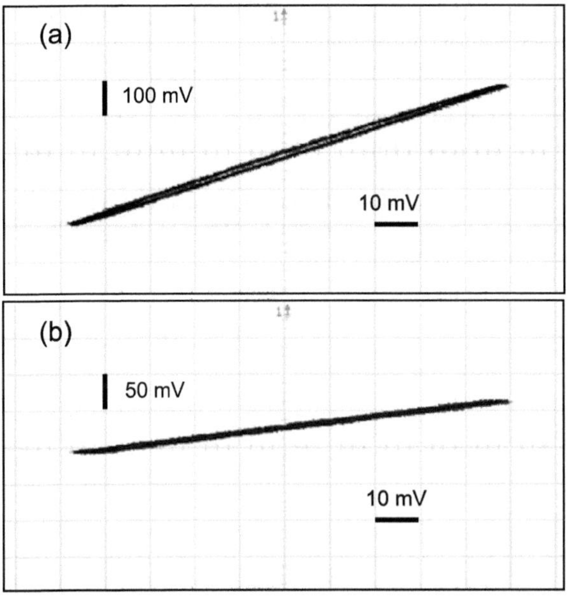

Fig. 5.6 Lissajous figures of the transfer function of the system measured with a sinusoidal input signal with an amplitude of 100 mV and a frequency of 8 kHz. Uncompensated system (a) and one compensated with the combined feedbacks (b) are compared. Input signal is on the horizontal axis whereas output on the vertical in both (a) and (b)

5.4 Conclusions

Wide-band and hysteresis-free regulation of a piezoelectric actuator was successfully implemented. Velocity of the actuator's displacement is detected via induced current, and converted to displacement by integration. By appropriately applying both velocity and displacement feedbacks, the fundamental resonance was completely eliminated and flat response was achieved with a bandwidth of ca. 300 kHz. Intrinsic hysteresis of the actuator was also greatly suppressed.

References

1. Kodera N, Yamashita H and Ando T (2005) Rev. Sci. Instrum. 76: 053708
2. Mizutani K, Kawano T and Tanaka Y (1990) Prec. Eng. 12:219
3. Ge P and Jouaneh M (1995) Precis. Eng. 17: 211
4. Newcomb CV and Flinn I(1982) Electron. Lett., 18: 442
5. Furutani K, Urushibata M and Mohri N (1998) Nanotechnology 9: 93

6. Furutani K, Urushibata M and Mohri N (1998) Proceedings of the 1998 IEEE International Conference on Robotics and Automation 1504
7. Ando T, Kodera N, Takai E, Maruyama D, Saito K and Toda A(2001) Proc. Natl. Acad. Sci. USA 98: 12468
8. Mertz J, Marti O and Mlynek J (1993) Appl. Phys. Lett. 62: 2344
9. Anczykowski B, Cleveland JP, Krügen D, Elings V and Fuchs H(1998) Appl. Phys. A, 66: S885
10. Humphris ADL, Tamayo J and Miles MJ(2000) Langmuir 16: 7891
11. Sulchek T, Hsieh R, Adams JD, Yaralioglu GG, Minne SC, Quate CF, Cleveland JP, Atalar A and Adderton DM (2000) Appl. Phys. Lett. 76: 1473

Chapter 6
PZT Driven Micro XY Stage

Takahito Ono[1], Mohd Faizul Mohd Sabri[2], and Masayoshi Esashi[3]

Abstract Novel piezo-driven XYZ and XY microstages have been developed. The design, fabrication and evaluation of the two kinds of microstages are presented. One of the microstages is fabricated from a monolithic PZT plate. Using dicing, electroplating of nickel, photolithography and laser machining, stacked PZT actuators are formed in the PZT plate, also the structure of XYZ microstage with 16×15 mm^2 are defined. This microstage has a capability of 6 degrees of freedom in motion. Using capacitive displacement sensor, precise motion control is demonstrated. However, it is difficult to form support spring with a low stiffness due to the difficulty in the structure formation of PZT. Another type of the microstage is hybrid of Si and PZT, a Si XY microstage structure is formed by deep reactive ion etching, and PZT stacked actuators are assembled in to the Si microstage. In order to amplify the displacements, Moonie amplification mechanism is chosen. The stage is supported by the Moonie mechanism and support springs. Over 80 μm of displacement is obtained at an application voltage of 70 V, also 18 times amplitude amplification is demonstrated.

6.1 Introduction

With the development of nanotechnologies, the importance of nanopositioning technology is increasing for various applications including nanoindentation, nanolithography, scanning probe microscopy and atomic force micrscopy (AFM) based high-density data storage devices, etc. The PZT($Pb(ZrTi)O_3$)-powered mirostage,

[1] Takahito Ono

Graduate School of Engineering, Tohoku University

[2] Mohd Faizul Mohd Sabri

Department of Mechanical Engineering, Faculty of Engineering, University of Malaya

[3] Masayoshi Esashi

The World Premier International Research Center Initiative for Atom Molecule Materials, Tohoku University one of the nanopositioning devices, has attracted considerable attention from many researchers owing to its merits such as small size, high stiffness, high resolution, fast response, low power consumption, relative large output force and frictionless actuation. The need for better nanopositioning devices will continue as the research and technology in the micro/nano world becomes more rapid. Therefore, quite a number of researches concerning micro/nano positioning for various applications such as data storage systems [1,2], micro-optical systems [3], and various measurement devices can be found [4,5].

The developed micro/nano positioning devices utilize different types of actuation methods. Pantazi et al. [6] and Golda et al. have developed a microstage driven by electromagnetic microactuators for probe-based data storage application [7]. Xu et al. have developed a microstage with PZT actuators [8], Sasaki et al. is working on electrostatically driven XY-stages [9]. A majority of the microfabricated XY-microstages are based on electrostatic and electromagnetic actuation. This is due to the long travel range offered by these two types of actuators compared to other type. However, the drawback of the electrostatic actuator is its low energy density [9]; therefore, a large array of comb drive actuators is required in order to produce a high output power, resulting in a large overall size of the XY-stage. In the case of electromagnetic actuation it is difficult to integrate actuators using microfabrication technology since it requires 3 dimensional coil structures and deposition of magnetic materials [6, 10].

There are few researches that have been executed using piezoelectric microactuators for nanopositioning, but none have succeeded to produce large enough displacements for practical use. The advantages of piezoelectric actuators are their relatively high operating frequency, high energy density and high accuracy with sub-nanometer resolution.

The aim of this research is to develop a microstage which is able to produce a large displacement, high accuracy, a large stage area and high resonant frequency with low cost.

The designs and fabrications of two kinds of PZT-driven XY microstages are reported. Laboratory-made stacked PZT actuators are employed for actuation. The actuation performances of the XY-microstage are evaluated.

6.2 Monolitic PZT XYZ Microstage

6.2.1 Design of Monolitic PZT XYZ Microstage

A monolithic PZT XYZ microstage is proposed and fabricated. Using a novel fabrication method for integration of PZT actuators into monolithic PZT, multi-degrees of freedom in the motion of microstage can be achieved. In addition, capacitive sensors for position monitoring can be integrated.

The displacement of piezoactuators themselves is very small, therefore, many kinds of mechanisms for enlarging displacement of *XY*-linear motion have been adapted. In order to achieve accurate *X* and *Y* linear motion of the stage, the *XY*-stage should have higher symmetric structure to minimize the cross-axis coupling. In this microstage, the stage corners or sides should be supported by above enlarging mechanisms. 'Parallelogram mechanism' has a high symmetry in their structure for enlarging the displacement, but most of them are driven by asymmetrically arranged piezoactuators.

A novel parallelogram mechanism for the stage with six degrees of freedom is designed, and the actuator is integrated into the parallelogram mechanism. This integration allows a high symmetric structure in the design. Fig.6.1 shows the schematic figure of the stage design and the parallelogram mechanism. Four arms with parallelogram mechanism support the center stage and the double-layered piezo-stack actuator is integrated in each center of the arm.

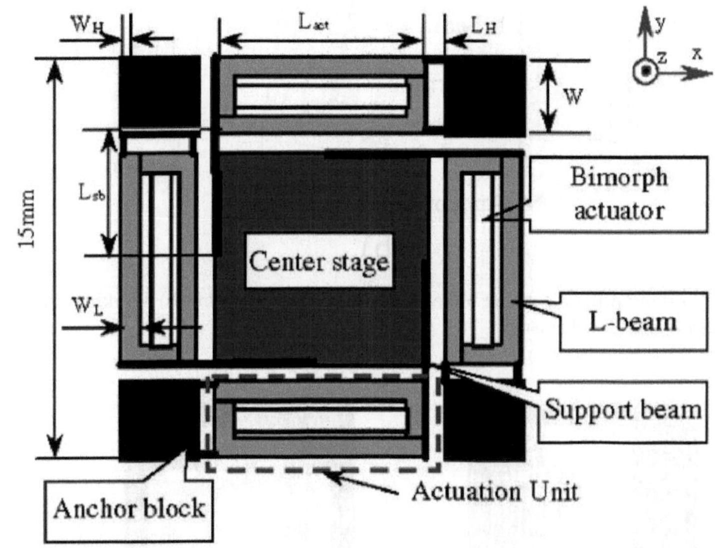

Fig. 6.1 Schematic layout of the XYZ microstage

The double-layered piezo-stack actuator, as the cross section is shown in the top of Fig.6.2(a), consists of two stacked piezoactuators, and the stacked piezoactuators can be individually driven by applying an appreciate voltage. The double-layered piezo-stack actuators can elongate to drive the stage in *XY*-plane, and can bend to move into out-of-plane. Individual operation of the four-parallelogram arms can drive the stage with six degrees of freedom in *X*, *Y*, *Z*, θ_x, θ_y and θ_z directions, as the typical motions are illustrated in Fig.6.3.

The enlarging ratio K_{XY} for *XY* displacement of the built-in parallelogram mechanism is given by:

$$K_{XY} = \frac{X}{L}. \tag{6.1}$$

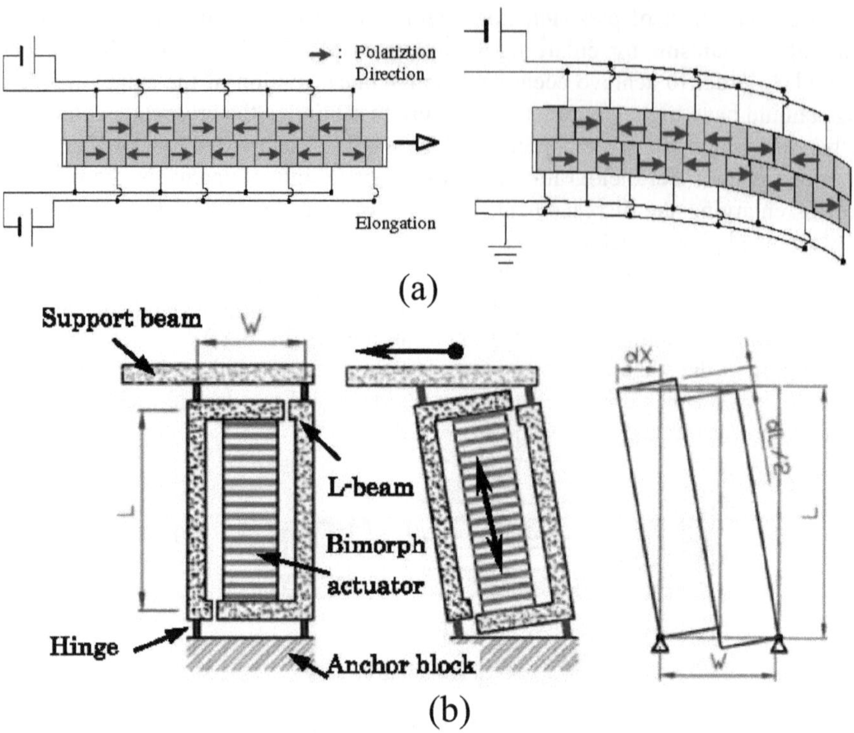

Fig. 6.2 (a) Working principle of the bimorph stacked piezo actuator. (b) Working principle of parallelogram mechanism

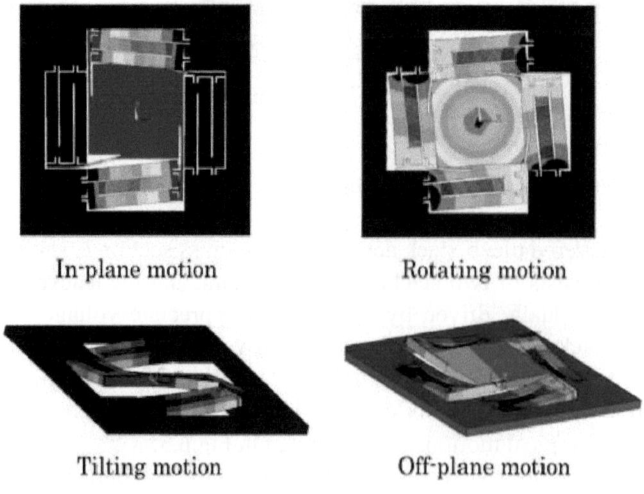

Fig. 6.3 Typical motion of the XYZ microstage

6.2.2 Fabrication of Monolitic PZT XYZ Microstage

The key elements of the microstage with six degrees of freedom are the double-layered piezo-stack actuators. The fabrication flow of the inner electrodes is shown in Fig.6.4. A 800 μm-thick PZT plate was prepared by polishing the both surfaces, and photoresist as a masking layer was spun on its one-side. The blind grooves with a depth of 400 μm were formed by dicing. In order to fabricate the double-layered piezo-stack actuators, the deep blind grooves were formed by a dicing saw and electroplated to form the inner electrodes for stacked piezoactuator. The complete filling of electroplated metal into the blind grooves was difficult due to the generation of voids or hollows if the seed metal layer for electroplating was remaining on the sidewall or top surface. Therefore, the electroplating process was performed as follows. First the blind grooves with 30 μm width and 130 μm pitch were formed for the inner electrodes of the stacked piezoactuators, as well as the grooves with 70 μm width were formed for the hinges and reinforced ribs. After sputtering an Cr-Au seed layer, the metals at sidewall were removed by the dicing saw. The Cr-Au on the photoresist surface was removed, resulting in leaving the seed layer only at the bottom of grooves. From the seed layers, Ni for forming the inner electrodes, hinges and reinforced ribs, were electroplated. It can be seen that the 30 μm ×400 μm grooves were diced. After polishing of the electroplated face of the PZT plate, the same processing was repeated on other side of the plate, and then the electroplating process was completed. Then, the photosensitive polyimide as an insulating layer with 8 μm thickness was patterned by photolithography. Cr-Au was deposited by sputtering and patterned by photolithography and by etching of Cr-Au Then the completed PZT plate was defined by machining with a femto-second laser, as the part of the structure is shown in Fig.6.4(d). Finally, the lead lines were bonded to the metal pads with conductive glue. The width of the cut grooves by the laser was about 150 μm, the hinge width was about 150 μm and the PZT actuator length is 6.5 mm. SEM images of the fabricated XYZ stage is shown in Fig.6.5.

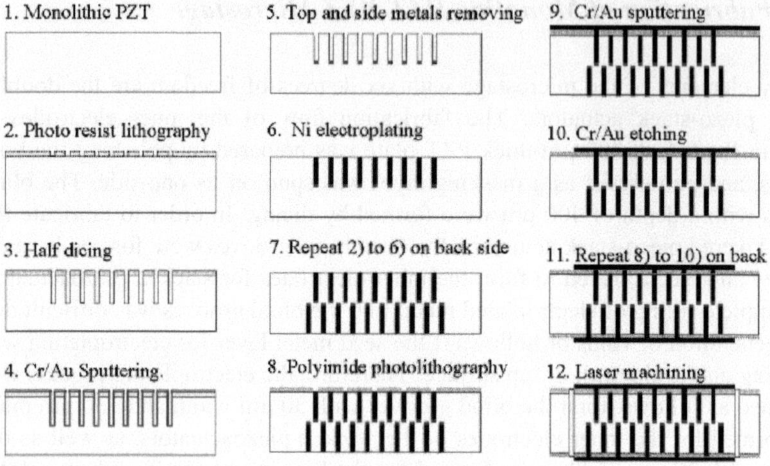

Fig. 6.4 Fabrication process of monolithic PZT XYZ microstage

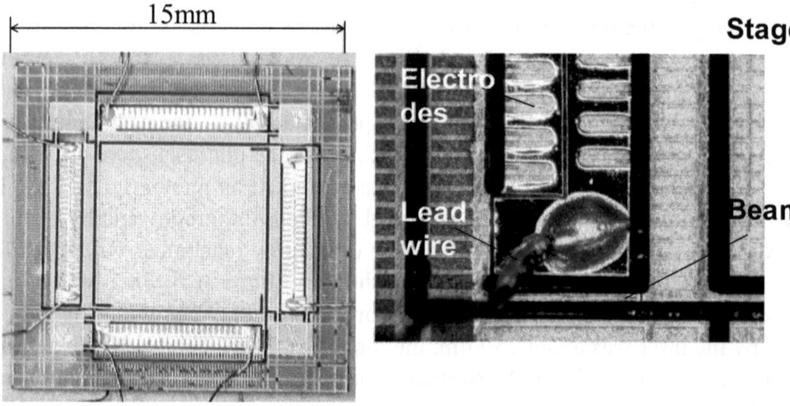

Fig. 6.5 Fabricated monolithic PZT XYZ microstage

The static responses of the successfully fabricated prototype have been evaluated using a laser displacement sensor. To evaluate the static response of the stage, driving voltages from the power sources are applied to the actuators of the microstage, the displacements are then detected by the laser displacement sensor. Displacement in x-direction is measured by applying the electric field to the four actuators of left and right side actuation units with the same direction as the poling direction of the PZT actuators. Displacement in z-direction is measured by applying electric field to all the four actuators on the top side with the same direction as the poling direction of the actuators, while the electric field applied to all the four actuators on the back side are opposite to the poling direction of the actuators. Tests on the prototype show that the displacements at the center of the stage are 2 μm in x-axis and 2 μm in z-axis under the applied voltage of 40 V. These values are approximately 10 times smaller than those estimated value from device design.

The reason of this discrepancy relies on the large stiffness of the beam supporting the stage.

6.3 Si-PZT Hybrid Microstage with Moonie Amplification Mechanism

6.3.1 Design of Si-PZT Hybrid Microstage

Recent deep reactive ion etching of Si is well developed; therefore, a Si-PZT hybrid XY-microstage with novel design are fabricated and evaluated. In this stage, Si is used as a base material and stacked PZT actuators are used to drive the stage.

The design of the proposed XY-microstage driven by stacked PZT actuators with Moonie amplification mechanisms [11] is shown in Fig.6.6. This design consists of two movable structures arranged so that movement in both the X and Y directions are controlled by actuators. The Moonie amplification mechanism will amplify the stroke of stacked PZT actuators for the respective directions. The center stage is supported by two sets of support beams, and can be actuated by the Moonie amplification mechanism. In addition, the center stage is placed in the movable outer frame, which is also supported by support beams and can be actuated by another Moonie amplification mechanism.

The stage is made of a Si substrate with a size of 20×20×0.4 mm. Silicon is chosen as a base material as it is preferable to metals due to its material properties such as inelasticity, hardness and small thermal expansion. Furthermore, the use of silicon will lower the cost on the basis of batch production.

As shown in Fig.6.6, the Moonie mechanism is based on four beams arranged in a "diamond" configuration in order to drive translation along one axis [12]. The four beams are connected by rounded hinge with radius *r*. The basic amplification factor of the Moonie mechanism can be calculated by the following equation.

$$AMP_{\text{factor}} = \left(\frac{L_{\text{M}}}{W}\right) = \left(\frac{\text{COS}\,\theta}{\text{SIN}\,\theta}\right) = \text{COT}\,\theta , \tag{6.2}$$

where AMP_{factor} is the amplification factor, L_{M} is the half length of the Moonie mechanism in the expansion direction, W is the half height of the Moonie mechanism and Moonie angle θ is the angle between the diamond rhombic direction and the longitudinal direction of the actuator. This equation implies that smaller values of Moonie angle will result in larger amplification factors. However, this calculation only employs geometric considerations in order to estimate the amplification factor. For actual factor, we have to consider blocking force, which is exerted onto

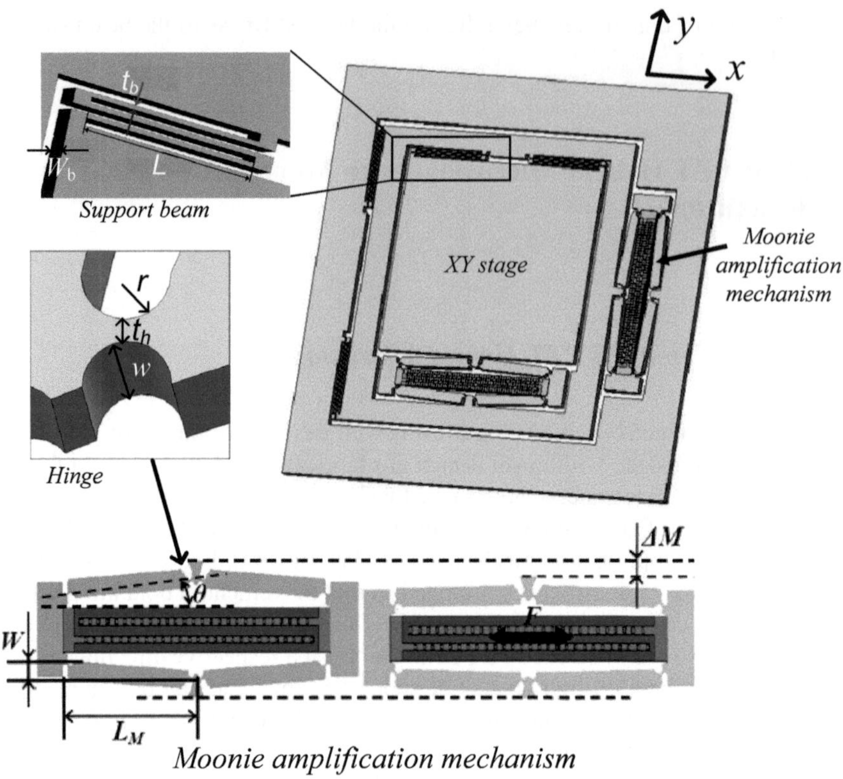

Fig. 6.6 Schematic of silicon XY-microstage containing Moonie amplification mechanism, and the working principle of the Moonie amplification mechanism

the actuator by the Moonie amplification mechanism. As the Moonie angle decreases, blocking force increases and maximum displacement is restricted.

FEM simulation of the microstage was then carried out for the design containing different Moonie angles, thus yielding amplification factor vs. Moonie angle, as shown in Fig.6.7. Figure 6.7 contains the curves of the amplification factors obtained through simulation. In addition, the geometrical amplification factor obtained from Equation (6.2) is plotted. It can be seen that there is a significant deviation in behavior between the simulation value and the geometrical amplification factor, especially at low Moonie angles. In practice restraining forces of the amplification mechanism restrict the displacement, as the Moonie angle decreases. From the simulation, it is predicted that the amplification factor will reach a maximum value at a Moonie angle of 1.5 degrees, and will decrease rapidly for lower values of Moonie angle. This prediction can be explained by the fact that the restraining forces are more significant for small Moonie angles. The experimental results of the amplification factors, for Moonie angles of 8.5, 2 and 1.5 degrees correspond well with the simulation curves, and serve to validate the simulation results. Important parameters for the microstage are listed in Table 6.1.

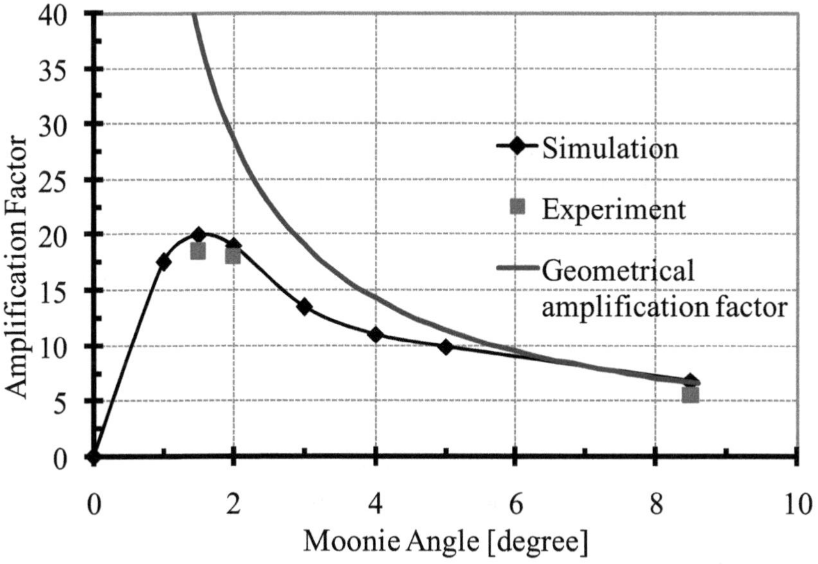

Fig. 6.7 Experimental and simulation results of the Moonie mechanism amplification factor as a function of Moonie angle

Table 6.1 Typical dimensions of fabricated XY microstage

Support spring	
Width, w_b	0.4 mm
Length, L	4 mm
Thickness, t_b	0.05 mm
Hinge	
Thickness, w	0.4 mm
Radius, r	0.1 mm
Thickness of most thin part, t_h	0.05 mm
Material properties (Silicon)	
Young`s modulus	190 GPa
Density	2.32 g/cm^3

6.3.2 Evaluation of Si-PZT Hybrid Microstage

A silicon microstage with dimensions of 20×20×0.4 mm was fabricated by conventional microfabrication processes using deep reactive ion etching of Si. The fabricated microstage has a Moonie angle of 2 degrees.

Evaluation of the XY-microstage was conducted using the following procedure. A commercially available stacked PZT actuator (type N-10 by NEC/Tokin) was used. These actuators were cut, and slotted into the Moonie amplification structure. In order to fix, the actuators were glued by an adhesive. Using a stroboscopic video microscope, in-plane motions of periodically moving structures were measured. A laser emitting diode is used as a pulsed light source that ensures constant illumination power of the strobe pulses.

The displacement of the Moonie amplification was also measured as a function of applied voltage. The applied DC voltage was swept in the range of 0 to 60 V with a step of 10 V at intervals of 50 msec. Thus, the measurement displacements are assumed to be under quasi-static conditions. From these results, the amplification factor 17 of the Moonie amplification mechanism can be obtained from displacements of the actuator and Moonie mechanism. These experimental results of the amplification factor are consistent with FEM simulation. In Fig.6.7, experimental results of the Moonie amplification factor for XY-stages with Moonie angle of 8.5, 2 and 1.5 degrees are also plotted. These results show that the simulation results are reasonably accurate, and useful in identifying the optimum parameters for the microstage.

The displacement of the microstage in the X and Y directions were measured as a function of applied voltage. An input voltage was applied to the microstage, from 5 V to 70 V. Figure 6.8 shows the displacement results of the fabricated microstage containing the Moonie amplification mechanism with a Moonie angle of 2 degrees. Typical of PZT materials and actuators, the displacements both in the X and Y directions exhibit hysteresis. As can be seen from the plot, significant displacement was obtained, where approximately 82 μm and 60 μm stage displacements were obtained at applied voltages of 70 V for X and Y directions, respectively. These magnitudes of the displacement are considerably large and comparable with reported achievements [2, 5, 7, 9, 10], and serves to illustrate the viability of the applications of this microstage in microsystems.

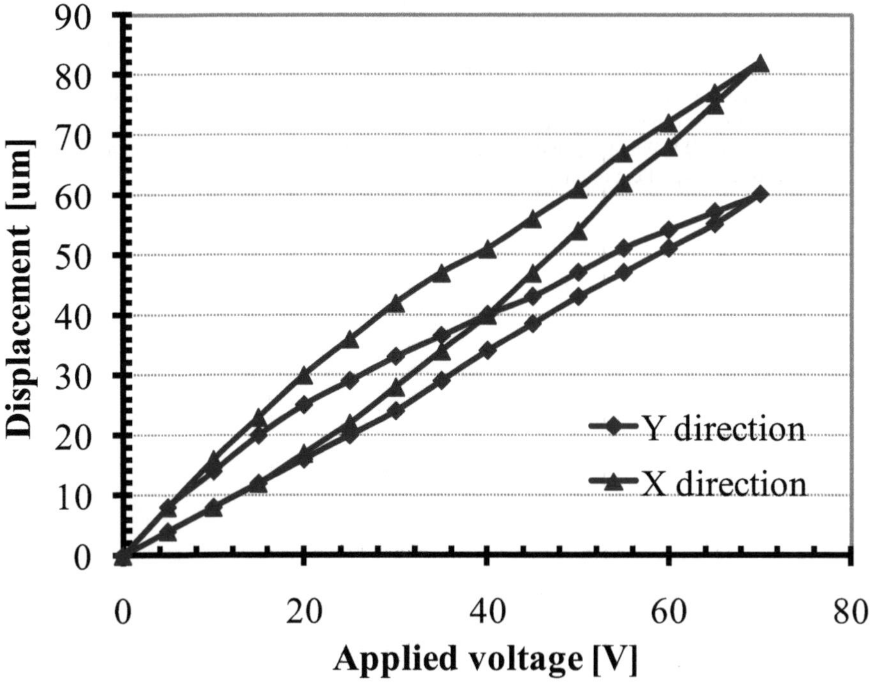

Fig. 6.8 Displacement of the XY microstage as a function of applied voltage

References

1. Eleftheriou E, Bächtold P, Cherubini G, Dholakia A, Hagleitner C, Loeliger T, Pantazi A and Pozidis H (2003) A Nanotechnology-based Approach to Data Storage Proc. of the 29th VLDB Conference 3-7
2. Faizul MS, Ono T, Kawai Y and Esashi M (2008) A Si-PZT Hybrid XY-microstage utilizing an Actuator Amplification Mechanism Proc. of the Actuator 2008 75-78
3. Debeda H, Freyholf TV, Mohr J, Walrabe U and Wengelink J (1999) Development of minituarized piezoelectric actuators for optical applications realized using LIGA technology IEEE J. MEMS 8 :258-263
4. Sarajlic E, de Boer MJ, Jansen HV, Arnal N, Puech M, Krijnen G and Elwenspoek M (2005) Bulk micromachining technology for fabrication of two-level MEMS in standard silicon substrate Proc. of the TRANSDUCERS '05 1404-1405
5. Takahashi T, Mita M, Fujita H and Toshiyoshi H (2006) A high fill-factor comb-driven XY-stage with topological layer switch architecture IEICE El. Ex 3 :197-202
6. Pantazi A, Lantz MA, Cherubini G, Pozidis H and Eleftheriou E (2004) A servomechanism for a micro-electromechanical-system-based scanning-probe data storage device Nanotechnology 15 :612–621
7. Golda DS and Culpepper ML (2008) Design of a six axis meso-scale nanopositioner driven by moving coil microactuators Proc. of Solid-state sensors, actuators, and Microsystems workshop 60-638

8. Xu H, Ono T, Zhang DY and Esashi M (2006) Fabrication and characterizations of a monolithic PZT microstage Microsyst Technol 12 :883-890
9. Sasaki M, Bono F and Hane K (2008) Large displacement micro-xy-stage with paired moving plates Japs J. of Applied Physics 47 :3226-3231
10. Choi JJ, Park H, Kim KY and Jeon JU (2001) Electromagnetic micro X-Y stage with very thick Cu coil for probe-based mass data storage device Proc. SPIE 4334 :363-371
11. Ma HW, Yao SM, Wang LQ and Zhong Z (2006) Analysis of the displacement amplification ratio of bridge-type flexure hinge Sensors and Actuators A 132 :730-736

Chapter 7
Precise Position Stages Using Pneumatically Driven Bellows Actuator and Cylinder Equipped with Air Bearings

Kenji KAWASHIMA[1] **and Toshinori FUJITA**[2]

Abstract In this chapter, two precise positioning stages using pneumatic actuators are introduced. One is for coarse movement and utilizes a pneumatic cylinder with air bearings. The other is for fine movement and utilizes an actuator driven by two pneumatic bellows. Novel servo valves, also developed by the authors, control the actuators. A high performance pneumatic servo valve having high dynamics up to 300 Hz is used for the coarse stage, while a slit type nozzle flapper servo valve that can minimize pressure fluctuations is used for the fine stage. As the proposed system is pneumatically driven, both stages have low heat generation and are non-magnetic, making them suitable for ultra precise positioning. Our investigations determined that, for the coarse stage, the maximum tracking error was less than 20 μm and the steady state error was just 0.4 μm. For the fine stage, the tracking error was less than 50 nm and steady state error was just 10 nm.

7.1 Introduction

Pneumatic servo systems are used in many fields including pneumatic robot systems, aspherical glass molding machines and vibration isolation systems. The use of air power has advantages such as compressibility, a high power ratio and low heat generation. Furthermore, air is nonmagnetic and can provide a clean energy source.

Pneumatic servo system control methods have been studied since the 1950s [1,2]. The performance of these systems improved significantly when pneumatic servo

[1] Kenji KAWASHIMA

Precision and Intelligence Laboratory, Tokyo Institute of Technology, Japan

[2] Toshinori FUJITA

Department of Mechanical Engineering, Tokyo Denki University, Japan valves became commercially available in the late 1980s [3]. Pneumatic servo valves are the key element of such systems. Because of this, the structure and parameters of such valves are important to good system performance [4,5,6,7,8]. Recently, with the development of precision manufacturing, the need for improved controllability of pneumatic servo systems has increased, highlighting the need for high-performance pneumatic servo valves. At present, one of the best pneumatic servo valves has a dynamic response of 100 Hz and a spool position accuracy of 2% to the full stroke. With the development of such pneumatic servo valves, precise position control of pneumatic servo systems has entered the practical stage.

Much research has gone into improving pneumatic positioning control [9,10], but the positioning accuracy of such systems does not exceed 0.01 mm, still inferior to electrical control. Because of this, highly precise pneumatic servo positioning systems often use piezoelectric actuators for fine strokes [11]. However, using electric actuators in hybrid systems nullifies some pneumatic system advantages, such as the lack of a magnetic field and reduced heat generation.

In this chapter, two proposed precise positioning stages that utilize pneumatic actuators are introduced. One is for coarse movement and uses a pneumatic cylinder with air bearings, while the other is for fine movement and uses an actuator driven by two pneumatics bellows. Novel servo valves developed by the authors control the actuators. Because the proposed system is pneumatically driven, these stages retain the advantage of low heat generation and are non magnetic, making them suitable for ultra precise positioning.

7.2 Coarse Movement Using Pneumatic Cylinder Equipped with Air Bearings

7.2.1 Pneumatic Cylinder Using Air Bearings

A photograph of the pneumatic cylinder table system and the experimental set up, including a schematic view of the pneumatic actuator, are shown in Fig.7.1 and Fig.7.2, respectively.

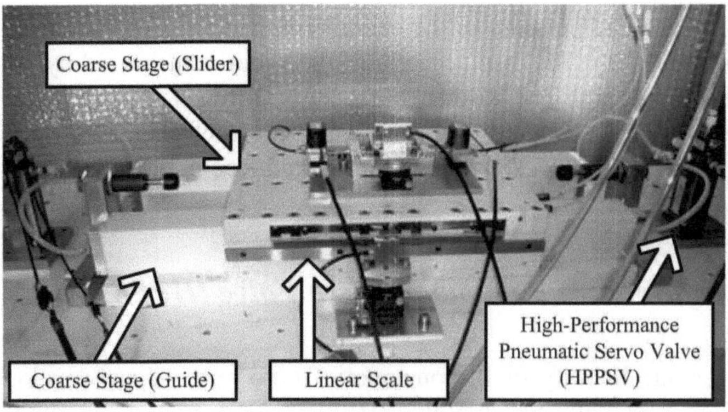

Fig. 7.1 Pneumatic servo table with air bearing

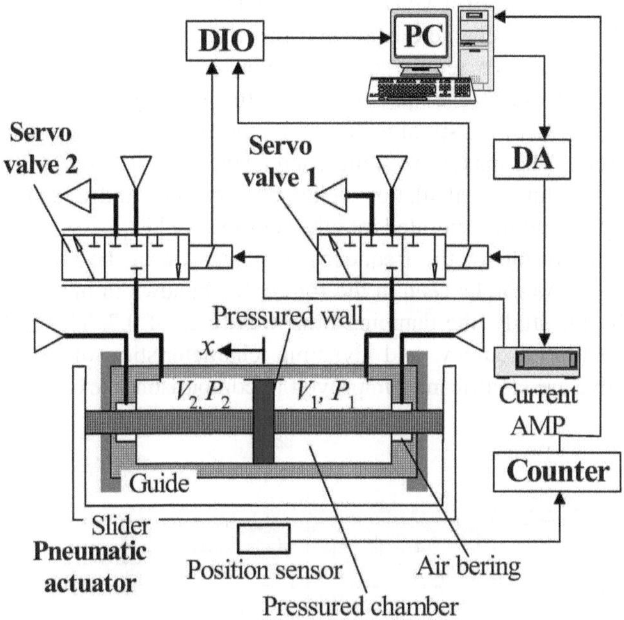

Fig. 7.2 Experimental setup of pneumatic servo table for coarse movement

The pneumatic actuator consists of a slider, which is a one degree of freedom (1-DOF) moving part, and a fixed guide. The slider is mounted with externally pressurized air bearings. The air bearings act through holes in the surface of the guide, thus preventing contact between the slider and the guide during movement. Use of this air bearing allows smooth acceleration and movement without stick-slip effects. The slider is driven by the pressure difference between both pressurized chambers.

The full stroke of the slider is 235mm. The inner area of the piston, the mass of the piston and the average pressure in the chamber are 12.2×10^{-4} m^2, 26 kg and 1.18×10^5 Pa, respectively. The displacement of the slider is measured by a linear scale of resolution 62.5pm. The displacement data are entered into a personal computer (PC) through a counter. The PC acts as a controller and sends the control signals to the servo valves, which will be introduced in the next section.

7.2.2 High-performance Pneumatic Servo Valve (HPPSV)

To improve the controllability of the pneumatic servo system, we developed a High-Performance Pneumatic Servo Valve (HPPSV) [10]. Figure 7.3 shows the schematic structure of this direct drive type three-port spool valve. The drive system of this valve consists of a voice coil motor (VCM) that has lightweight moving parts, high dynamic characteristics and provides good linearity between the coil current and thrust. To reduce spool friction, an air bearing is attached between the spool and the sleeve that maintains the spool under non-contact conditions. The full stroke of the spool is ±0.75mm and spool position accuracy is 0.1% to the full stroke. The maximum sonic conductance is 9.9×10^{-9} m^{4}s/kg. A digital controller, including a calculation delay compensator and a disturbance observer, was designed and incorporated into the servo valve [12].

Figure 7.4 shows the results of the frequency response test of the HPPSV with 10% of the full-stroke. As can be seen in the figure, the bandwidth of the HPPSV is up to 300 Hz. In this study, the damping ratio was fixed at 0.7. This indicates that the spool position accuracy and dynamic characteristics of the newly developed servo valve are significantly improved in comparison to existing valves.

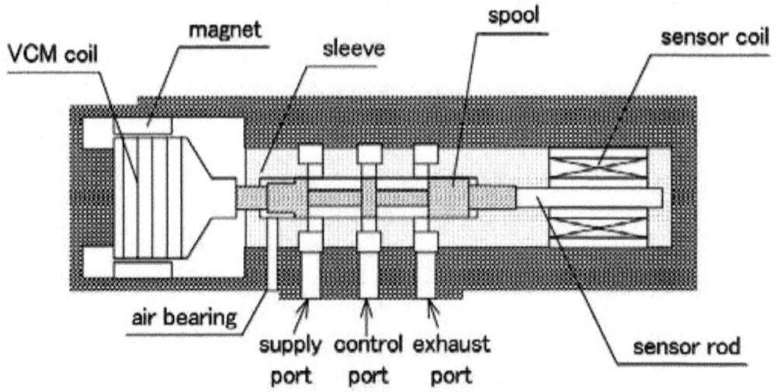

Fig. 7.3 Schematic diagram of high-performance pneumatic servo valve with digital control

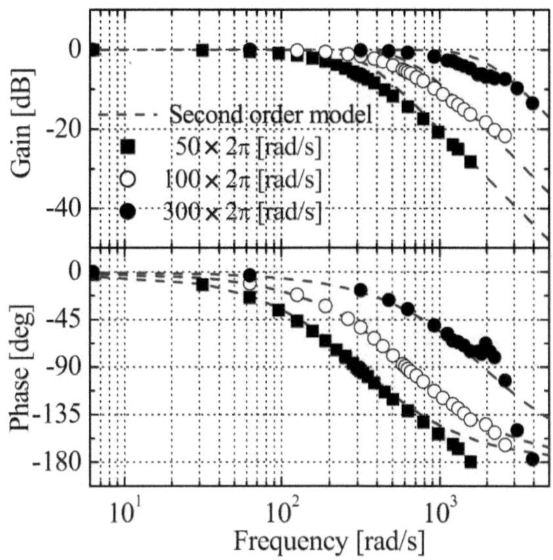

Fig. 7.4 Frequency response of the servo valve

7.3 Fine Movement with Bellows Actuator

7.3.1 Fine Stage

Figure 7.5 shows a photograph of the newly developed precision positioning stage driven by two pneumatic bellows [13]. The base, elastic hinges, and the mobile plate of the stage are all-in-one part that is cut from a steel plate. The size of the moving plate is 100 × 50 × 24 mm. The elastic hinges guide the mobile plate at four points. The two pneumatic bellows are attached between the base and the flexible region, facing each other. The pressure differences between the two bellows cause the mobile plate to move. The maximum displacement of the plate is 263 μm when the differential pressure between the bellows is 300 kPa. The natural frequency of the mass-spring system, composed by the plate and hinges, is approximately 60 Hz.

Figure 7.6 shows a schematic diagram of the positioning control system for the stage drive. Two slit-type nozzle flapper servo valves, which will be introduced in the next section, are used to generate the differential pressure. A linear scale displacement sensor with a resolution of more than 62.5 pm is used for stage position detection, and the servo valves are connected to the appropriate bellows.

The supply pressure is set to 300 kPa to maintain a constant pressure. Measurements are fed to a PC via a 16-bit analog/digital (A/D) converter the control signal are sent from the PC to the servo valves via a digital/analog (D/A) converter. The sampling time of the controller is 0.2 ms.

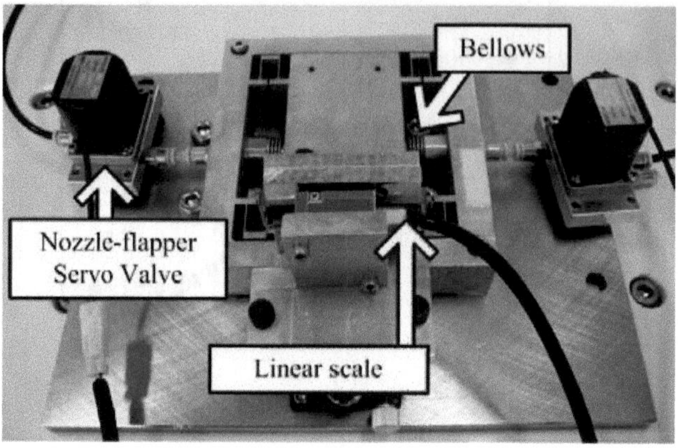

Fig. 7.5 Developed fine stage driven by pneumatic bellows actuators

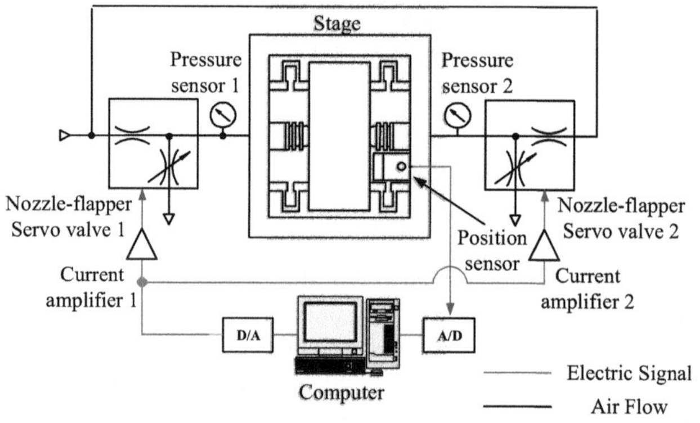

Fig. 7.6 Schematic diagram of the stage control system

7.3.2 Nozzle-flapper Type Servo Valve Using Slit Structure

When pressurized air passes through servo valves, noise and pressure fluctuations that adversely affect the precise position control of the system are often experienced at the downstream side.

In response to this, we developed a novel nozzle-flapper type servo valve that utilizes a slit structure instead of an orifice [14]. The slit structure shown in

Fig.7.7 maintains a laminar flow condition that that minimizes noise and pressure fluctuations.

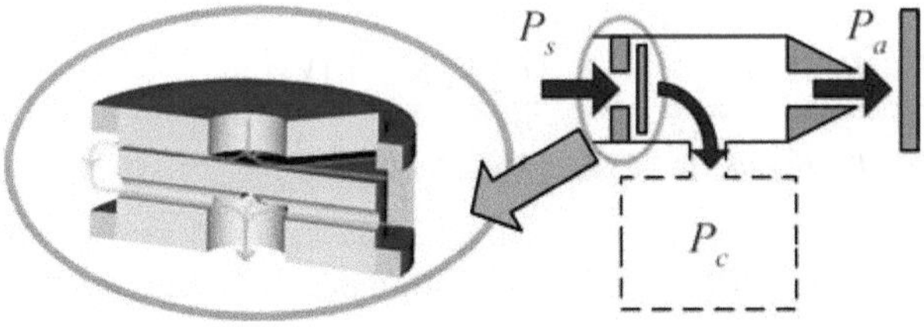

Fig. 7.7 Developed nozzle flapper type servo valve using slit structure

The slit structure is fabricated using etching technology. We investigated the flow characteristics of the slit theoretically and experimentally to evaluate the design specifications and characteristics of the valve.

Our experimental results indicated that the noise level decreased by approximately 15 dB and pressure fluctuations could be reduced by 75% in comparison with an ordinal novel nozzle-flapper type servo valve that uses an orifice plate. This indicates that the newly developed valve is more effective than many existing valves.

7.4 Control Method

7.4.1 Coarse Movement

Since the dynamics of the servo valve is much higher than that of the table system, the servo table system is given as a 3rd order system, as shown in Fig.7.8. Therefore a PDD^2 controller is used to control the table system. The position, velocity and acceleration are used to calculate controller input u_c:

$$u_c = K_{cp}\left(x - x_{ref}\right) - K_{cv}\dot{x} - K_{ca}\ddot{x} \tag{7.1}$$

where, K_{cp} K_{cv} and K_{ca} are the proportional gain, velocity gain and acceleration gain respectively. The velocity and acceleration are estimated by an observer. The transfer function of the control system can be written as follows using a normalized Laplace operator s_0 and dimensionless parameters α and β,

$$G(s)=\frac{x(s)}{x_{ref}(s)}=\frac{1}{s_0{}^3+\alpha s_0{}^2+\beta s_0+1} \quad (7.2)$$

Here, $\alpha=\dfrac{K_{ca}K_nK_{sv}\omega_n^2}{\sqrt[3]{\left(K_{cp}K_nK_{sv}\omega_n^2\right)^2}}$, $\beta=\dfrac{\left(K_vK_nK_{sv}+1\right)\omega_n^2}{\sqrt[3]{\left(K_{cp}K_nK_{sv}\omega_n^2\right)^2}}$

K_n : gain of the stage m/(kg/s), K_{sv} : proportional gain of the servo valve kg/(s m), ω_n : natural frequency of the stage (rad/s).

The velocity gain K_{cv} and the acceleration gain K_{ca} can be expressed as functions of α , β and the proportional gain K_{cp}. The input signal to the servo valve u_c is the reference spool displacement of HPPSV. The output of HPPSV is the flow rate G. If K_{cp} =1.0 μm / m, α =2 and β =3, K_{cv} and K_{ca} were calculated as 5.54 × 10^{-2} s μm /m and 1.66 × 10^{-3} s^2μm /m, respectively.

Furthermore, feed-forward compensation is realized by mounting inverse model of the feedback system, as shown in Fig.7.8. The feed-forward compensation has an s^3 term in the numerator. That is, the position reference trajectory x_{ref} is required to have 3rd order differentiable curve. We selected minimum jerk trajectory as a reference input to the system [15].

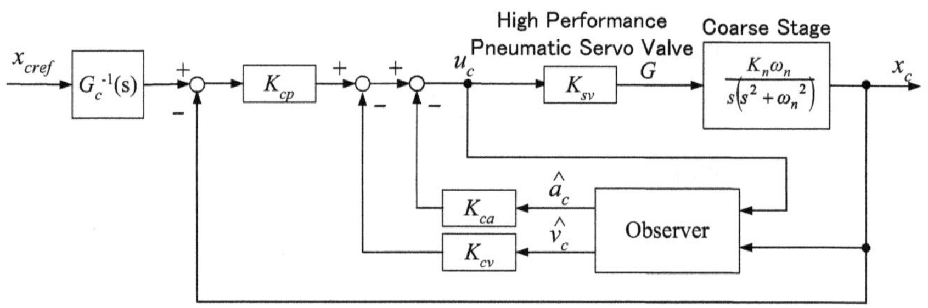

Fig. 7.8 Block diagram of the coarse stage control system

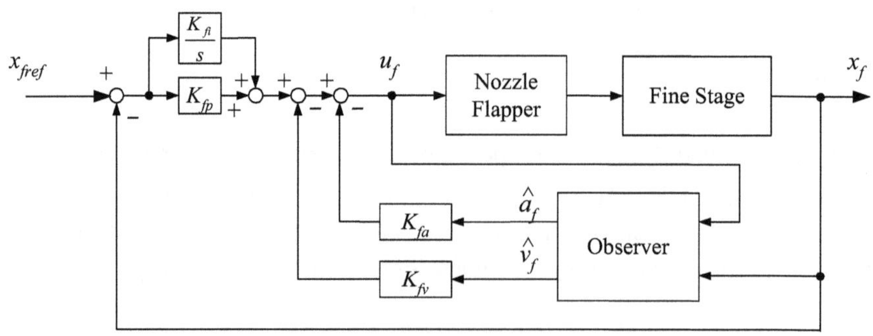

Fig. 7.9 Block diagram of the fine stage control system

7.4.2 Fine Movement

Figure 7.9 shows a block diagram of the positioning control system for the fine stage. Normally, the transfer function of the pneumatic servo system using a cylinder as an actuator is a third order time lag system that includes an integral element. This positioning system is a third order time lag system; however, it does not include an integral element. When proportional control is applied to such a system, an offset in displacement occurs. Hence, proportional-integral (PI) control is applied. Furthermore, the velocity and the acceleration feed-forward are added to improve performance. The velocity and the acceleration of this stage are also estimated by an observer. The gain values of controller are obtained by trial and error, and they are the proportional gain, K_{fp} : 3.0 × 10^3 V/m, the integration time, K_{fi} : 5.0 × 10^6 V/(m s), the velocity gain K_{fv} : 30 Vs/m, the acceleration gain K_{fa} :0.3 vs^2/m.

7.5 Experimental Results

7.5.1 Coarse Movement Using Pneumatic Cylinder Equipped with Air Bearing

Minimum jerk trajectory control was implemented with 10 kg payload at the 20 mm amplitude. To improve dynamic characteristics and position controllability, a larger proportional gain value is required. In an actual system, the proportional gain is limited by the nonlinearity of the system, the servo valve dynamics and the maximum sonic conductance of the valve.

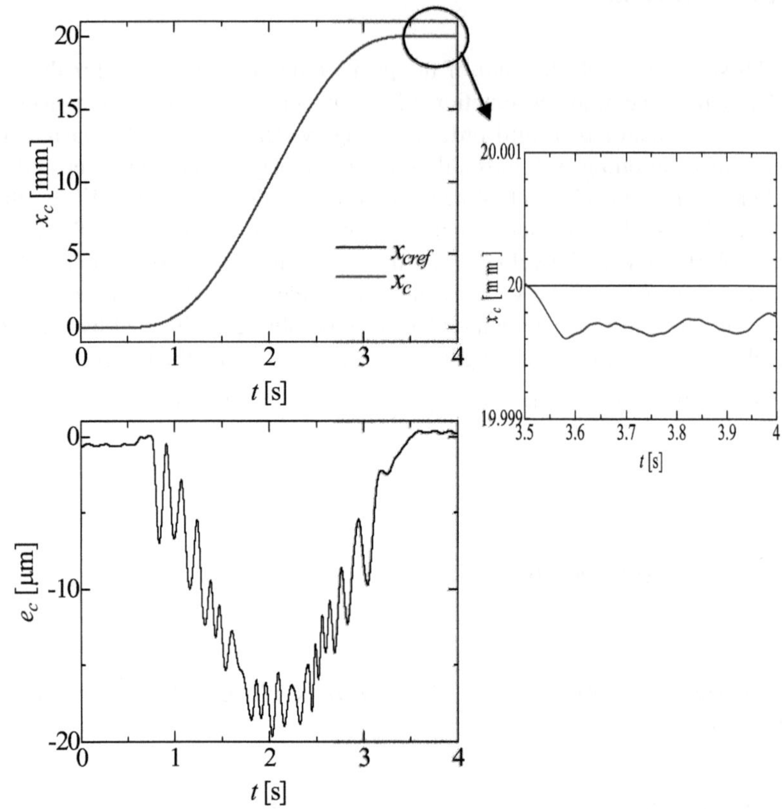

Fig. 7.10 Experimental results of trajectory tracking (coarse stage)

Figure 7.10 shows experimental results of trajectory tracking when the natural frequency of the servo valve was set at 250 Hz. The upper, lower and right figures show the position, trajectory errors and the enlarged steady state error after 3.5 sec, respectively. These show that the maximum tracking error is less than 20 μm and steady state error is just 0.4 μm.

7.5.2 Fine Movement with Bellows Actuator

The trajectory tracking of the fine stage with feedback control is shown in Fig.7.11. The movement width is 20 μm. The upper, lower and right figures show the position, trajectory errors and the enlarged steady state error after 3.5 sec, respectively.

These results indicate that the tracking error is less than 50 nm and steady state error is just 10 nm. When the stage is mounted on a pneumatic isolation table, 7 nm fluctuations can be observed. This indicates that the accuracy of the stage could be improved if it were used in a better working environment.

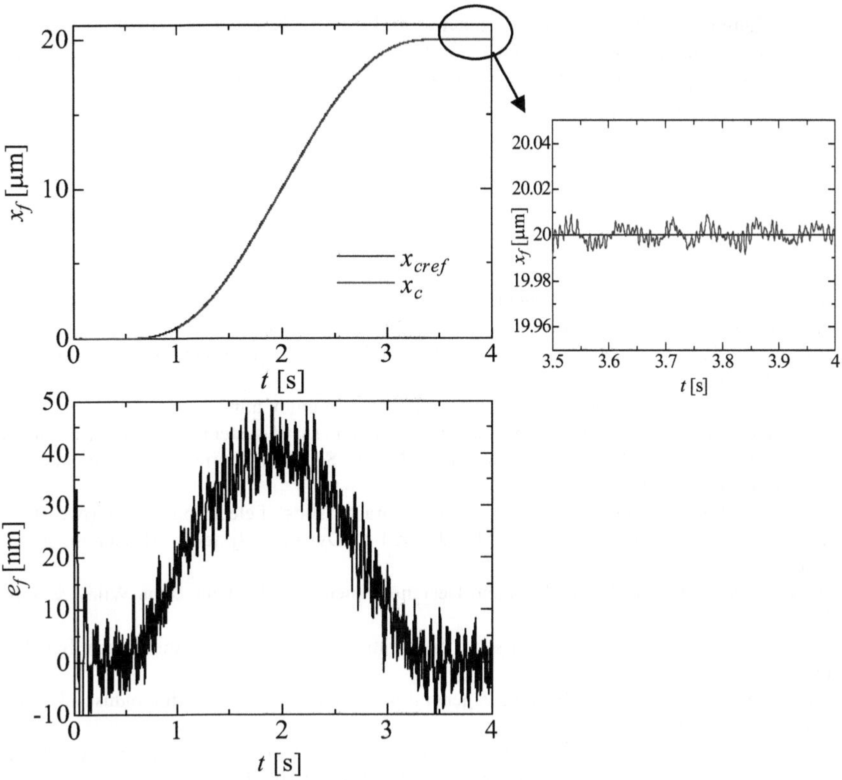

Fig. 7.11 Experimental results of trajectory tracking (fine stage)

7.6 Conclusion

In this chapter, two newly developed, pneumatically driven, precise positioning stages were introduced. One stage is for coarse movement and uses a pneumatic cylinder with air bearings while the other is for fine movement and uses an actuator driven by two pneumatic bellows. Our novel servo valves control the actuators. A high performance pneumatic servo valve having high dynamics up to 300 Hz is used for the coarse stage. A slit type nozzle flapper servo valve that can minimize pressure fluctuation is used for the fine stage. Our investigations clarified that the maximum tracking error was less than 20 μm and the steady state error was just 0.4 μm for the coarse stage. For the fine stage, the tracking error was less than 50 nm and steady state error was just 10 nm.

Our future work will be the development of a coarse/fine dual stage utilizing these two actuators.

Acknowledgments The authors are grateful to Prof. T. Kagawa of Tokyo Institute of Technology and Mr. K. Sasaki of Pneumatic Servo Controls LTD., for their advice and support.

This work is supported by a Grant in Aid from the Ministry of Education, Culture, Sports, Science and Technology (MEXT) of Japan.

References

1. Shearer LJ (1956) Study of pneumatic processes in the continuous control of motion with compressed air- I, II. Transactions of the ASME, Feb: 233–249
2. Burrows RC and Webb RC (1966) Use of the root loci in design of pneumatic servo-motors. Control: 423–427
3. Festo. http://www.festo.com/INetDomino/coorp_sites/en/index.htm
4. Liu S and Bobrow EJ (1988) An analysis of a pneumatic servo system and its application to a computer controlled robot. Transactions of the ASME, Series G, Journal of dynamic systems, measurement and control 110: 228–235
5. Krivts LI (2004) Optimization of performance characteristics of electropneumatic (two-stage) servo valve. Transactions of the ASME, Journal of Dynamic Systems, Measurement, and Control 126: 416–420
6. Andersen WB (1967) The Analysis and Design of Pneumatic Systems. John Wiley & Sons, Inc.
7. Backe W (1986) The application of servo pneumatic drives for flexible mechanical handling techniques. Robotics 2 (1): 45–56
8. Pu J and Weston HR (1988) A new generation of pneumatic servo for industrial robot, Robotica 7: 17–23
9. Wang J, Pu J and Moore P (1999) Accurate position control of servo pneumatic actuator systems: An application to food packaging, Control Engineering Practice, 7: 699–706
10. Hamiti K, Voda-Besan A and Roux-Buisson H (1996) Position control of a pneumatic actuator under the influence of stiction, Control Engineering Practice, 4 (8): 1079–1088
11. Chiang M, Chen C and Tsou T (2005) Large stroke and high precision pneumatic-piezoelectric hybrid positioning control using adaptive discrete variable structure control, Mechatronics 15: 523-545
12. Miyajima T, Fujita T, Sakaki K, Kawashima K and Kagawa T (2007) Development of a Digital Control System for High-performance Pneumatic Servo Valve, Precision Engineering 31: 156-161
13. Fujita T, Sakaki K, Kawashima K and Kagawa T (2007) Ultra Precise Positioning Control of a Stage Driven by Pneumatic Bellows, FLUCOME, CD-ROM
14. Kawashima K, Youn C and Kagawa T (2007) Development of a Nozzle-flapper Type Servo Valve using a Slit Structure, Trans. ASME Journal of Fluid Engineering, 129 (5): 573-578
15. Fujita T, Kawashima K, Miyajima T, Ogiso T and Kagawa T (2008) Effect of Servo Valve Dynamic on Precise Position Control of a Pneumatic Servo Table, Int. J. of Automation Technology, 2 (1): 43-48

Chapter 8
Development of a New Nano-Micro Solid Processing Technology Based on a LIGA Process and Next-Generation Microactuators

Daiji NODA[1] and Tadashi HATTORI[1]

Abstract The demand of micro-fabrication such as microactuators, microcoils, smart sensors is continually increasing. Actuators can occupy a large part of the volume and the weight of an overall system, and therefore required to be reduced in size. However, there has been little progress in fabricating microactuators using existing technologies. Micro-fabrication processing and new technologies are needed in order to form three-dimensional electromagnetic type microactuators. The LIGA process could be used to fabricate nano- and micro-scale parts for many applications. Consequently, we fabricated spiral microcoils with a narrow pitch and high aspect ratio coil lines for an electromagnetic type microactuator using the LIGA process. We have fabricated coil lines with a width of 10 μm and an aspect ratio of 5. We have also estimated the suction force of actuators using these microcoils. It is very expected that these high aspect ratio microcoils would be capable of delivering high performance in spite of their miniature size.

8.1 Introduction

Actuators are finding increasing use in a variety of fields and many applications. Therefore, they are one of the most important components in various machines because the operation of the machine depends on their performance. Recently, actuators can constitute a large part of the weight of a system, and although demands have been made for reductions in size and greater sophistication, very little progress have been fabricate so far. However, the miniaturization of actuators has made little progress since it requires micro-fabrication, micro-processing, and other new technologies that are not compatible with traditional machining techniques.

[1] Daiji NODA and Tadashi HATTORI

Laboratory of Advanced Science and Technology for Industry, University of Hyogo

Typical driving power sources for actuators are electrostatic, piezoelectric, electromagnetic, shape memory alloy (SMA), etc. Among these actuators, we are focusing on the electromagnetic type actuators driven at a low voltage, with high power, high efficiency, and low cost. However, the current carrying capacity of miniature coils is small when current paths of coil lines are microscopic, making it difficult to obtain sufficient output power. In addition, it is also very difficult to fabricating process microscopic current paths by means of conventional machining techniques.

On the other hand, the LIGA (German acronym for Lithographite, Galvanoformung, and Abformung) process [1] could be used to fabricate nano- and micro-scale parts and devices. The LIGA is a total process for fabricating the master mold for micro-structured parts using X-ray lithography, electroforming a micro pattern mold, and molding plastic micro-structure parts [2,3]. For X-ray lithography, the NewSUBARU synchrotron radiation facility at our university [4] was used. This was operated at an energy of 1.0 or 1.5 GeV modes. The X-ray exposure at BL11 of NewSUBARU was carried out with the workpiece held in a specially manufactured nine parts operation exposure stage [5]. Thus, this X-ray exposure stage makes it feasible to form three-dimensional (3D) structure such as spiral coil patterns [6-8]. With this technique, it was possible to fabricate high aspect ratio coil line structures.

8.2 Design and Simulation of Electromagnetic Actuator

An electromagnetic type actuator including a magnetic circuit was designed with the aid of calculated by simulation. The simulation was carried out varying the aspect ratio of the coil lines.

8.2.1 Design of Electromagnetic Actuator

For the design of the magnetic circuit, we used the type known as an "*open frame solenoid*", which is open at the sides as shown in Fig.8.1 [6,7]. For the material of the magnetic core (fixed core and plunger) and the shield parts (yoke) we used the nickel iron alloy Permalloy 45, because it has the largest permeability of the soft magnetic metals. Therefore, it can generate a strong magnetic field with a very small electric current. When a voltage is applied to the coil, a magnetic flux is formed in a gap, which deforms the magnetic field and produces a suction force on the plunger.

An acrylic pipe with an outside diameter of 5 mm and an inside diameter of 3 mm was used as the base material for coil lines fabrication. The pipe material is PMMA (polymethylmethacrylate) which has the properties of a positive type photoresist. Therefore, it could be directly exposed to X-ray lithography to form high aspect ratio structures on the acrylic pipe surface.

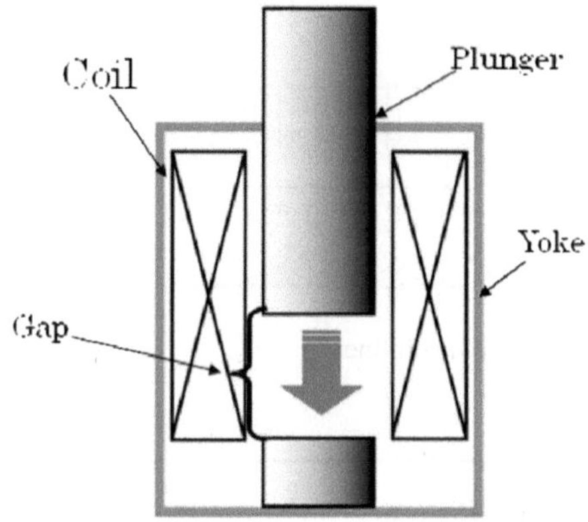

Fig. 8.1 Designed model of actuator operation with magnetic circuit

8.2.2 Simulation of the Suction Force of the Electromagnetic Actuator

We proposed a spiral microcoil with high aspect ratio coil lines. Figure 8.2 shows images of the coil lines. Conventional wire type coils are limited to coated copper wire of a few ten of micrometers and aspect ratio of 1. However, in this research, high aspect ratio type was fabricated using the X-ray lithography technique. In this model, a magnetomotive force is proportional to squares of current paths. If the aspect ratio of coil is increased, the cross sectional area of coil lines is also increased allowing a greater current flow. Figure 8.3 shows the calculated results of the suction force and permitted currents in coils with different aspect ratios. Here, we used coil parameters as the coil line width of 10 μm and the number of coil turns of 675. The gap between the plunger and the fixed core was 1 mm. When the aspect ratio is 5, the suction force may be about 25 times greater than for a coil with an aspect ratio of 1.

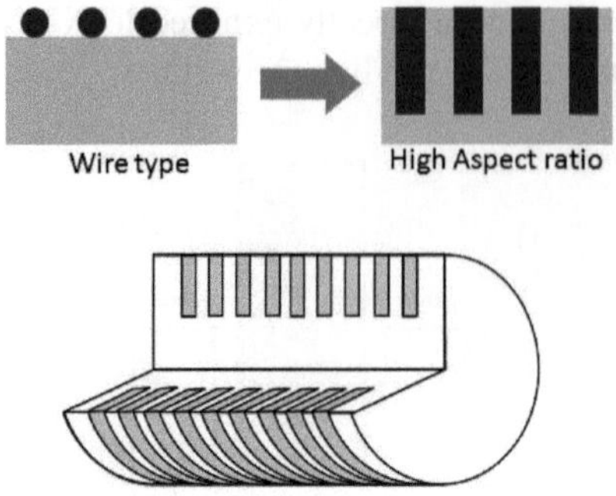

Fig. 8.2 Image of high aspect ratio coil lines

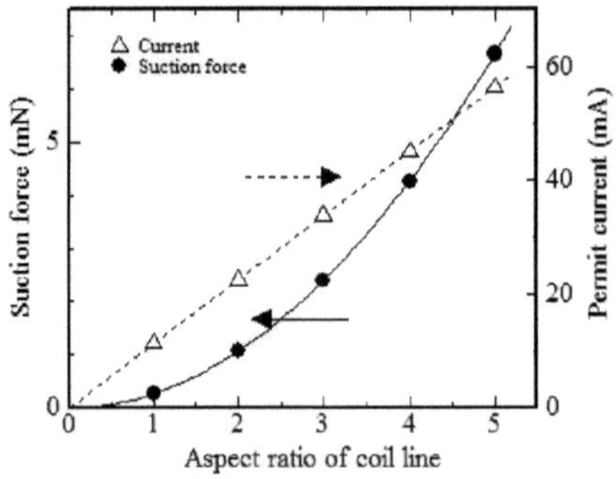

Fig. 8.3 Calculation of suction force and permit current in different aspect ratio

8.3 Fabrication Process for Coil Lines

A spiral microcoil was formed on the surface of the acrylic pipe using X-ray lithography and metallization techniques. Fabrication process for coil lines is shown in Fig.8.4. First, a thread structure was formed on the pipe surface using X-ray lithography. Next, a thin seed layer of copper to be used as an electrode in electroforming was deposited on the pipe by spattering. The pipe was then immersed into a copper plating bath for electroforming and electroforming carried out until the spiral groove was filled with copper film. Finally, the plated copper was chemically etched to remove copper from the surface, but leave copper remaining in the spiral groove thus forming a coil. The following sections give detailed descriptions of each of the process steps.

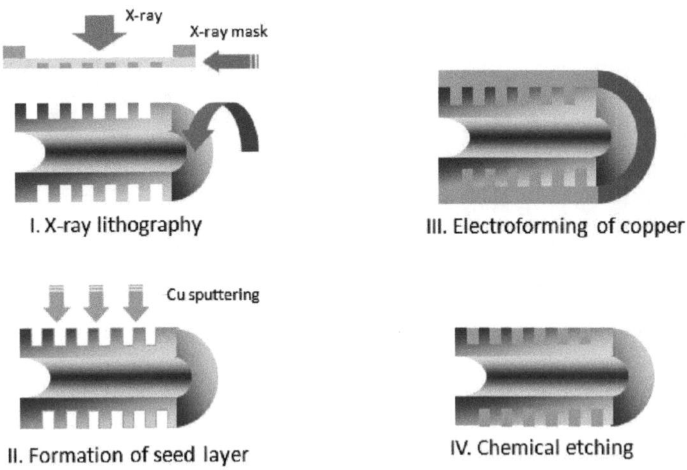

Fig. 8.4 Fabrication process for coil lines by X-ray lithography

8.3.1 X-Ray Lithography

In this experiment, we used an X-ray mask with feature widths of 10 μm and 30 μm. Therefore, screw thread structures for 10 μm and 30 μm line width were formed. To make a 3D micro coil line structures, the acrylic pipe was rotated using a stepping motor and movement of the X-ray mask was controlled by piezoelectric elements. To expose on the pipe surface, X-ray exposure strategy was implemental, in which the process was divided into 60 steps that is close to continuity exposure. Thus, the pipe was rotated through an angle of 6 degrees while the X-ray mask was advanced by just 1/60 of the pitch for each X-ray exposure cycle [7].

After X-ray exposure, the PMMA was developed in GG developer (diethyleneglycolmonobutyether: 60 vol.%, morpholine: 20 vol.%, ethanolamine: 5 vol.%, distilled water: 15 vol.%) at room temperature to form the screw thread structures on the pipe surface. The spiral structure of coil lines was observed using a scanning electron microscope (SEM). Figure 8.5 shows the spiral coil lines. In the case of Fig.8.5a, the aspect ratio of coil lines was about 5 with a width of 10 μm. In the case of 30 μm lines and spaces pattern, an aspect ratio of 2 was obtained, as shown in Fig.8.5b. From these figures, we were able to confirm that the joints between each section of the groove pattern were perfectly aligned. The processing depth, which determined the aspect ratio of the coil lines, was controlled by the X-ray exposure dose and the development time as shown in Fig.8.6.

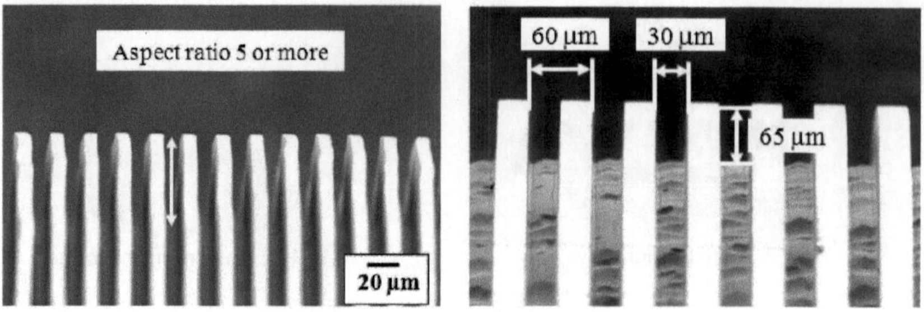

(a) 10 μm lines and spaces pattern (b) 30μm lines and spaces pattern

Fig. 8.5 SEM images of coil lines with high aspect ratio structure

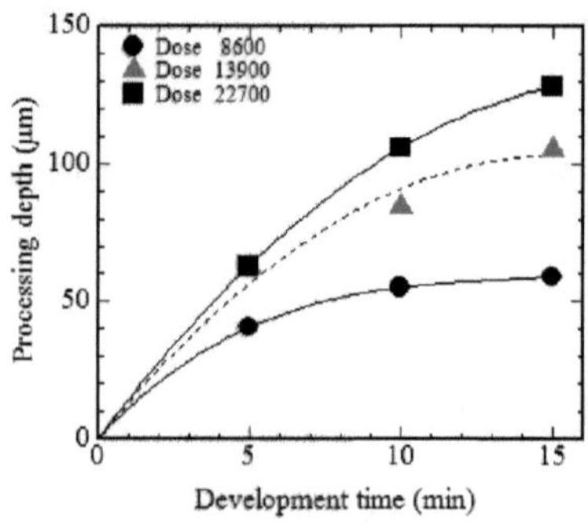

Fig. 8.6 Relationship between processing depth and development time

8.3.2 Formation of Seed Layer

Since the acrylic pipe is nonconductive, a conductive seed layer is required for electroforming. A 300 nm thick seed layer was formed on the surface of the pipe by sputtering. In order to obtain a low resistance seed layer, we considered that the pipe was moved along its axis and sputtering carried out with the pipe in three different positions. As a result, the resistance around the circumference of the pipe was sufficiently low.

8.3.3 Electroforming of Copper

Following deposition of the seed layer, the acrylic pipe was immersed in an electroplating bath of copper sulfate solution, which included a leveling agent to promote uniform growth by reducing the electric field strength at the edges of coil line structures. Therefore, the thickness of copper at the flute was thicker than on the convex lines. Figure 8.7 shows a SEM image of a coil lines after copper electroforming. This figure shows that the copper layer was grown up from the bottom of grooves, completely filling the high aspect ratio structures.

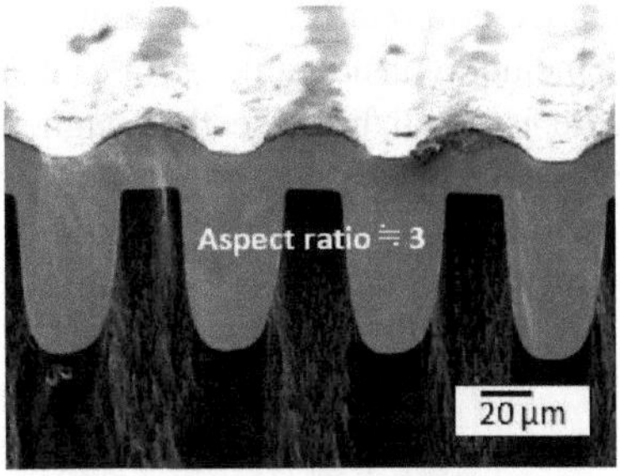

Fig. 8.7 SEM image of coil lines after electroforming

8.3.4 Isotropic Chemical Etching

Isotropic chemical etching of copper using E-process-W etchant was performed until only the copper in the grooves remained, thus forming the coil lines. The pipe rotation mechanism was also used to rotate the acrylic pipe in the etchant to ensure uniform pipe surface etching. From this result, we produced coil lines by copper etching until the protrusions of groove structures were exposed, as shown in Fig.8.8.

Fig. 8.8 SEM image of coil lines after isotropic etching

8.4 Measurements of Suction Force

We also built a measurement system, as shown in Fig.8.1, in order to measure the suction force of the designed electromagnetic type actuator. This system is a very simple structure and it is easy to change the coil [7], as shown in Fig.8.9. The gap between the plunger and the fixed core was adjusted by an XY stage. Figure 8.10 shows a comparison of the theoretical values by simulation and actual measurement of the suction force generated by a coil with 30 μm width and an aspect ratio of 2. The measured results were in relatively good agreement with the theoretical values. Here, the results include considerable errors where the gap between the plunger and the fixed core is small because the magnetic flux assumed in the simulation might be much different from the actual flux. Currently, we have been carrying out measurements of the suction force by fabricating spiral microcoils with higher aspect ratio structure produced by X-ray lithography and metallization techniques.

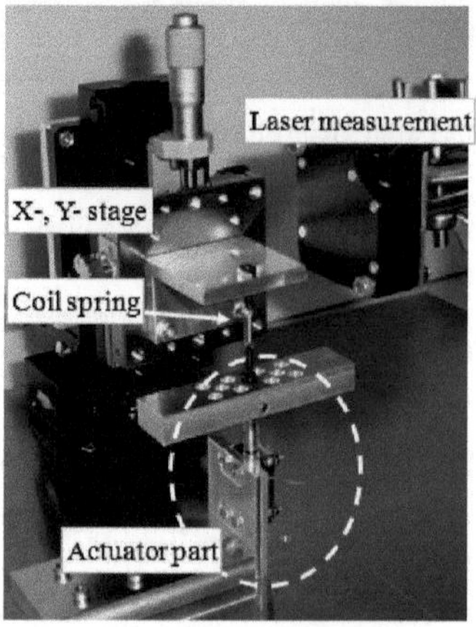

Fig. 8.9 Measurement system for suction force of actuator

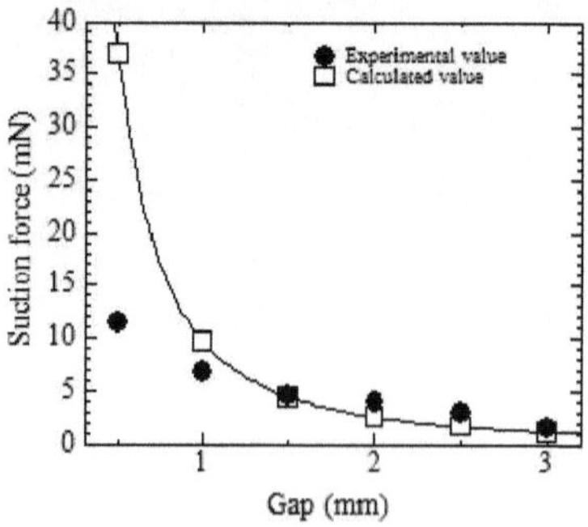

Fig. 8.10 Suction force comparison between measurement and simulation values

8.5 Development of 1 mm Diameter Microcoil

The outside diameter of the acrylic pipe used was 5 mm. Therefore, the size of the coil is too big for a microcoil and microactuator. So, we used metal master bars with diameters of 0.5 to 1 mm for microcoil fabrication. PMMA was applied onto the master bar using a dipping method [9,10]. The thickness of PMMA determines the structure of the coil line depth. Thus, this is a very important factor in microcoil fabrication.

8.5.1 Dipping Method

The dipping method was used in order to obtain a thick layer of photoresist on the metal bar. Figure 8.11 shows the fabrication process for metal bar and dipping process. The fabrication process is largely identical to that used for the acrylic pipe, expect the final etching step. The dipping method comprises four steps: dipping, recovery, air drying, and baking. A highly viscous photoresist solution and control over the centrifugal force were important factors to obtain a thick uniform coating, and thus enable the production of high aspect ratio coil lines.

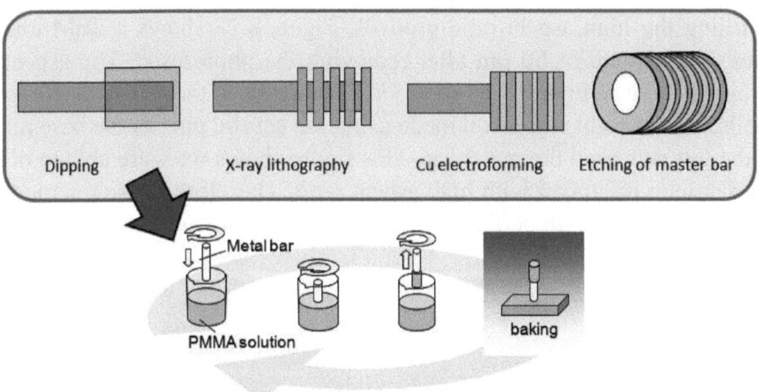

Fig. 8.11 Fabrication process and dipping method

8.5.2 Results and Discussions

We were able to control the thickness of PMMA on metal bar by the speed of rotation and concentration of PMMA [9]. In these results, PMMA thickness of more than 100 μm was obtained on metal bar in single coating. Thus, the aspect ratio of coil lines achieved for 30 μm width grooves was greater than 3.

A spiral coil structure was formed in the PMMA on the metal bar using X-ray lithography technique. In this case, we used an X-ray mask with 30 μm lines and spaces patterns. The diameter of the metal bar was 0.5 mm. Figure 8.12 shows a SEM image of coil line structures with a pitch of 60 μm. This figure shows that the aspect ratio realized was about 6 because the grooves were narrower than the designed width of the coil.

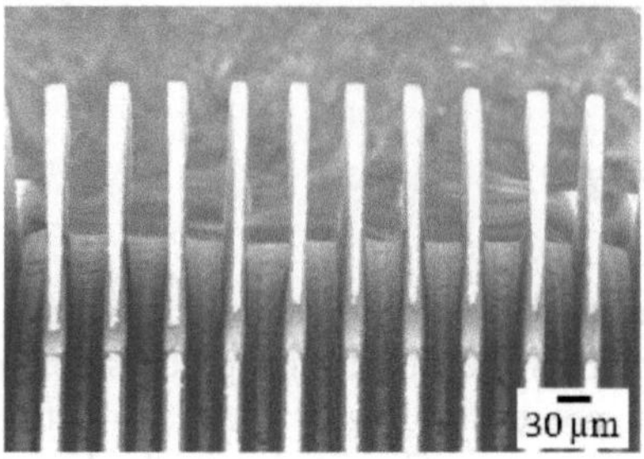

Fig. 8.12 SEM image of coil lines

Next, we performed a metallization process, including electroforming and photoresist etching. In this case, the metal bar acts as the seed layer for electroforming. Therefore, electroforming layer was grown up from the bottom completely filling the high aspect ratio grooves. Figure 8.13 shows a SEM image of coil lines with a pitch of 60 μm after removing the photoresist. The aspect ratio was obtained about 2. Figure 8.14 shows a comparison of the size of the fabricated microcoils. On the right was a coil made using the acrylic pipe as the base material and on the left was used the metal bar. This figure shows we were able to obtain a 0.5 mm diameter microcoil with high aspect ratio. Therefore, these microcoils are very expected to have high performance despite their miniature size.

Fig. 8.13 SEM image of coil lines after resist etching

Fig. 8.14 Fabricated microcoils comparison metal bar with acrylic pipe

8.6 Conclusions

We have fabricated narrow pitch and high aspect ratio spiral microcoils for an electromagnetic type actuator using 3D deep X-ray lithography technique and metallization process. Using these techniques, we succeeded in producing a grooved structure with 10 μm in coil line widths with a maximum aspect ratio of about 5. We also succeeded in electroforming copper in the high aspect ratio structure and forming a coil line by isotropic copper etching. Therefore, we could obtain microcoils with high aspect ratio coil lines structures.

In addition, we developed a measurement system to measure the suction force produced by these electromagnetic type actuators. The results of suction force measurements enabled us to confirm the results of simulation. These measurement results were in relatively good agreement with the simulated ones.

We also attempted to fabricate microcoils with diameters of less than 1 millimeter. Using a dipping method, photoresist thickness of over 100 μm were achieved using a highly viscous solution and controlling the centrifugal force. We succeeded in producing a spiral microcoil with 30 μm coil lines width with an aspect ratio of about 2 using X-ray lithography and metallization techniques.

Using these techniques, we were able to fabricate microcoils with high aspect ratio coil lines. Thus, it is very expected that electromagnetic type microactuators with high suction force could be manufactured despite their miniature size.

Acknowledgments This research was partially supported by the Grant-in-Aid for Scientific Research on Priority Area, No. 438, "Next-Generation Actuators Leading Breakthroughs", from the Ministry of Education, Culture, Sports, Science and Technology, Japan

References

1. Becker EW, Ehrfeld W, Hagmann P et al (1986) Fabrication of microstructures with high aspect ratios and great structural heights by synchrotron radiation lithography, galvanoforming, and plastic moulding. Microelectron Eng 4:35-56
2. Mekaru H, Kusumi S, Sato N et al (2005) Development of three dimensional LIGA process to fabricate spiral microcoil. Jpn J Appl Phys 44:5749-5754
3. Mekaru H, Kusumi S, Sato N et al (2007) Fabrication of a spiral microcoil using a 3D-LIGA process. Microsyst Technol 13:393-402
4. Ando A, Amano S, Hashimoto S et al (1998) Isochronous storage ring of the New SUBARU project. J Synchrotron Rad 5:342-344
5. Mekaru H, Utsumi U, Hattori T (2001) Beam line BL11 for LIGA process at the NewSUBARU. Nucl Instrum Methods A 467-468:741-744
6. Mochizuki H, Mekaru H, Kusumi S et al (2007) Design of solenoidal electromagnetic microactuator utilizing 3D X-ray lithography and metallization. Microsyst Technol 13:547-550
7. Matsumoto Y, Setomoto M, Noda D, Hattori T (2008) Cylindrical coils created with 3D X-ray lithography and metallization for electromagnetic actuators. Microsyst Technol 14:1373-1379
8. Setomoto M, Matsumoto Y, Yamashita S et al (2008) Fabrication of Spiral Micro Coil Lines for Electromagnetic Actuators. J Adv Mech Des Syst Manuf 2:238-245
9. Noda D, Yamashita S, Matsumoto Y et al (2008) Fabrication of High Aspect Ratio Microcoil Using Dipping Method. J Adv Mech Des Syst Manuf 2:174-179
10. Noda D, Matsumoto Y, Setomoto M, Hattori T (2008) Fabrication of Microcoils Using X-ray Lithography and Metallization. IEEJ Trans SM 128:181-185

Chapter 9
New Microactuators Using Functional Fluids

Shinichi Yokota[1], Kazuhiro Yoshida[1], Kenjiro Takemura[2] and Joon-wan Kim[1]

Abstract An electro-conjugate fluid (ECF) generates jet flow in inhomogeneous electric field. Electro-rheological fluid (ERF) and magneto-rheological fluid (MRF) change their apparent viscosities in electric or magnetic field. Such functional fluids are expected to realize new promising actuators with simple and miniaturizable structure without sliding part. The paper describes objective and outline of our research on new microactuators using functional fluids. Then, as some results of our research, ECF micromotors, ECF gyroscopes, soft actuators using pressure due to ECF jet, forced liquid cooling systems using ECF jet, micro-actuators using ERF/MRF, and high output power piezoelectric micropumps are described briefly.

9.1 Introduction

Functional fluids are liquids that change their physical properties under external stimulus. As an example of functional fluids, an electro-conjugate fluid (ECF) generates jet flow called ECF jet in high inhomogeneous electric field. Electro-rheological fluid (ERF) / magneto-rheological fluid (MRF) changes its apparent viscosity in electric / magnetic field. In application of such functional fluids, high density fluid power can be controlled with simple and miniaturizable structure without sliding part.

MEMS (Micro Electro Mechanical Systems), micromachine and microactuator technologies have realized many new micro devices to handle information signals with many efforts. However, researches on practical millimeter-sized microactuators with high power densities have still been insufficient. New microactuators using functional fluids are thought to be promising candidates for such applications.

[1] Shinichi Yokota, Kazuhiro Yoshida and Joon-wan Kim

Precision and Intelligence Laboratory, Tokyo Institute of Technology

[2] Kenjiro Takemura

Faculty of Science and Technology, Keio University

Our research is to realize new microactuators using functional fluids. In this paper, objective, outline and some results of our research are described briefly.

9.2 Objective and Research Outline

We have been researching on actuators using functional fluids, e. g., ECF micromotors.

In a cylindrical chamber (stator) that has some pairs of rod type electrodes on the cylindrical inside surface and is filled with ECF, rotational flow is generated due to the ECF jets [1]. A rotor with blades rotates in the stator. One of the authors has proposed this type of SE (Stator Electrodes) type ECF micromotor and another type of RE (Rotor Electrodes) type that has a rotor with some pairs of rod type electrodes and rotates due to the reaction forces of the ECF jets. The ECF micromotors are also classified by the shape of the rotor: C type has a cylinder rotor and realizes simple and miniaturizable structure; DP type has a disk plate rotor and realizes very thin structure and high density structure by stacking.

Several ECF micromotors were fabricated in sizes of 20 - 1 mm inner diameter and the characteristics were experimentally investigated. As a result, the followings have been confirmed: 1) Small inner diameter increases output power density; 2) in DP type micromotors, small inner height increases the output power density.

The objective of the research is to realize new practical microactuators by using functional fluids such as ECF, ERF, and MRF. The research executed the following studies:

1. Development and improvements of ECF micromotors

Higher power density ECF micromotors were developed through investigations on dense stacking of DP rotors, optimization of the structure including electrodes, selection of ECF, and so on. Also, new ECF micromotors were proposed and developed for sensors and so on.

2. Development of soft actuators using pressure due to ECF jet

By using the flexibility of the fluid actuators and the compactness of the ECF jet generators, soft actuators using pressure due to ECF jets were proposed and developed without bulky tanks.

3. Development of forced liquid cooling system using ECF jet

In application of the ECF jet, a simple and compact pump using ECF jet were proposed and developed for forced liquid cooling system for electronic chips in advanced notebook computers and so on.

4. Development of microactuators using ERF/MRF

To improve performance of an ER microactuator that is a fluid microactuator with ER microvalves, the mathematical model of the ER valve, flexible ER valve, and so on were proposed and developed. Also, microvalves using MRF were proposed and developed.

5. Development of high output power piezoelectric micropumps

As a micro fluid power source, high output power piezoelectric micropumps using fluid inertia effect in a pipe and the one using resonance drive were proposed and developed.

6. Applications of the developed actuators

9.3 Some Research Results

9.3.1 ECF Micromotors [2]-[8]

To realize high output power DP-RE type ECF micromotors (Fig.9.1), the several electrode configurations are investigated as in Fig.9.2. The experimental results implies the ECF micromotor possibly generates 130 W/kg or higher in millimeter scale. In addition, we are also developing micrometer scale motors using MEMS technologies.

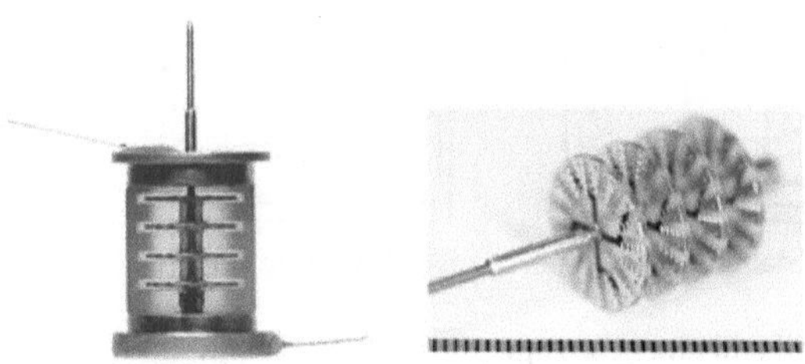

(a) Inner structure (b) Rotor

Fig. 9.1 Fabricated four-layer DP-RE type ECF micromotor

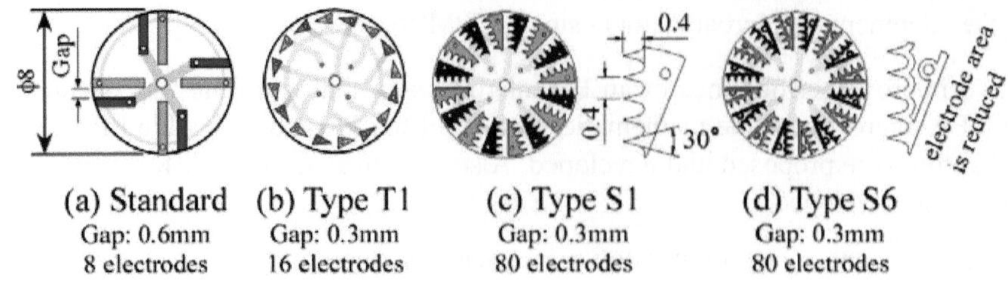

Fig. 9.2 Electrode patterns of rotor plates

9.3.2 ECF Gyroscopes [9]-[11]

We are developing a novel gyroscopes using ECF. This is called an ECF liquid rate gyroscope shown in Fig.9.3. This gyroscope measures a drift flow of the ECF jet, which is occurred due to Coriolis force. Hence there need no mechanically moving parts inside. This type may have advantages like long life time, high resistance to vibration/impact, resulting in making this gyroscope to be a candidate for next generation micro rate gyroscope, which may be used for car navigation systems etc. Figure 9.4 shows an inner structure of prototype with very thin structure. The prototype with the best configuration could detect the angular rate with input voltage of 640 V. If the angular rate is changed with 1 Hz, the prototype could follow it by increasing the input voltage to 1 kV.

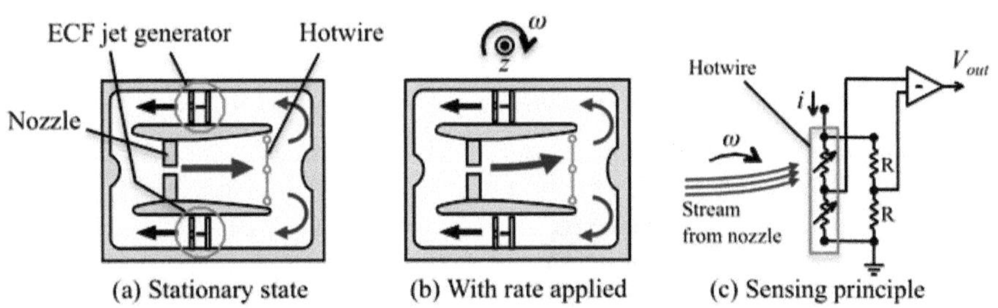

Fig. 9.3 ECF liquid rate gyroscope

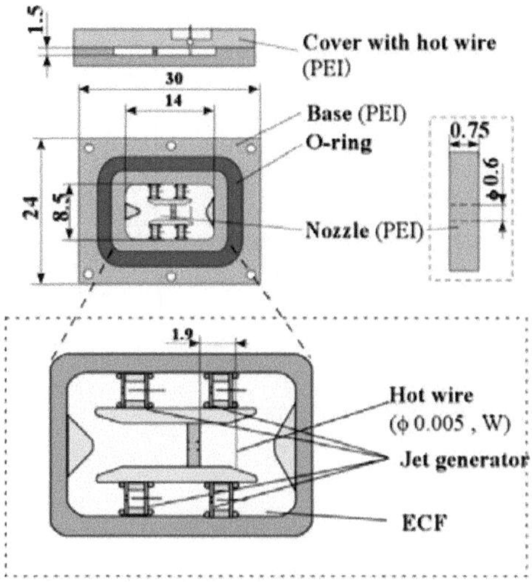

Fig. 9.4 Prototype of the ECF micro gyroscope

9.3.3 Soft Actuators Using Pressure due to ECF Jets [12]-[19]

The ECF jet can possibly control the pressure of soft actuators. Namely, we can develop many kinds of micro soft actuators integrated with fluid power source.

One application is an ECF micro artificial muscle cell shown in Fig.9.5. This artificial muscle mainly consists of an ECF tank membrane, a fiber-reinforced rubber tube and an ECF jet generator. The tank membrane and the fiber-reinforced tube are arranged to be a bicylindrical structure. Then the actuator could be a cell, easy to be integrated. When the ECF jet is generated, the inner pressure of the fiber-reinforced tube increases, resulting in making the artificial muscle cell contract. The fabricated cell with ϕ13.5 mm × 14 mm generates contraction of 1.2 mm. The ECF micro artificial muscle cell developed here is suitable for integrating in combinations in series and parallel as shown in Fig.9.6, to obtain desired output force and displacement.

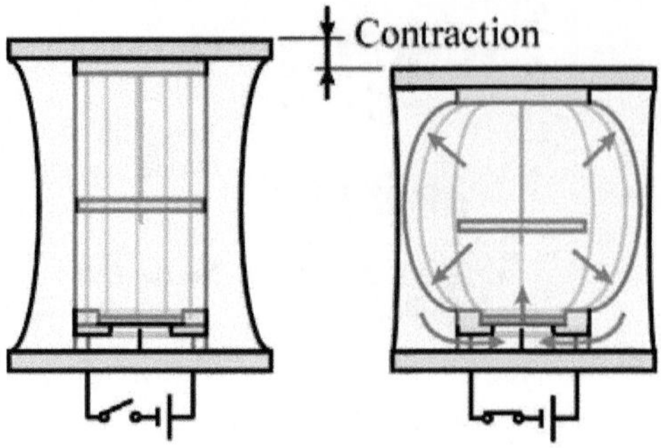

Fig. 9.5 ECF micro artificial muscle cell

Fig. 9.6 Integrated cells with serial and parallel

Another application is an ECF micro finger shown in Fig.9.7. A finger tube has three chambers as in the figure, and each of them is connected through an ECF jet generator. Namely, the ECF can actively be moved from one chamber to the others. Then, the finger tip moves in 3D space. A large model prototype with 4.5 mm in diameter was developed and the above-mentioned principle was experimentally confirmed.

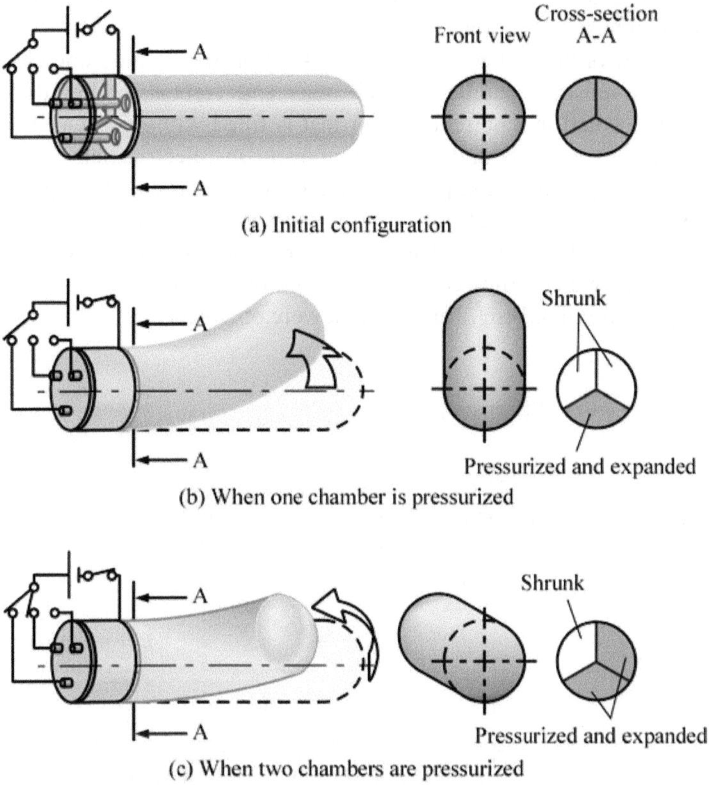

Fig. 9.7 ECF micro finger

9.3.4 Forced Liquid Cooling Systems Using ECF Jets [20][21]

Recently, as increasing the performance and density of CPUs, the heat generation of CPUs has become a serious problem especially in notebook computers that have limited space for a heat sink and cooling fan. As for a solution of this problem, a novel forced liquid cooling system for advanced notebook computers is developed. The system has a simple structured, ultra-thin planar ECF pump with A4 size and thickness of less than 1 mm.

Fig.9.8 shows the structure and the working principle. Fine line electrodes are located on a substrate parallel to each other and vertical to flow direction. The substrate is located in a pump chamber that is filled with ECF. Outflow is generated due to the ECF jets under the inhomogeneous electric fields. The proposed pump has a simple ultra-thin planer structure without any moving parts. The proposed pump plays roles not only of a power source of the forced liquid cooling system but also of a radiator. Through experiments, the electrode sizes and flow channel height were optimized to have a high output power. The maximum flow rate and pressure of 5.5 cm^3/s and 7.2 kPa were obtained.

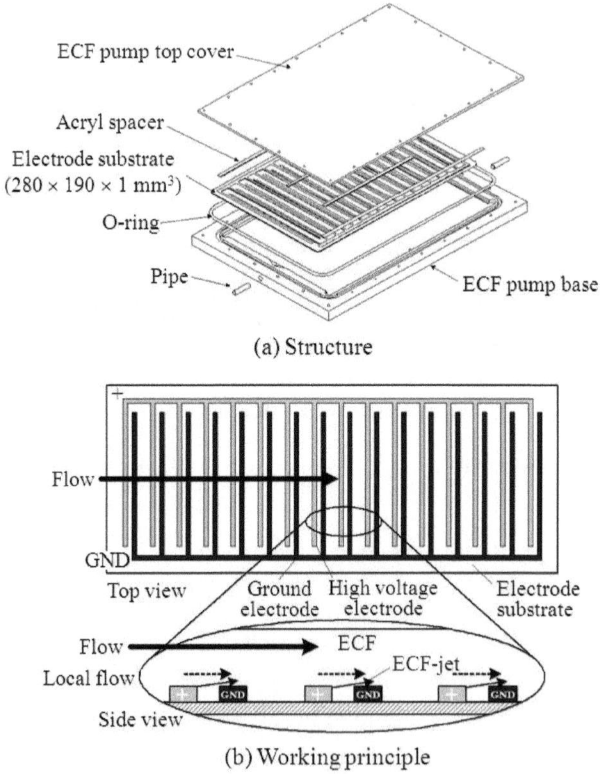

Fig. 9.8 Ultra-thin planar ECF pump for forced liquid cooling system

9.3.5 Microactuators Using ERF/MRF [22]-[30]

We are developing microactuators using ER microvalves. The ER microvalve controls the ERF flow by the apparent viscosity change due to the applied electric field through fixed electrodes.

To realize soft microactuators for medical applications and so on, an FERMA (Flexible ER MicroActuator) that has an FERV (Flexible ER Valve) was proposed and developed as shown in Fig.9.9. An FERV based on flexible SU-8 cantilever structures and the novel MEMS fabrication process was proposed. The FERV shown in Fig.9.9(b) was successfully fabricated with 5 mm length, 2.4 mm width and 0.2 mm thickness and the sufficient flexibility was confirmed. It was experimentally confirmed that the static characteristics are independent from the bending. Furthermore, an FERMA using the FERV was fabricated and tested.

Also, we have proposed an inherently robust ER microactuator against disturbance as shown in Fig.9.10. The microactuator consists of a pair of movable and fixed parallel plate electrodes with variable gap length and an upstream restrictor. The microactuator can not only generate displacement but also suppress the displacement due to external force by the inherent position feedback mechanism. The microactuator was fabricated and the validity was confirmed experimentally.

As for microactuators using MRF, we have developed an MR microvalve whose differential pressure is controlled by the applied magnetic field with a permanent magnet, a thermosensitive ferrite and Peltier elements. Also, we have proposed and developed a microvalve using MRF column as the valve-body.

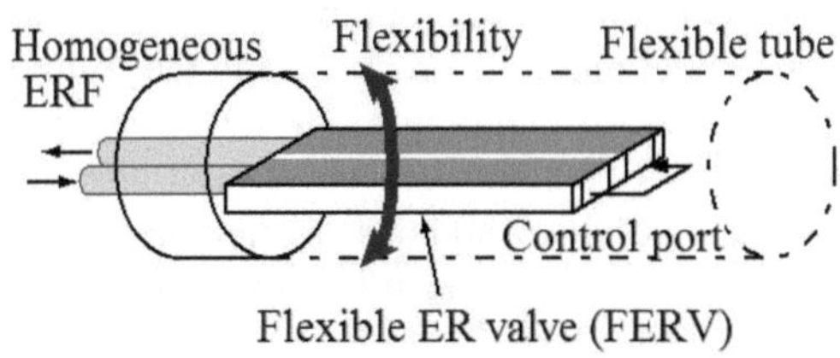

(a) Basic structure

(b) Fabricated FERV

Fig. 9.9 FERMA (Flexible ER MicroActuator)

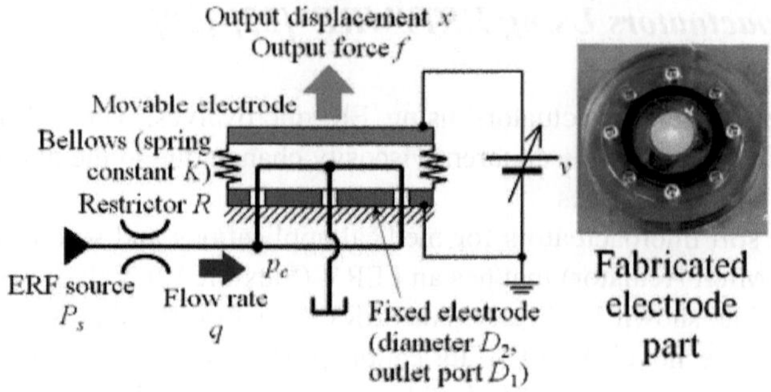

Fig. 9.10 Inherently robust ER microactuator against disturbance

9.3.6 High Output Power Piezoelectric Micropumps [31]-[33]

As micro fluid power sources, we are developing high output power piezoelectric micropumps.

A micropump that has an outlet pipe element in place of a check valve was proposed and fabricated. Utilizing the fluid inertia effect in the outlet pipe, the micropump flows out in both pumping and suction periods and obtains output flow larger than the displacement. The fabricated micropump with 2.3 cm^3 in volume can generate fluid power up to 0.22 W in water pumping. Also, for higher output fluid power, a full-wave rectifying micropump that has two sets of pump chamber, inlet check valve and outlet pipe was proposed and the validity was confirmed experimentally. Furthermore, for pumping higher viscosity fluid like homogeneous ERF, the micropump using a multi-reed valve was proposed and fabricated as shown in Fig.9.11. As a result of silicone oil pumping, 1.6 times higher output fluid power density was realized than the pump with a conventional reed type check valve.

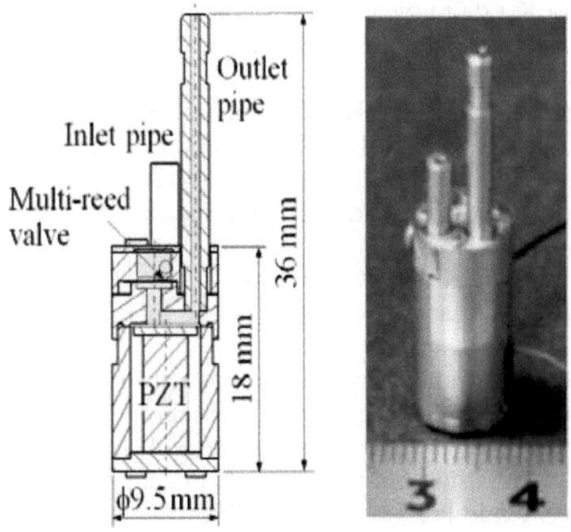

Fig. 9.11 Piezoelectric micropump using fluid inertia effect in pipe

9.4 Conclusions

The objective, research outline and some research results of our research to realize new microactuators using functional fluids were briefly introduced. We are convinced of the validity of the new microactuators using functional fluids.

Acknowledgments A part of the research was supported by Grant-in-Aid for Scientific Research in Priority Areas No. 16078205 of the Ministry of Education, Culture, Sports, Science and Technology of Japan.

References

1. Raghavan R, Qin J, Yeo LY, Friend JR, Takemura K, Yokota S, Edamura K (2009) Electrokinetic Actuation of Conductivity Dielectric Liquids. Sensors and Actuators B 140-1: 287-294
2. Yokota S, Nakada T (2004) Micro Actuators Using ECF-Jets. Fluid Power Systems 35-6: 368-374 (in Japanese)
3. Yokota S, Kuwajima S, Edamura K (2004) Realization of a Higher Integrated Multi-layered DP-RE-type ECF Micro-Motor. Trans of JSME(C) 70-693: 1463-1469 (in Japanese)
4. Yokota S (2004) Micro Actuators Using Electro-Conjugate Fluids (ECF). Proc of Int Conf on Mechatronics Technology (ICMT2004): 19-21 {Invited}
5. Yokota S, Kawamura K, Takemura K, Edamura K (2005) High-Integration Micro Motor Using Electro-Conjugate Fluid (ECF). J of Robotics and Mechatronics 17-2: 142-148

6. Yokota S, Edamura K (2006) New Construction of an Electro-Conjugate Fluid-jet Driven Micromotor with Inner Diameter of 2mm. Proc of IMechE Part I J of Systems and Control Engineering 220-4: 251-256
7. Yokota S, Kawamura K, Takemura K, Edamura K (2006) Improvement of Output Power of ECF Micromotor. Trans of IEEJ(E) 126-4: 137-143 (in Japanese)
8. Takemura K, Kozuki H, Edamura K, Yokota S (2007) A Micromotor using Electro-conjugate Fluid - Improvement of motor performance by using saw-toothed electrode series -. Sensors and Actuators A 140: 131-137
9. Yokota S, Nishizawa R, Takemura K, Edamura K (2006) Concept of a Micro Gyroscope using Electro-conjugate Fluid and Development of an ECF Micro Gyro-motor. J of Robotics and Mechatronics 18-2: 114-120
10. Takemura K, Yokota S, Suzuki S, Edamura K, Imamura T, Kumagai H (2009) Micro Liquid Rate Gyroscope using Electro-conjugate Fluid. Sensors and Actuators A 149-2: 173-179
11. Takemura K, Imamura T, Edamura K, Kumagai H, Yokota S (2009) The Practical Design of a Liquid Rate Gyroscope using Electro-conjugate Fluid. Proc of IMechE Part I J of Systems and Control Engineering: In Press
12. Takemura K, Yokota S, Edamura K (2005) Micro Artificial Muscle Actuator using Jet Flow of Electro-conjugate fluid. Trans of JSME(C) 71-708: 2571-2577 (in Japanese)
13. Abe R, Yokota S, Takemura K, Edamura K (2006) A Tube-type ECF Microactuator Using Pressure due to ECF-jet. Trans of Japan Fluid Power System Society 37-5: 55-60 (in Japanese)
14. Yokota S, Hong Y-P, Takemura K, Edamura K (2007) Earthworm Type Micromachine Driven by Electro-conjugate Fluid. J of Robotics Society of Japan 25-6: 140-145 (in Japanese)
15. Yokota S, Abe R, Takemura K, Edamura K (2007) Proposition of a Micro Finger Using ECF-jet and Characteristics Evaluation of Large Model. Trans of Japan Fluid Power System Society 38-5: 65-70 (in Japanese)
16. Takemura K, Yokota S, Edamura K (2007) Development and Control of a Micro Artificial Muscle Cell using Electro-conjugate Fluid. Sensors and Actuators A 133-2: 493-499
17. Abe R, Takemura K, Edamura K, Yokota S (2007) Concept of a Micro Finger using Electro-conjugate Fluid and Fabrication of a Large Mode Prototype. Sensors and Actuators A 136-2: 629-637
18. Takemura K, Yajima F, Edamura K, Yokota S (2008) Integration of Micro Artificial Muscle Cells using Electro-conjugate Fluid. Sensors and Actuators A 144-2: 348-353
19. Yokota S (2008) Micro-actuators by making use of jet flows due to Electro-conjugate Fluid. Int J of Mechanics Based Design of Structures and Machines 36-4: 330-345
20. Yokota S, Seo W-S, Yoshida K, Edamura K (2005) A Thin-Planar Pump Using Electro-Conjugate Fluid (ECF) for Liquid Cooling of Electronic Chips. Trans of JSME(C) 71-709: 2798-2804 (in Japanese)
21. Seo W-S, Yoshida K, Yokota S, Edamura K (2007) A high performance planar pump using electro-conjugate fluid with improved electrode patterns. Sensors and Actuators A 134-2: 606-614
22. Yoshida K, Soga T, Yokota S, Kawachi M, Edamura K (2004) A Valve-Integrated MR Cylinder Using Magneto-Rheological Fluid. Proc of Int Conf on New Actuators (ACTUATOR2004): 609-612
23. Yoshida K, Takamatsu T, Yoneda Y, Yokota S (2005) A Microvalve Using Magneto-Rheological Fluid. Trans of JSME(C) 71-704: 1355-1360 (in Japanese)
24. Yoshida K, Park J-H, Yano H, Yokota S, Yun S-N (2005) A Study of Valve-Integrated Microactuator Using Homogeneous Electro-Rheological Fluid. Sensors and Materials 17-3: 97-112
25. Yoshida K, Jung Y-O, Yokota S (2005) A Micro Fluid Power System Using Magneto-Rheological Fluid Valve-Body. Proc of Int Conf on Manufacturing, Machine Design and Tribology (ICMDT2005) (CD-ROM): MAA-101

26. Yoshida K, Takahashi H, Park J-H, Yokota S, Kawachi M, Edamura K (2006) A Control Valve Using Ferrite Particle-Dispersed Magneto-Rheological Fluid. Trans of JSME(C) 72-714: 457-463 (in Japanese)
27. De Volder M, Yoshida K, Yokota S, Reynaerts D (2006) The use of liquid crystals as electrorheological fluids in microsystems: model and measurements. J of Micromechanics and Microengineering 16:612-619
28. Yoshida K, Ogiso T, Yokota S (2007) A Flexible ER Microactuator Using Homogeneous Electro-Rheological Fluid. Trans of JSME(C) 73-733: 2508-2513 (in Japanese)
29. Kim JW, Yoshida K, Kouda K, Yokota S (2007) Flexible Electro-rheological Microvalve (FERV) Based on SU-8 Cantilever Structures and its Application to Microactuators. Proc of 6th Int Symp on Linear Drives for Industrial Applications (LDIA 2007) (CD-ROM)
30. Yoshida K, Kamiyama K, Kim JW, Yokota S (2008) Proposition of an ER Microactuator with Inherent Position Feedback Mechanism. Proc of 7th Int Symp on Fluid Power Toyama 2008: 551-554
31. Park J-H, Yoshida K, Ishikawa C, Yokota S, Seto T, Takagi K (2004) A Study on High-Output Resonance-Driven Piezoelectric Micropumps Using Active Check Valves. J of Robotics and Mechatronics 16-2: 171-177
32. Yoshida K, Jung Y-O, Seto T, Takagi K, Park J-H, Yokota S (2006) Optimal Structure of a Micropump Using Inertia Effect of Pipeline Element. Trans of JSME(C) 70-697: 2668-2673 (in Japanese)
33. Yoshida K, Sugiura K, Yokota S (2006) A Full-Wave Rectifying Piezoelectric Micro Fluid Power Source Using Fluid Inertia Effect In Pipes. Proc of Int Conf in Mechatronics Technology (ICMT2006)(CD-ROM): MN04

Chapter 10
Two-Axial Piezoelectric Actuator and Its Motion Control Toward Development of a Tactile Display

Masahiro OHKA[1]

Abstract Since tactile displays are enhanced, a parallel typed two-axial micro actuator is composed of two bimorph piezoelectric elements and two small links connected by three joints. A control system for the two-axial actuator is designed on the basis of a multi-layered artificial neural network to compensate hysteresis of piezoelectric elements. The output neuron emits time derivative of voltage; two bits signal expressing increment or decrement condition is generated by two input neurons; one of the other two input neurons and the other calculate current values of voltage and displacement, respectively. The neural network is featured with a feedback loop including an integral element to reduce number of neurons. In experiment, if the result of the left piezoelectric element is compared to that of right element, the displacement amplitudes and the inclinations coincide on the right and left piezoelectric elements. Although precise hysteresis characteristics such as loop width are considerably different, the present neural system can follow the difference.

10.1 Introduction

In the realm of virtual reality technology, several display mechanisms have been tentatively presented for tactile displays in addition to such already established visual and auditory displays as head-mounted displays and 5.1-ch surround-sound systems. Examples of tactile displays adopted so far include mechanical vibratory pin arrays [1], surface acoustic waves [2], and pin arrays driven by pneumatic actuators [3] and piezoelectric actuators [4].

Since distributed pressure must be applied to the human skin to stimulate human tactile receptors distributed over it, very high-level actuator technology is required to realize high-density distributed actuator arrays. Despite many attempts, developing tactile displays that can satisfy requirements for practical use continues to be difficult.

[1] Masahiro OHKA

Graduate School of Information Science, Nagoya University

Haptic displays like PHANToM, on the other hand, are commercially available and are attracting the attention of researchers because they can be applied to manufacturing tasks such as CAD/CAM and remote operations. Haptic means a sensation relating to touch, caused by signals from both the tactile receptors distributed across the skin and muscle receptors in the muscles. In a haptic display, a tactile sensation is restricted to the points of touch that resemble sensations accepted through a pencil when we touch an object with it. Thus, it seems reasonable to regard a haptic display as a force display. Tactile displays, which are required for presenting surface conditions such as texture, are essential for any handling task.

Although a tactile display requires actuator arrays composed of a number of micro actuators, realizing such actuator arrays is very difficult because relatively large displacement and force are required for a tactile display. So far, a tactile display has generally suggested a pressure display that generates normal force distribution on the finger surface. Recently, shearing force distribution plays an important role in tactile sensation. For instance, comb illusion [5], a tactile illusion, causes feelings of unevenness to the finger surfaces on the teeth tips of a comb when scratching the teeth. If the illusion is applied to a tactile display, we can enhance the reality of the virtual surface feeling demonstrated by it.

Thus, we are studying a two-axial micro-actuator to develop a tactile display pad comprised of two-axial actuators for a tactile display. The two-axial actuator is composed of two bimorph piezoelectric elements and two small links connected by three joints. We formulated its kinematics with the two-dimensional control of its center joint (manipulable joint, hereafter) in the two-dimensional coordinate system.

The two-axial actuator's manipulatable joint does not follow the same route in the increment and the decrement of the applied voltage because the present two-axial actuator utilizes the piezoelectric elements possessing a hysteresis phenomenon. Although feedback control effectively compensates the hysteresis phenomenon so that the manipulatable joint follows the desired trajectory, additional sensors for measuring its displacement are required to realize feedback control. However, feedback control is not suitable for the present actuator because the actuator array requires a huge number of actuators to apply it to the tactile display.

A new control method for piezoelectric actuators is established based on a multi-layered artificial neural network to achieve sensor-less control of the actuator. Since the network scale becomes larger without a new idea for learning the hysteresis loop, we apply causality to formulate a modified neural network. In causality, a certain current time derivative of a physical variable is determined by the current physical variables if all of them can be measured at the moment.

10.2 Tactile Display Based on Comb Illusion

The comb illusion, which is a tactile illusion, causes a feeling of unevenness to the finger surfaces touching a comb's teeth tips by scratching the teeth. We can experience this illusion using a comb and a ballpoint pen. After preparing a comb and a ballpoint pen and putting a finger on the comb's teeth tips, scratch the teeth (Fig.10.1). Although the horizontal vibration of the teeth tips is very small, a relatively strong unevenness feeling will be experienced.

Using this illusion, we are developing a tactile display (Fig.10.2) in which the finger's contour is shown by a dotted line. An operator put his/her finger on the top of the display surface. 12 tactile pins were connected to the operator's finger surface. Each tactile pin moves in a two-dimensional direction to generate virtual unevenness. Each pin is driven by a two-axial actuator described in the next section.

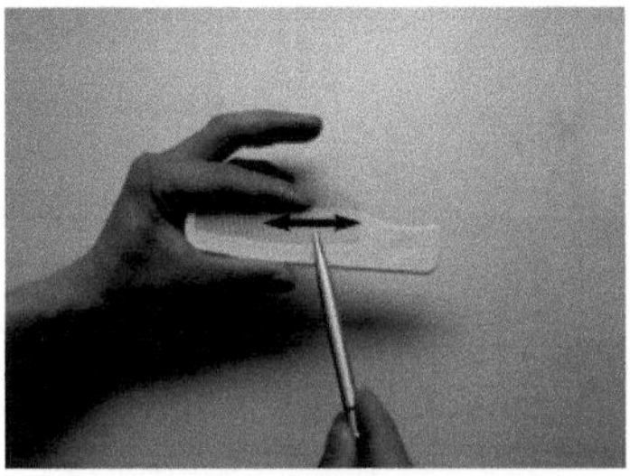

Fig. 10.1 Comb illusion

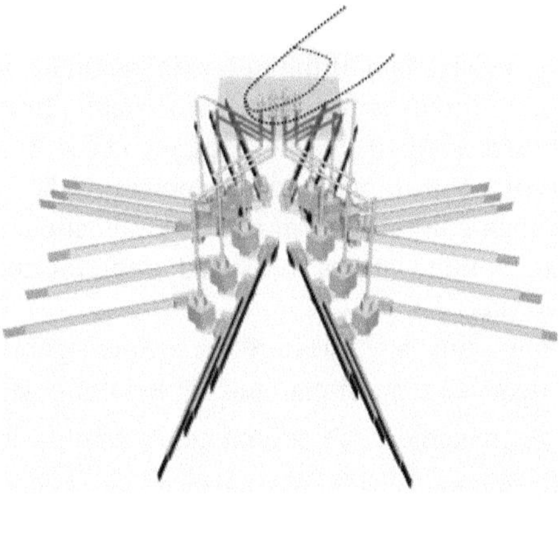

Fig. 10.2 Tactile display

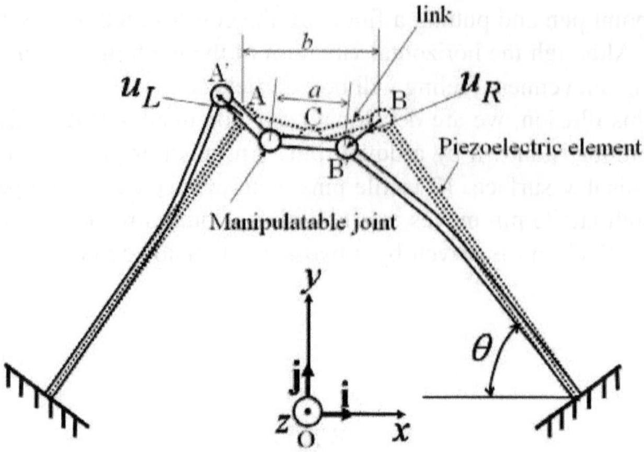

Fig. 10.3 Parallel type two-axial actuator

10.3 Parallel Type Two-Axial Actuator

Figure 10.3 shows the mechanism of the two-axial actuator. It is composed of two piezoelectric elements, three joints, and two small links to generate two-dimensional displacement (Fig.10.3). The displacement of manipulatable joint position C is controlled by controlling the displacement of the left and right bimorph type piezoelectric elements. First, to formulate a kinematic equation of a two-axial actuator, the following nomenclatures were used in the formulation:

a : length of small link

b : distance between joints A and B in reference configuration

u_x : x-directional displacement of manipulatable joint position C

u_y : y-directional displacement of manipulatable joint position C

u_R : bending displacement of right piezoelectric element

u_L : bending displacement of left piezoelectric element

V_R : applied voltage of right piezoelectric element

V_L : applied voltage of left piezoelectric element

θ : establishment angle of piezoelectric element

$\dot{x}$: time derivative of variable x

Coordinates of points A' and B' were calculated from the geometrical relationship shown in Fig.10.3 as follows:

Point A': $\left(-\frac{1}{2}b+u_L\sin\theta,\ \ \sqrt{a^2-\frac{b^2}{4}}-u_L\cos\theta\right)$

Point B': $\left(\frac{1}{2}b-u_R\sin\theta,\ \ \sqrt{a^2-\frac{b^2}{4}}-u_R\cos\theta\right)$

where $a>b/2$.

When the displacement vector of the manipulatable joint position is expressed as $\mathbf{u}=u_x\mathbf{i}+u_y\mathbf{j}$, the following expressions were obtained because the condition of the link length is constant:

$$\left(u_x+\frac{1}{2}b-u_L\sin\theta\right)^2+\left(u_y-\sqrt{a^2-\frac{b^2}{4}}+u_L\cos\theta\right)^2=a^2 \tag{10.1}$$

$$\left(u_x-\frac{1}{2}b+u_R\sin\theta\right)^2+\left(u_y-\sqrt{a^2-\frac{b^2}{4}}+u_R\cos\theta\right)^2=a^2. \tag{10.2}$$

The following equations were obtained from differentiating Eqs. (10.1) and (10.2):

$$2\left(u_x+\frac{1}{2}b-u_L\sin\theta\right)(\dot{u}_x-\dot{u}_L\sin\theta)+2\left(u_y-\sqrt{a^2-\frac{b^2}{4}}+u_L\cos\theta\right)(\dot{u}_y+\dot{u}_L\cos\theta)=0 \tag{10.3}$$

$$2\left(u_x-\frac{1}{2}b+u_R\sin\theta\right)(\dot{u}_x+\dot{u}_R\sin\theta)+2\left(u_y-\sqrt{a^2-\frac{b^2}{4}}+u_R\cos\theta\right)(\dot{u}_y+\dot{u}_R\cos\theta)=0 \tag{10.4}$$

The simultaneous equation composed of Eqs. (9.3) and (9.4) is written as the following matrix expression:

$$\begin{pmatrix}\dot{u}_L\\ \dot{u}_R\end{pmatrix}=\begin{bmatrix}\dfrac{A_{11}}{A_{11}\sin\theta-A_{12}\cos\theta} & \dfrac{A_{12}}{A_{11}\sin\theta-A_{12}\cos\theta}\\ -\dfrac{A_{21}}{A_{21}\sin\theta+A_{22}\cos\theta} & -\dfrac{A_{22}}{A_{21}\sin\theta+A_{22}\cos\theta}\end{bmatrix}\begin{pmatrix}\dot{u}_x\\ \dot{u}_y\end{pmatrix} \tag{10.5}$$

$$A_{11} = u_x + \frac{1}{2}b - u_L \sin\theta \tag{10.6}$$

$$A_{12} = u_y - \sqrt{a^2 - \frac{b^2}{4}} + u_L \cos\theta \tag{10.7}$$

$$A_{21} = u_x - \frac{1}{2}b + u_R \sin\theta \tag{10.8}$$

$$A_{22} = u_y - \sqrt{a^2 - \frac{b^2}{4}} + u_R \cos\theta \,. \tag{10.9}$$

10.4 Neural Network Including Feedback Loop

In the present research, we are attempting to achieve sensor-less control of an actuator with a new piezoelectric actuator method using a neural network model. Sensor-less control is necessary for a tactile display because it is comprised of many actuators. The present network structure features causality as a basic idea in which the time increment of the physical variable is determined at a certain instant, if all the current physical variables can be measured simultaneously.

Generally, the piezoelectric actuator slope of the voltage-displacement curve is noticeably different between the increment and decrement sequences of the applied voltage. If the above causality is applied to the control method of the piezoelectric actuator, the voltage's time increment at a certain instant can be determined by the current voltage and displacement, and information related to whether it is the increment or the decrement condition.

Figure 10.4 shows the structure of the present neural network that is comprised of four neurons in the input layer, ten in the hidden layer, and one in the output layer. The output neuron emits the voltage's time derivative, and the two bits signal expressing the increment or decrement condition is generated by two input neurons; one of the other two input neurons and the other show the current voltage and displacement values, respectively. After the time integration of the output time derivative of the displacement, it is not only used for the voltage that should be applied to the piezoelectric element but also fed to the neuron in the input layer through a feedback loop. The neural network is featured with the feedback loop including an integral unit to reduce the number of neurons.

In the learning process, synaptic connection weight $w_{ij}^{(s)}$ is adjusted by the back propagation algorithm for neural networks after the feedback loop shown in Fig.10.4 is removed. Here, suffix s shows the layers, and $s = 0$ and 1 are the input layers to the hidden layer and the hidden layer to the output layer, respectively. Suffixes i and j denote the number of neurons and depend on the layer; if $s = 0$, then $i = 0, 1, 2, 3$ and $j = 0,1, \ldots,9$; if $s = 1$, then $i = 0,1, \ldots,9$ and $j = 0$. Since the error back propagating method [6] for adjusting the synaptic connection's weight is introduced in many other references, expressions related to it are abbreviated in the present paper.

Voltage history including the increment and decrement process is applied to the piezoelectric element to obtain data for network learning. In network learning, we used data composed of about 100 patterns of two bits that express increment or decrement, applied voltage V, displacement u, and the voltage's displacement derivative dV/du. Therefore, the value of displacement derivative dV/du of the voltage will be output from the output neuron. Since du/dt is given from the trajectory planning, the time derivative of the voltage is calculated by multiplying du/dt by dV/du.

In addition, error back propagation is applicable to the present neural network by removing the feedback loop from the present network. This is different from the RTRL [7] of recurrent neural networks.

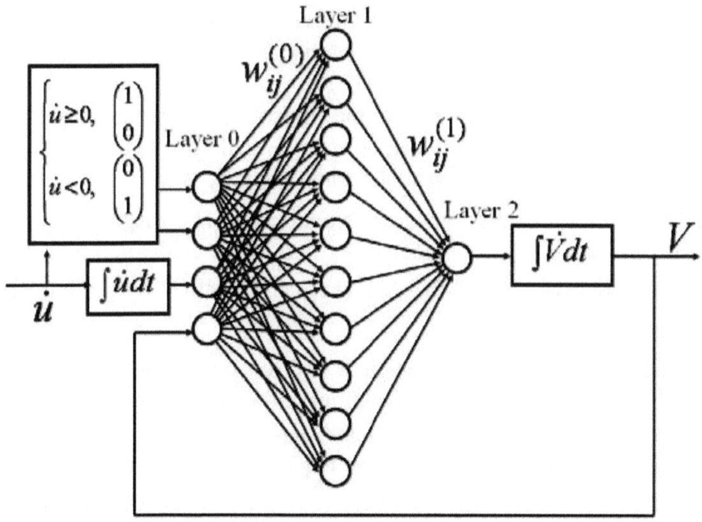

Fig. 10.4 Neural network with feedback loop

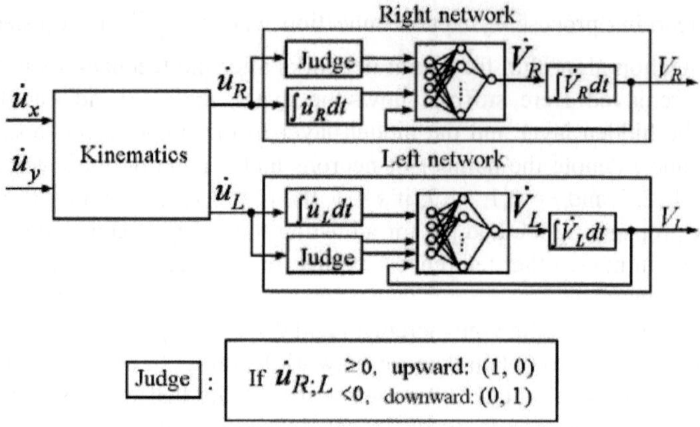

Fig. 10.5 Block diagram for controller equipped with neural networks

10.5 Control System

In the present two-axial actuator, positional error is caused by individual differences, not only about the inclination of a linear portion in the hysteresis loop but also by its non-linearity and width. The influence of the individual difference is modified by individually adding the characteristics of the left and right actuators based on the neural network model.

Figure 10.5 shows a block diagram of the control system whose design is based on the above idea. The neural network is incorporated into this control system to control the left and right piezoelectric elements. First, $\dot{u}_x$ and $\dot{u}_y$ are decided from the designed trajectory in the two-dimensional area. The displacement rates of right and left type piezoelectric elements $\dot{u}_R$ and $\dot{u}_L$ are calculated by substituting $\dot{u}_x$ and $\dot{u}_y$ into kinematic Eq. (9.5). u_L and u_R are obtained by the numerical integral of $\dot{u}_L$ and $\dot{u}_R$.

The 0th and the 1st neurons of the right and left neural networks in the input layer accept binary number 1 or 0 for judging either the increment or decrement condition. The increment or decrement condition is simply decided based on the signs of $\dot{u}_L$ and $\dot{u}_R$. If the sign is positive or negative, the judgment is increment or decrement. For example, if the input vector of the left piezoelectric element is $(1 \;\; 0 \;\; u_L \;\; V_L)^T$ or $(0 \;\; 1 \;\; u_L \;\; V_L)^T$, the condition is specified as increment or decrement.

Since the voltage's displacement derivative is output from the output neuron of the neural network, as previously mentioned, the voltage's time derivative is calculated by multiplying the time derivative of the displacement by it. In addition, after integration, the voltage is applied to the piezoelectric element and fed to the input layer through the feedback loop.

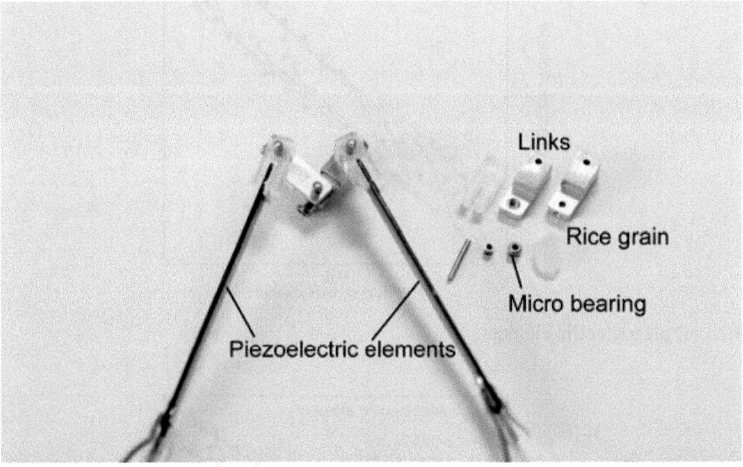

Fig. 10.6 2D micro actuator and mechanical elements

10.6 Experiment and Discussion

The parallel two-axial actuator is comprised of two bimorph type piezoelectric elements, two small links, and three joints (Fig.10.6). The piezoelectric element (31 mm long, 2.0 mm wide, 0.50 mm thick) was disassembled from a Braille cell SC9 developed for the visually handicapped by KGS Ltd. Small 5-mm links made of aluminum alloy were used in this actuator.

Since the two links or the link and the piezoelectric element end are connected with an aluminum alloy joint, they can be rotated mutually. The center of the three joints functions as the manipulatable joint. In these joints, micro-sliding bearings are used to remove the gap between the shaft and the bearing, which is about the same size as a grain of rice (Fig.10.6).

We used an OLYMPUS, IX71, inverted research microscope to measure the displacements of the three joints. A 1.25-power PlanApo×1.25 for the objective lens and 0.35-power U-TV0.35×C-2 for the camera adaptor were used. The movement trajectories of joints A, B, and C were measured by image data processing of the image retrieved by a CCD camera mounted in the microscope. The centroid coordinates of these joints were obtained with noise reduction and circular regression. The above operation was executed in each stepwise voltage variation to measure the trajectories of the three joints. Multifunctional universal image analysis software (Library Ltd., cosmos 32 Ver4.6) was used.

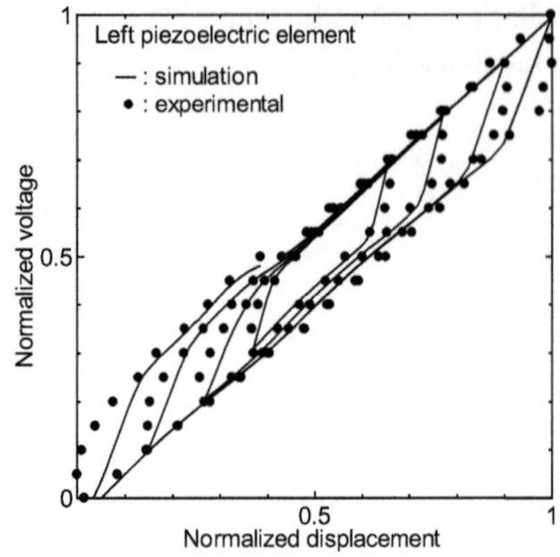

Fig. 10.7 Left piezoelectric element

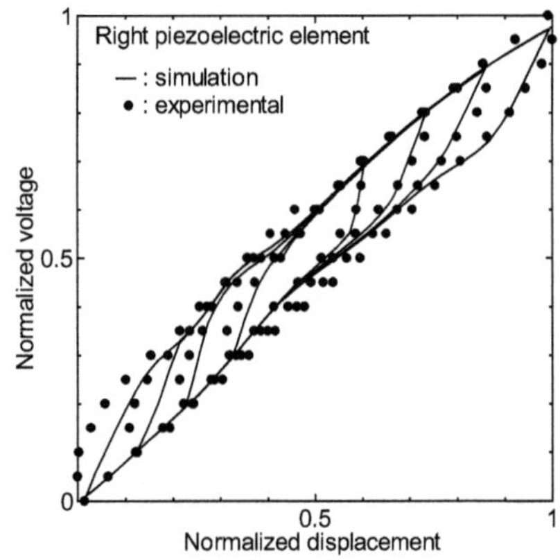

Fig. 10.8 Right piezoelectric element

To adjust weight $w_{ij}^{(s)}$ of the synaptic connection, the following experiment was performed to examine the relationship between the applied voltage and the displacements of the left and right piezoelectric elements. 100 V datum voltage was applied to the left and right piezoelectric elements at first to make a starting point. In the discussion in subsequent sections, voltage 100 V will be expressed as 0 V. In the beginning the voltage was sequentially increased to 40 V at 10-V intervals, and then the applied voltage was decreased to -40 V at 10-V intervals. Subsequently, it was decreased to -60 V after being increased to 60 V at 10-V intervals; it was decreased to -80 V after being increased to 80 V at 10-V intervals; it was decreased to –100 V after being increased to 100 V at 10-V intervals. In addition, it was increased to the starting point of 0 V at 10-V intervals. Centroid displacements of joints A, B, and C were measured at every stepwise variation.

The hysteresis loops of the left piezoelectric element obtained by the present experiment are shown in Fig.10.7 by ●. The solid line in Fig.10.7 is the output result of the control system shown in Fig.10.5. The output result from the neural network almost coincides with the experimental result not only for a large loop of ± 100 V but also for small inside loops of ± 40, ± 60, and ± 80 V.

Next, the hysteresis loops of the right piezoelectric element obtained by the present experiment are shown in Fig.10.8 by ●. The solid line in Fig.10.8 is the output result of the control system. The right element's inclination is the same as that of the left element. The output result from the neural network almost coincides with the experimental result. Therefore, a high accuracy learning result is obtained, as mentioned above.

If the result of the left piezoelectric element is compared to that of right element, the displacement amplitudes and the inclinations coincide on the right and left piezoelectric elements. Although precise hysteresis characteristics such as loop width are considerably different, the present neural system can follow the sight difference.

In addition, 100,000 iterations were needed to obtain the learning result of Figs. 10.7 or 10.8. The learning calculation, executed on a notebook computer (Panasonic, CF-R4), required about 15 minutes to complete.

10.7 Conclusion

A control method of a two-axial actuator was presented to enhance the positioning accuracy and applied to the tactile display. To realize sensor-less control of piezoelectric actuators possessing obvious hysteresis characteristics, we established a new neural network model that included a feedback loop based on causality where the applied voltage's time derivative was determined by either the increment or decrement condition, the current voltage, and the displacement.

The two-axial micro actuator was composed of left and right bimorph piezoelectric elements, two links, and three joints. The control system was also developed by incorporating the neural network for compensation of the hysteresis characteristics. The learning was terminated within reasonable calculation time; after the learning process, we could reproduce the hysteresis characteristics including several minor loops with high accuracy. In the future, software accuracy must be enhanced by increasing the amount of learning data and the accuracy of the joint bearings.

Acknowledgments This study was supported by fiscal 2008 grants from the Ministry of Education, Culture, Sports, Science and Technology (Grant-in-Aid for Scientific Research in Priority Areas, No. 1607807).

References

1. Ikei Y et al(1999) Tactile Texture Presentation by Vibratory Pin Arrays Based on Surface Height Maps. International Mechanical Engineering Conference and Exposition: 51-58.
2. Takahashi M et al(2000) A Tactile Display Using Surface Acoustic Wave, 2000 IEEE International Workshop on Robot and Human Interactive Communication: 364-367.
3. Tanaka Y et al(2002) Wearable Haptic Display for Immersive Virtual Environment. Fifth JFPS International Symposium: 309-310.
4. Ohka M et al (2006) A Two-axis Bimorph Piezoelectric Actuator Controlled by a Multi-layered Neural Network. 10th International Conference. on New Actuator: 503-506.
5. Hayward V(2008) A Brief Taxonomy of Tactile Illusions and Demonstrations that Can Be Done in a Hardware Store. Brain Research Bulletin 75-6: pp. 742-752.
6. Rumelhart D E et al(1986) Learning Representations by Back-propagating Errors. Nature 323-9: 533-536.
7. Williams R and Zipser D(1989) A Learning Algorithm for Continually Running Fully Recurrent Neural Networks. Neural Computation. 1: 270-280.

Chapter 11
High-Performance Electrostatic Micromirrors

For Accuracy, Low-Voltage driving, Temperature Stability, and High frequency

Minoru SASAKI[1]

Abstract The electrostatically driven micromirror is the attractive device for steering the optical beam. The basic performance is expected to be improved in addition to the size reduction. The required performances which often appear in applications can be categorized into common technical issues. Some of them are accuracy of tilt angle, decreasing the driving voltage, temperature stability, and high speed response. Dividing the complicated technology into these issues, researches are described. The accuracy is the attempt for integrating the sensor. The low-voltage driving is realized by decreasing the rotational spring constant. The temperature stability relates to the material selection. The high frequency is realized by the light-weighted planer mirror plane.

11.1 Introduction

The definition of the micromirror may be the small device including some new mechanisms compared to the bulk mirror devices. At present (2009), Digital Micromirror Devices, product from Texas Instruments, is well-known because of the mass-production and the commercial success. There are many other variations and potentials of micromirrors. The continuing growth of optical applications (e.g., display technologies or optical-fiber telecommunication infrastructure) are stimulating the progress of miniaturized, high-contrast, high-efficient, power reduction, portability, reliable, inexpensive optical devices for beam steering.

There are many issues about micromirrors. Although the exact specification depends on the application, the micromirrors share some of fundamental requirements. The research topics are described mainly from our results.

[1] Minoru SASAKI

Dept. of Advanced Science and Technology, Toyota Technological Institute

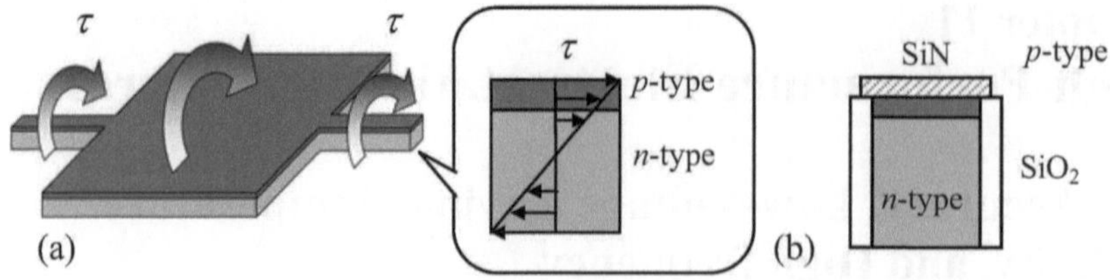

Fig. 11.1 Vertical design of the micromirror. Design with the passivation films. The inset shows the distribution of the shear stress inside the torsion bar

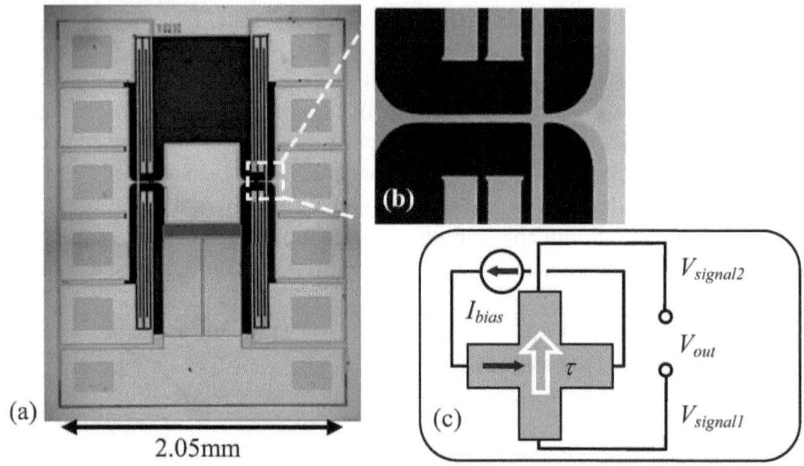

Fig. 11.2 Micromirror device. (a) Whole view. (b) Sensor in the torsion bar. (b) Simplified sensor circuit

11.2. Issues for Higher Performances

11.2.1 Integration of Sensor

In general, the applications require the accuracy of the mirror rotation angle. The galvanometric scanner is combined with the rotation angle sensor. Consider an example of the laser printer. Supposing 600 dpi and the optical arm of 10 cm, the neighboring dot corresponds to 0.012° mirror rotation. The accuracy will be on the order of 0.001°. The feedback control of the rotation angle based on the sensor signal, and integrating the sensor inside the micromirror are attractive.

The shear gauge measures the transverse voltage generated by the shear piezoresistance effect [1]. Figure 11.1(a) shows the vertical design of the micromirror. The Si layer has *p-n* junction. The sensor is in *p*-type region at the top surface. When the mirror rotates, the shear stress is generated having the maximum magnitude at the surface as shown in the inset. The shear stress is proportional to the mirror rotation angle. The reverse bias is applied for confining the bias current in *p*-type region. Since the leak current will decrease the sensitivity, the passivation film is prepared as shown in Fig.11.1(b). Top and side surfaces of the sensor are covered with SiN and SiO_2 films, respectively.

Figure 11.2(a) shows a fabricated device. The total size including electrode pads is 2.05×2.7 mm^2. The center mirror is 550×570 μm^2. The vertical comb drive actuator rotates the mirror. Figure 11.2(b) shows the magnified sensor. Figure 11.2(c) shows the simplified sensor circuit. The sensor consists of an element and 2 pairs of terminals. One pair is for flowing the bias current I_{bias}. The other is for detecting the signal of the transverse voltage. The signal V_{out} is expressed as follows.

$$V_{out} = \frac{f\,\rho_0 \pi\,\tau}{t} I_{bias} \tag{11.1}$$

f is a geometric factor. ρ_0 and t are the isotropic resistivity and thickness of the sensing layer, respectively. τ is the shear stress, which is proportional to the mirror rotation angle. π is the shear piezoresistance coefficient. The *p*-type Si has the larger value of π. The position of the sensor is near to the electrode pad for suppressing the temperature increase. This meandering structure has a rotational spring constant whose magnitude is ~1/10 of that of the torsion bar. The meandering Si spring has the resistance of 6.4 kΩ. The signals V_{signal1}, V_{signal2} are the input to the lock-in amplifier amplifying their difference.

The short-term noise is improved from ~0.1 to 0.01°, compared with our previous study [2]. The main improvement is the modulation frequency of the bias current setting higher value (50 kHz) than the mechanical resonant frequency of the micromirror (~10 kHz). 5 kHz is used in the previous device. With the electrical design for obtaining the lower impedance, the electrical band width of the sensor becomes larger than that of the mechanical one enabling the larger modulation frequency. The leak current is reduced with the passivation film.

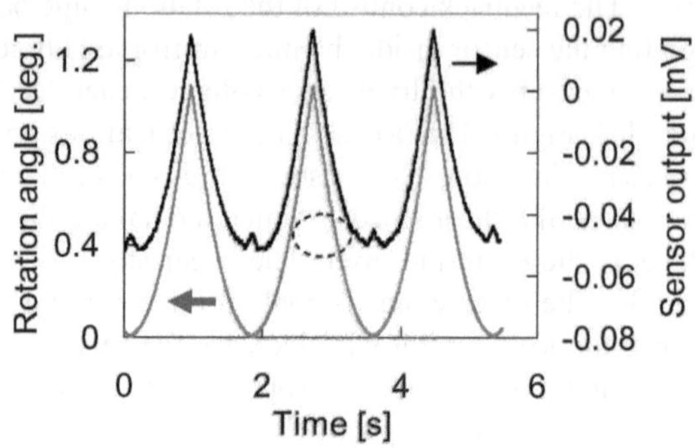

Fig. 11.3 Time responses of 3 cycles of mirror movement

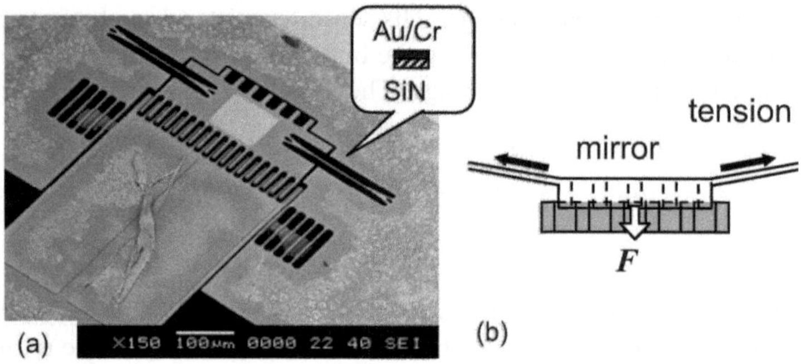

Fig. 11.4 Fabricated micromirror. Schematic drawing of the micromirror for illustrating the effect of the tension inside the torsion bar for suppressing the vertical displacement

Figure 11.3 shows the typical time responses of 3 cycles of the mirror movement. The triangular voltage is applied for driving the electrostatic actuator. The black and gray curves are the sensor signal and the mirror rotation angle, respectively. The average bias current is 0.82 mA modulated at 50 kHz. The sensitivity and the time constant of the lock-in amplifier is 20 mV/V and 10ms, respectively. When the mirror rotates, the sensor output increases. Comparing two signals, the signal shapes are approximately same. The difference appears when the mirror rotation angle is less than 0.1°. The small peak appears. The reason of this peak is not clear at now. The magnitude of these peak changes time to time implying the relation to the contact resistance between the sensor and the probe wire. There is a possibility that the residual stress introduced by the passivation films (top tensile SiN and side compress SiO_2 films). The sensitivity is 0.066 mV/$^{\circ}$ for this sensor. At now, the main noise is the drift. In this experiment, the drift during 1 cycle is 0.02°.

11.2.2 Low-Voltage Driving of Electrostatic Actuator

Especially for the electrostatic actuators, one effort is realizing the larger rotation angle with the lower driving voltage. This is because the established system usually works with dc 5, 12, or 24 V sources. There are two methods for decreasing the driving voltage. One is increasing the driving force. Another is decreasing the spring constant. Adopting both methods, Hah et al. reports the surface micromachined micromirror using the foundry service (Sandia's ultraplanar multilevel MEMS technology-V) [3]. This supposes the use of stepper.

The thin film torsion bar can decrease the spring constant [4, 5]. The tension is included for increasing the rigidity against the pull-in instability which limits the maximum operation region. Supposing the rectangular cross-section of the torsion bar, the rotational spring constant k_θ is as follows.

$$k_\theta \approx \frac{Gwt^3}{3l}\left(1-\frac{192}{\pi}\frac{t}{w}\tanh\frac{\pi w}{2t}\right) \tag{11.2}$$

G is shear modulus. l, w and t are length, width and thickness of torsion bar, respectively, supposing that $w>t$. The factor t^3 shows that decreasing t is effective for decreasing k_θ. The proximity patterning without using the stepper will give the minimum size of a few μm. As for the deposited film, the thickness of ~100 nm can be used as the structure. The spring constant from eq. (11.2) is $1.4\text{x}10^{-11}$ Nm/rad by setting values of G, l, w, and t as 80 GPa, 200, 4, and 0.3 μm, respectively. For obtaining larger spring constants for other displacement, tension is applied to the torsion bar. As seen from the inset of Fig.11.4, the tension works against the vertical displacement of the mirror. The tensile stress σ_0 inside SiN film can be 760 MPa [6]. The tension $\sigma_0 wt$ is 910 μN. This value is quite large compared to the driving force of about 7 nN in our actuator design. The thin film torsion bar is softest in the vertical direction. The vertical spring constant k_z is expressed as follows [7].

$$k_z \approx \frac{1}{\sum_{n=1,odd}^{\infty}\frac{1}{l}\frac{2}{\frac{Ewt^3}{12}k_n^{\ 4}+\sigma_0 wtk_n^{\ 2}}}, \quad k_n=\frac{n\pi}{l} \quad (n=1,3,5\cdots) \tag{11.3}$$

There are terms correspond to the factors due to the elasticity E (290 GPa) and the stress σ_0. k_z are estimated to be 0.016 and 19 N/m when σ_0 is 0 and 760 MPa, respectively. The increase of the rotational spring constant $\Delta k_{\theta\ stretch}$ generated by the beam stretching can be estimated as follows.

$$\Delta k_{\theta\ stretch} \approx \frac{1}{8l}\left\{\frac{\sigma_0(w^3t+wt^3)}{3}+\frac{E}{2l^2}\left(\frac{w^5t}{10}+\frac{w^3t^3}{9}+\frac{wt^5}{10}\right)\theta^2\right\} \tag{11.4}$$

The first term corresponds to stretching against the stress σ_0. The second nonlinear part corresponds to the elastic stretching. The dominant part is the first term having the value of $3.1x10^{-12}$ Nm/rad when the mirror rotation is small (<6.6 rad). This is 22% of the spring constant estimated from eq (11.2). The tension is fundamentally perpendicular to the rotational displacement inside the torsion bar. The increase of the spring constant is minimized.

Figure 11.4 shows the fabricated micromirror. The dimensions of the torsion bar is 200x4x0.3 μm^3. The structures of the thin film torsion bar and the vertical comb drive actuator are realized by combining the isotropic and the anisotropic Si plasma etching.

Figure 11.5 shows the mirror rotation angle as a function of the driving voltage. The thickness of the comb finger is 10 μm. The rotation angle reaches 7.3° at 5 V. The comb gap is 4 μm in the lateral direction. The curve is smooth showing a nearly linear relation. The data is for a round trip. The hysteresis (defined by the maximum difference of the rotation angle at the same driving voltage in the round trip motion) is 0.1°. This corresponds to 1.4% of the full scale.

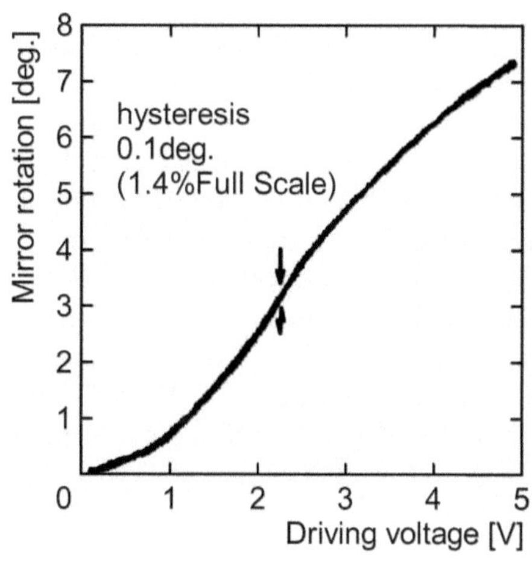

Fig. 11.5 Mirror rotation as a function of the driving voltage

11.2.3 Temperature Stability

A performance at high temperatures becomes eventually important when developing micromirrors for high optical power applications, and for product reliability. Electrostatic actuators have the lower risk of temperature increase, since they consume little power and generate the negligible heat. As for a disadvantage, the magnitude of the electrostatic force is rather small compared to other actuation forces. As described before, the thin film torsion bar with the tension can realize the low-voltage driving. Although SiN film generates a large tensile stress, it is an insulating material. A metal overlayer is necessary for constructing the electrostatic actuator. Au/Cr metals are deposited as shown in the inset of Fig.11.4(a). The temperature characteristics are found to be poor. For example, the mirror rotation angle is 4.3^{o} at 5 V at room temperature (17 ^{o}C). At 60 ^{o}C and 5 V, the mirror rotation angle decreases to 1.3^{o}. This can be attributed to the differing CTEs of the materials used in the torsion bar.

Realizing the tense thin film torsion bar with Si material having the same CTE with that of Si substrate will be ideal. Many actuators have been fabricated using crystalline (c-) Si or polycrystalline (poly-) Si on Si substrate. The performances at high temperatures have been discussed in limited literatures. It is valuable to confirm the temperature characteristics of the thin film torsion bar, since the thin film torsion bar is smaller in its thickness and has lower spring constant which strongly affects the total performance of the micormirror. The intrinsic nature of the tension is also important.

The tense poly-Si film is introduced to the thin film torsion bar for improving the temperature characteristics. The principle for realizing the low-voltage driving is the use of compliant spring which is the same with that of the previous device using SiN. The difference is the use of poly-Si. The tensile stress is generated by the crystallization of amorphous (a-) Si.

Figure 11.6 shows the fabricated micromirror using the tensile poly-Si film [8].

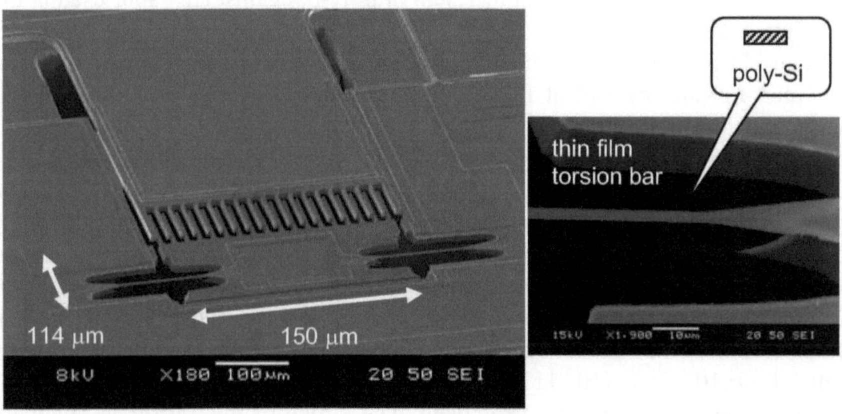

Fig. 11.6 Fabricated micromirror and the magnified image of the torsion bar. The inset shows the lateral cross-section of the thin film torsion bar. The width of the torsion bar is 5 μm

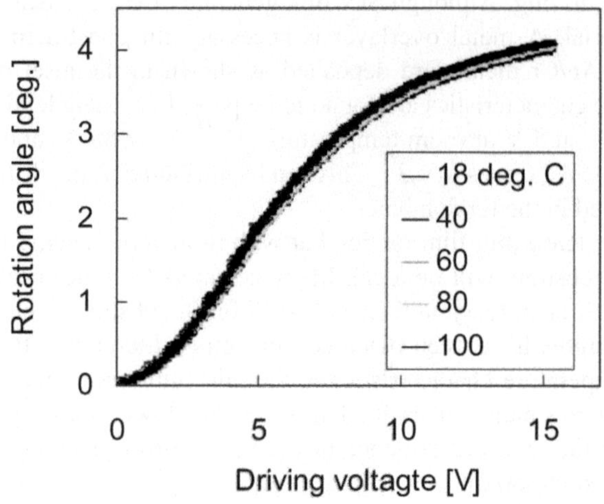

Fig. 11.7 Mechanical mirror rotation angle as a function of driving voltage at different temperatures. The initial comb-to-comb height distance of this mirror is 6.1 μm

The surface roughness is 0.006-0.055 μm Ra. The device surface is smoother compared to that (0.10-0.58 μm Ra) of the previous device shown in Fig.11.4. This is attributed to the chemical inertness of poly-Si against HF vapor used in the sacrificial SiO_2 layer etching.

Figure 11.7 shows the mirror rotation angle measured at different temperatures. This mirror rotates by 4.0° at 15 V. The curves almost overlap each other. When the temperature increases up to 100 $^{\circ}C$, the fluctuation of the rotation angle increase is 0.14°. This corresponds to 3 % of the stroke and the sensitivity of 1.7×10^{-3} $^{\circ}/^{\circ}C$. The rotational spring becomes softer at higher temperature. When the measurement is carried out at the same temperature, the fluctuation of 0.08° is observed in our experiment. Miner et al. examined the high temperature performance of the micromirror of single c-Si. Their micromirror has the stroke of 4.4 or 6.0° for two axes. The sensitivity of the rotation angle is $0.6\text{-}1.2 \times 10^{-3}$ $^{\circ}/^{\circ}C$ [9]. The micromirror for variable optical attenuator can be estimated to have the sensitivity of $\sim 1 \times 10^{-4}$ $^{\circ}/^{\circ}C$ [10, 11]. This micromirror is fabricated from single c-Si with the metal layer on the mirror plate. The stroke of this mirror is relatively small (e.g., 0.4°). The thin film torsion bar of tensile poly-Si is considered to have similar performance to that of c-Si.

11.2.4 Lightening Mirror Plate Keeping Rigidity

The fast response is necessary especially for the high-resolution laser display. The light-weighted mirror is ideal. There have been efforts realizing the flat mirror with the stretched thin film over the drum frame. Satisfying $<\lambda/10$ is not easy when the thin film is used, because the film bends due to the stress and its distribution inside. The large tension inside the film is considered to flatten the mirror plate. Figure 11.8 shows the illustration. The tensile stress increases the stiffness without increasing the mass. Figure 11.9 shows the calculation of the inertia normalized by the value of the solid disk. The lateral axis is the inner radius of the drum relative to the outer radius. The supposed values are 250 μm for outer radius, 50 μm for drum thickness, and 0.5 μm for the film thickness. When the inner radius increases, the inertia decreases. The mirror based on the same concept

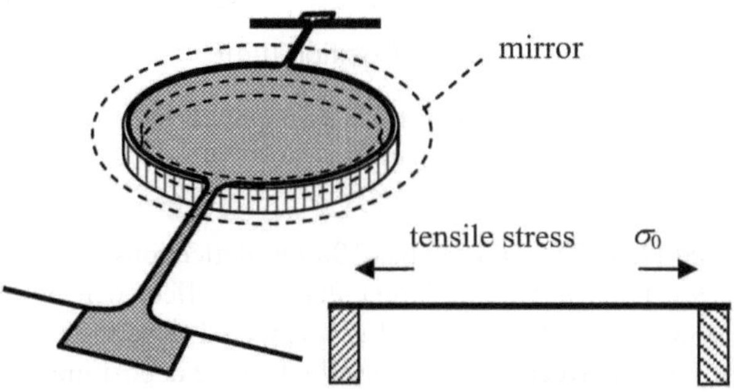

Fig. 11.8 Schematic drawing of the mirror, which consists of a rigid c-Si drum and the thin film

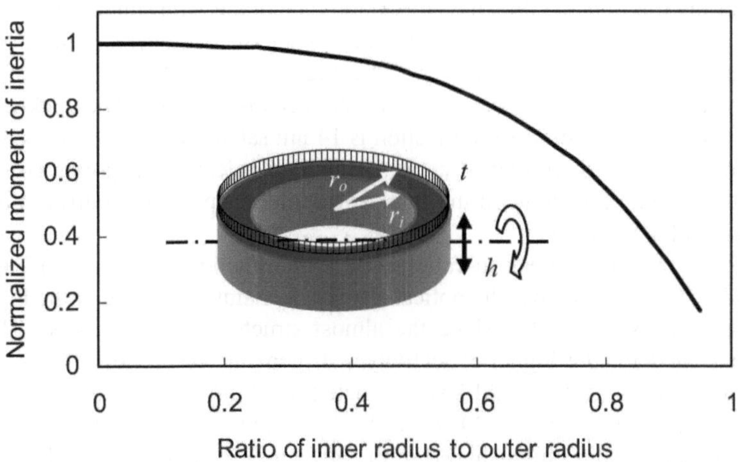

Fig. 11.9 Schematic drawing of the mirror, which consists of a rigid c-Si drum and the thin film

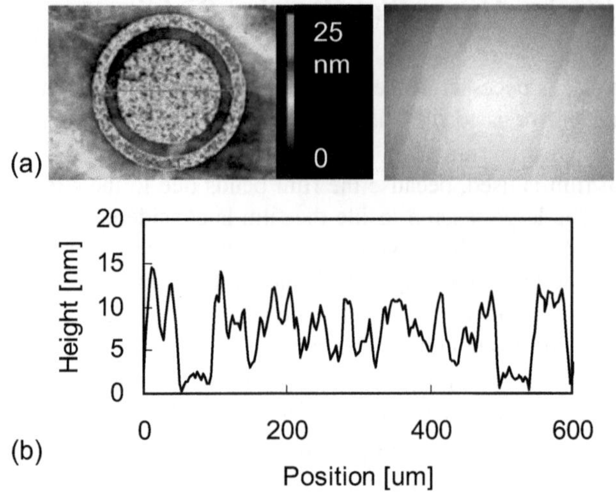

Fig. 11.10 (a) The profile and the optical image of the mirror when the membrane is single poly-Si film. (b) Typical cross-section of the micromirror

has been prepared [12]. Their mirror has 100 nm deflections in peak-to-valley. If the tensile stress can be larger, the larger flattening effect will be obtained. The film prepared is deposited at 590 ^{o}C and is considered to be the mixture of a- and poly-Si phases. The improvement is expected when the deposition is carried out at pure a-Si phase [13]. After the annealing, the crystal matrix formation generates the strong tensile stress shrinking the volume and removing hydrogen.

The outer radius r_o and inner radius r_i are 250 and 190 μm, respectively. h is the height of the ring. t is the thickness of the top membrane. The supposed values are 200 μm for h and 0.5 μm for t. The moment of inertia can decrease to 62 %. The resonant frequency will increase by 27 %. Figure 11.10(a) shows the profile of the fabricated mirror. The tensile stress of ~600 MPa is obtained. The thickness of a-Si film is 500 nm. The mirror deflection is 14 nm satisfying $<\lambda/10$ for visible light. Figure 11.10(b) shows typical corss-section. Two depression points corresponds to the pinning positions on the ring. The center shape of the mirror is almost flat within the noise level.

As for the Si mirror, the peak-to-valley surface deformations are 20 and 15 nm at 20 and 110 ^{o}C, respectively. The optical flatness is maintained even at the high temperature. This is reasonable since the almost structural material is Si. The metal layer coated for the higher reflectivity will generate the bending moment due to the different CTE. When Al layer is coated by 100 nm, the peak-to-valley deformations are 16 and 18 nm at 20 and 110 ^{o}C, respectively. Although Al/Si bilayer generates the bending moment due to the different CTEs, the optical flatness is maintained. This mirror flatness satisfies $<\lambda/10$ supposing the visible light.

As for the surface roughness of the films, as-deposited a-Si shows 0.75 nm Ra. In generall, the amorphous film gives the smooth surface suitable for the mirror surface. Even after the crystallization, the surface roughness is still 0.80 nm Ra. The slight increase of the surface roughness is observed due to the crystallization.

11.3 Summary

There are specifications for the micromirror elements. The topics discussed in many micromirror especially driven by the electrostatic actuator are described. They are about rotation sensor, highly compliant suspension spring, and lighted mirror plate. They are all fundamental but can be an answer to the requirements. The further research and the systematic organization of the technology will clear the future direction.

Acknowledgments There were supports from MEXT grant-in-aid for scientific research on priority areas (no. 19016003, 17040003), support program for forming strategic research infrastructure from 2008, Knowledge Cluster Initiative (the Second Stage)~Tokai Region Nanotechnology Manufacturing Cluster, electro-mechanic technology advancing foundation 2007-2008, JST seeds finding and testing program 2005. Main collaborations are with K. Hane, H. Miura, Tohoku University, S. Kumagai, Toyota Technological Institute. Facilities used for the device fabrication include the Micro/Nano-Machining Research and Education Center at Tohoku University.

References

1. Sasaki M, Tabata M, Haga T, Hane K (2006) Piezoresistive Rotation Angle Sensor Integrated in Micromirror, Jpn J Appl Phys 45, 4B: 3789–3793
2. Aonuma T, Kumagai S, Sasaki M, Tabata M, Hane K (2009) Characteristics of Improved Piezoresistive Rotation Angle Sensor Integrated in Micromirror Device, Jpn J Appl Phys 48, 4: 04C191
3. Hah D, Huang S T, Tsai J, Toshiyoshi H, Wu M C (2004) Low-Voltage, Large-Scan Angle MEMS Analog Micromirror Arrays With Hidden Vertical Comb-Drive Actuators, J Microelectromechanical Systems 13: 279–289
4. Sasaki M, Yuki S, Hane K (2006) Large-Rotation and Low-Voltage Driving of Micromirror Realized by Tense Thin Film Torsion Bar, IEEE Photon Technol Lett 18, 15: 1573–1575
5. Sasaki M, Yuki S, Hane K (2007) Performance of Tense Thin Film Torsion Bar for Large-Rotation and Low-Voltage Driving of Micromirror, IEEE J Selected Topics in Quantum Electronics 13, Issue 2: 290–296
6. Sekimoto M, Yoshihara H, Ohkubo T, Saitoh Y (1981) Silicon Nitride Single-Layer X-Ray Mask, Jpn J Appl Phys 20: L669–L672
7. Senturia S D (2000) Microsystem Design, Chapter 9, Springer Science+Business Media, Inc, New York
8. Sasaki M, Fujishima M, Hane K, Miura H (2009) Simultaneous Realization of Stabilized Temperature Characteristics and Low-Voltage Driving of Micromirror Using Thin Film Torsion Bar of Tensile Poly-Si, IEEE J Selected Topics in Quantum Electronics, 15, September/October accepted
9. Miner A, Milanovic V (2007) High temperature operation of gimbal-less two axis micromirrors, Proc IEEE/LEOS Int Conf on Optical MEMS & Nanophotonics, TuP19: 91–92

10. Isamoto K, Kato K, Morosawa A, Chong C, Fujita H, Toshiyoshi H (2004) A 5-V operated MEMS variable optical attenuator by SOI bulk micromachining, IEEE J Selected Topics in Quantum Electronics, 10: 570–578
11. Tei M (2005) Development of optical MEMS devices in Santec, technological trend, case study, and future, Seminar textbook, No.170422, Technological Media Center (in Japanese), 22 April
12. Nee J T, Conant R A, Lau K Y, Muller R S (2000) Stretched-film micromirrors for improved optical flatness Proc IEEE 13th Annual Int Conf Micro-Electro-Mechanical Systems: 704–709
13. Sasaki M, Sasaki T, Hane K, Miura H (2008) An Optically flat micromirror using stretched membrane with crystallization-induced stress, J Optics A, 10: 044004 (8pp)

Chapter 12
Non-Contact On-Chip Manipulation of 3-D Microtools and Its Applications

(Magnetically Modified Soft Microactuators for Particle Manipulation)

Fumihito ARAI[1] and Yoko YAMANISHI[1]

Abstract This chapter describes a novel non-contact manipulation system using an on-chip, magnetically driven micro-tool (MMT) rather than conventional bio-manipulation by hand, which has a higher risk of contamination and lower success rate and repeatability. MMTs can sort particles individually, and have the unique feature that they can be installed directly in a microchannel (width = 150 μm), unlike conventional cell sorting systems. The drive unit was significantly improved by focusing magnetic field on-chip and reduced the magnetic interaction region. Also, we have investigated an active size controlled droplet generation system by using magnetically driven microtool (MMT). With a lateral motion of the MMT in microchannels, the continuous phase can be pinched off by the movement of MMT to obtain size-controlled droplets actively.

12.1 Introduction

Improvement in microchip production driven by advances in micro/nano fabrication in chemistry and biology have led, for example, to an integrated microchip for chemical analysis that effectively conducts chemical reactions in microchannel microspaces [1], [2]. A microchip that can culture cells has been developed to maintain the desired environment by controlling oxygen, nutrients, and waste materials in the microchannel, where the transportation medium can be single-phase or multiphase flow, typically containing microbeads, chemicals, or cells. Sorting

[1] Fumihito ARAI and Yoko YAMANISHI

Department of Bioengineering and Robotics, Tohoku University microparticles in microchannels offers great potential in biology, chemistry, and environmental analysis.

In general, when microparticles are sorted in microchannels, the effects of centrifugal force and gravity on the particles of the order of several tens of microns in size are relatively small, compared to large-scale sorting. Therefore, large-scale sorting techniques are not applicable to sorting of microparticles in a microchannel. Instead, a fluid force, electrostatic force, and optical force are commonly used to sort microparticles in microchannels.

Cell sorters based on flow cytometry, for example, sort cells in a continuous cell-laden flow, collecting sorted cells in cell suspensions [3]. Cell sorting is generally conducted either electrically (by charging cells) or by mechanically moving receiving dishes after a shot of laser light. The laser pulse is controlled to place a single cell in a single droplet by pushing the diluted cell suspension into an air phase. Such systems tends to be large and expensive, however, so low-cost cell sorter have been developed on a microchip. There are many methods of microscale cell sorting, such as dielectrophoretic sorting [4], laser trap [5], magnetic isolation, and switching in microfluidic channels using microvalves [6]. Systems are designed to fit the characteristics of the sorting object being sorted – e.g., by size, carrier liquid condition, and the sorting speed.

The most popular microactuators that can be applied in the confined space of microchannels are electrostatic microactuators, optical tweezers [7], and magnetic microactuators [8]. Coulomb force is often used in manipulating cells of the order of 10 μm, while a high voltage must be applied to manipulate particles of the order of 100 μm, which risks damaging cells by a heat generation. Dielectrophoretic force is adjusted by varying the gradient of the squared value of the electric field, but this is controllable only in the limited region adjacent to the electrodes, and requires higher voltages to sort larger objects. Optical tweezers manipulate cells indirectly using microtools to reduce the risk of damaging cells during manipulation, however, the force generated is on the order of several pN, which is not suitable for manipulating cells on the order of 100 μm. On the other hand, the magnetic sorting limits the risk of cell contamination [9], and it has been used in many studies because of its low cost.

12.2 Magnetically Driven Microtools Actuated by a Focused Magnetic Field for Separating of Micro-Particles

12.2.1 On-Chip Sorting by MMT

We had developed magnetically driven microtools (MMT) in a previous research [10]-[14]. The basic concept of MMT was proposed in 2004 for noncontact manipulation of microscopic particles such as oocytes, cells and microbeads. The developed MMT can be applied to cell sorting involving the risk of contamination, and the MMT and microchannel were made of disposable materials. The development of polymer-based flexible and biocompatible MMT using magnetite powder and photolithography [12], [13], [14] provides many microchannel functions, such as valves, stirrer, and loader [14]. Microdevices actuated by magnetic force are a simple in structure, low-cost, and easy to integrate with the microchips. Non-contact sorting provides an especially robust, reliable solution unaffected by the properties of sorted objects, and can be used to sort relatively large particles on the order of 100 μm. In MMT sorting, chip capabilities should be enhanced by combining with other functioned MMTs. This promises to help automate complex on-chip microparticle manipulation processes. To do so, however, it is crucial that we integrate as many functions as possible in limited on-chip space. We thus focused on magnetic field concentration.

First prototype developed is shown in Figure 12.1 (a) and (b). The system consisted of two modules -- an upper module containing a disposable microchannel and a lower actuation module. Permanent magnets reportedly provide significant advantages in microscale systems effectively [15]. To actuate the soft magnetic MMT, our first system used a drive unit and permanent magnet unit. The density of magnetic flux generated by the electromagnetic coil was amplified by a permanent magnet unit mounted between the microchannel and the magnetic circuit, and a MMT was moved by non-contact actuation. The maximum actuation frequency of MMT was about 18 Hz. However, in terms of the integration of chip, the concentration of the magnetic field was restricted by the size of the permanent magnet. Therefore it was required to produce a drive unit which can concentrate the magnetic field in the limited region without using the permanent magnet unit.For the current study, we have solved these problems by focusing magnetic field of drive unit by using a couple of pins as shown in Figure 12.1 (c) and (d). To integrate MMT, we fabricated magnetized MMT using a composite of neodymium powder and PDMS. Also, we assembled a magnetic pin unit under a biochip to focus density of magnetic flux and to decrease magnetic field interference area.

In our experimental setup (Figure 12.2), the direction of the current in the coil of the magnetic circuit can be switched to reverse the electromagnet's polarity, and the density of magnetic flux generated by the electromagnetic coil is focused by the magnetic pin unit fabricated under a biochip, and the MMT is moved by non-contact actuation. The MMT was installed in the microchannel of the PDMS chip before bonding the chip and cover glass, applying 3.85 V and 0.15 A to electromagnetic coils. The density of magnetic flux at the tip of pin was 6.8 mT.

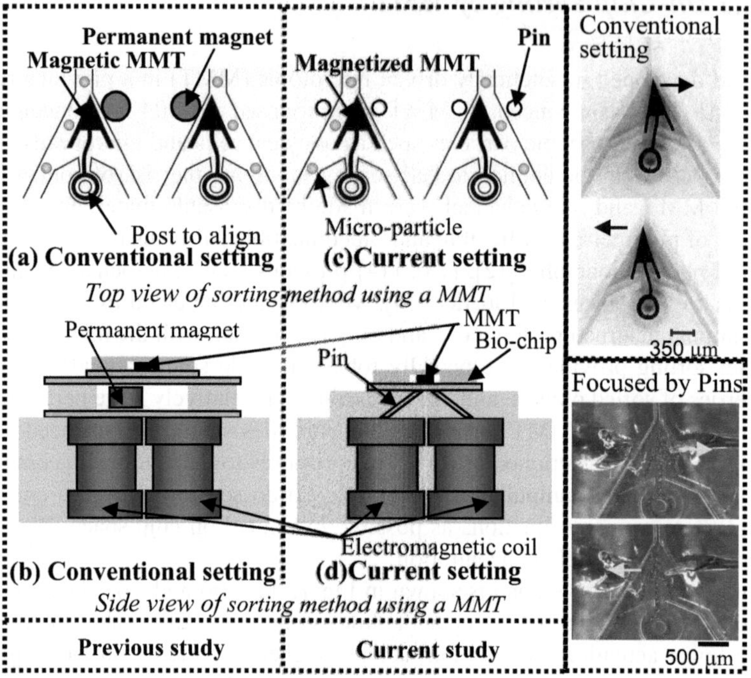

Fig. 12.1 Schematic of experimental setup

12.2.2 Fabrication of MMT and Microchannels

Figure 12.3 shows the fabrication process of the MMT, which is basically involves: (1) patterning SU-8 over the silicon substrate, and producing MMT mold by photolithography; (2) mixing PDMS and neodymium ($Nd_2Fe_{14}B$, 50 wt%) and spreading it over the patterned mold and baked it on a hotplate (100℃, 15 min); (3) spin-coating the PDMS over the layer fabricated in (2) and baking in an oven (90℃, 10 min); (4),(5) separating MMT from the mold with PDMS, and peeling MMT from PDMS; and (6) magnetizing MMT. The surface of MMT was Teflon coated with CF_4 gas by plasma ashing method (discharge power: 130 W) for 30 minutes to avoid any stiction in microchannels. The use of magnetic powder in the polymer has great advantages in arbitral shape during fabrication. The baked PDMS, from which the catalyst evaporates, is biocompatible, enabling it to be widely used as a biochip material. It is difficult to fabricate fine, complex forms from ferromagnetic materials. To manipulate cells, the shapes of actuator tend to be complex to accurately manipulate and protect sensitive cell material. The current shape of MMT sorters is simpler, but the MMT itself can be fabricated in more complex shapes for more complex functions.

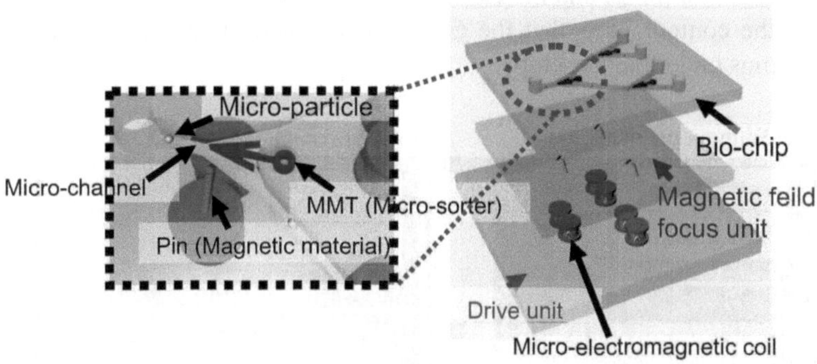

Fig. 12.2 Schematic of experimental setup

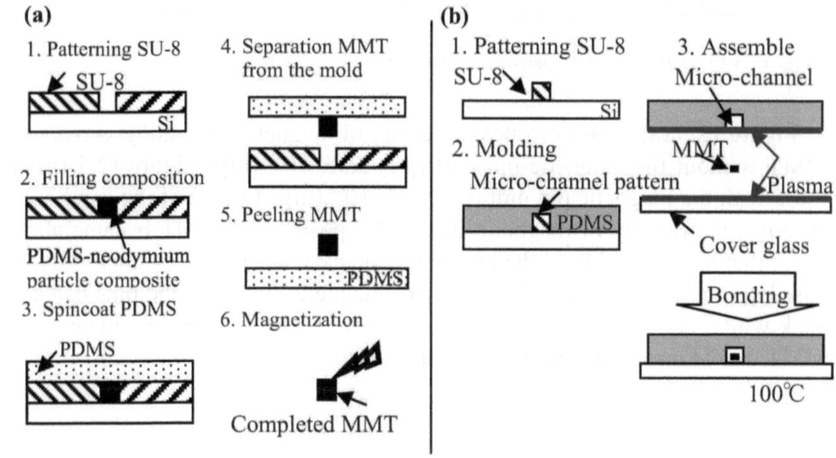

Fig. 12.3 Process flows of (a) MMT, (b) MMT installed microchip

12.2.3 Finite Element Analysis (FEM) of Magnetic Flux Density

After magnetization of the MMT, we have tried to actuate the MMT without using a permanent magnet unit. The MMT could be actuated only by the drive unit of electromagnetic coils, however the actuation power was insufficient. To strengthen magnetic flux density, we used a magnetic pin unit for focusing magnetic field and for actuation of the MMT. The actuation module consists of two electromagnetic coils and two magnetic pins under a bio-chip. The density of magnetic flux generated by the electromagnetic coils is focused by the magnetic pins under a biochip, and the MMT is moved by switching the polarity of the electromagnet's polarity.

For the setting with and without pins, the distribution of density of magnetic flux near the MMT was calculated by FEM using COMSOL software, and shown in Figures.12.4 (a) and (b) respectively. The total number of turns of electromagnetic coil was 500 times, and the electric current of the coil was 0.12 A. It is clear to see in the contour plots that the generated magnetic flux was focused by the magnetic pins under the bio-chip.

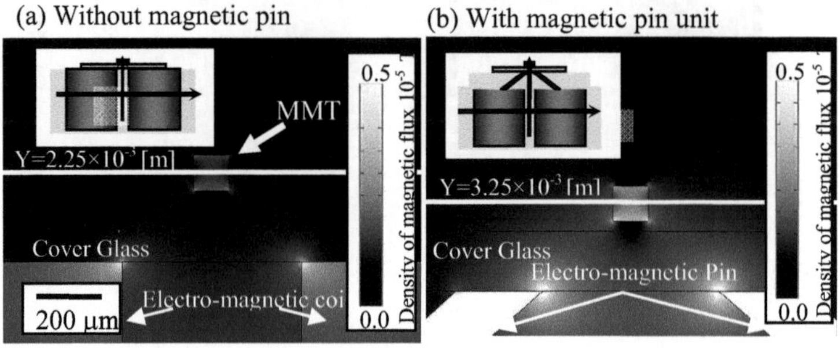

Fig. 12.4 FEM analysis of density magnetic flux

Figure 12.5 (a) shows profiles of density of magnetic flux along x-axis near the MMT without the magnetic pin unit (y = 2.25×10^{-3} [m]). Figure 12.5 (b) shows those with the magnetic pin unit (y = 3.25×10^{-3} [m]). In case of the condition with the pins, the density of magnetic flux at the position of MMT became about 1.8 times that of setup without the pin unit. It is clear to see that magnetic field is focused at the tip of the pin. The density of magnetic flux with the pins had a single peak while that without the pin had three peaks. By focusing the magnetic field, MMT can be installed within limited spaces.

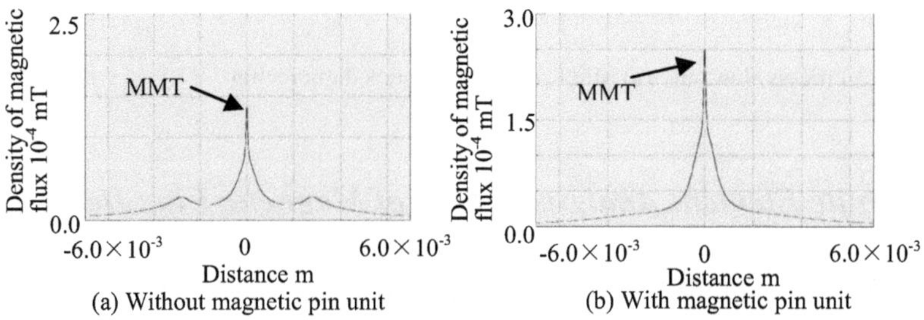

Fig. 12.5 Profiles of density of magnetic flux

12.2.4 Magnetic Flux Interaction

Figure 12.6 shows the magnetic interaction region of the previous setting with a permanent magnet and current pin setting. The magnetic interaction region was obtained by measuring the region, where other MMTs were moved by the applied current and voltage to the main MMT which was located between two electromagnetic coils. The measurement was carried out on the glass substrate and a circle drawn by several measured distances of magnetic interference (Figure 12.6, dashed line). The interference region was formed by moving a circular interference region of a permanent magnet (for pins, the region was equivalent to two circles). We confirmed that the size of interaction for the current setting was one twentieth of previous setting.

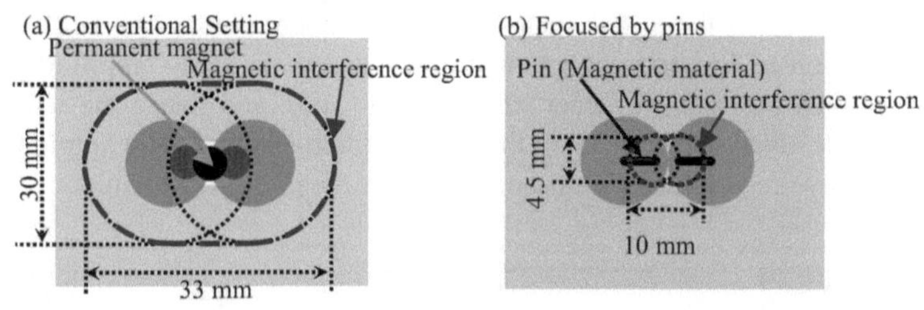

Fig. 12.6 Region of the interferences of the density of magnetic flux

12.3 On-Chip Production Droplets Using Magnetically Driven Microtool

12.3.1 Droplet Dispensing by MMT

Another function of MMT was developed for dispensing droplets. The MMT with valve function was directly installed to the microchannel. The MMT operated with a lateral motion by non-contact magnetic actuation, then MMT can act as "chopper" to disintegrate the multiphase-flow and the size of the dispensing droplet was actively controlled. The microchannel of dispersed phase was designed to be located at different height from the main microchannel for continuous phase, therefore only the dispersed phase microchannel can be switched selectively to be opened or closed. We have successfully control the size of dispensing droplet actively by using the MMT. This system can change the droplet size on-demand, and hence the size-classified droplets can be produced without any separation and collection of droplets after the production.

Figure 12.7 shows the three-dimensional plan view of the microchannel. We have fabricated two columns made of PDMS in the microchannel so that a couple of ring of MMT can fit them. These columns contribute to the simple and accurate installation of MMT and provide stable lateral actuation of MMT.

Figure 12.8 shows the photos of microchannel after the installation of MMT. Figure 12.8(b) shows the cross section at A-A' in Figure 12.8(a). As shown in the figure, the microchannel of dispersed phase is located in the middle height of the microchannel of continuous phase. The magnified image in Figure 12.8 shows the area of droplet dispensing in the microchannel. The MMT has a characteristic of softness (Young's Modulus ≈ 5 MPa) , and we have fabricated a square hole (200×500 μm) in the MMT which helps to reduce the drag force when the MMT is in the lateral motion by reducing the contact area of MMT and microchannel. The narrow arched-shaped microchannel located above the MMT was fabricated to remove any bubbles at the initial stage of the experiment whose exit is closed during the operation of the experiment of droplet dispensing. The amplitude of MMT is larger than the width of microchannel (200 μm), and hence the lateral motion of MMT is always restricted by the wall of microchannel. The amplitude of MMT is fixed by the width of the microchannel for the present experiment.

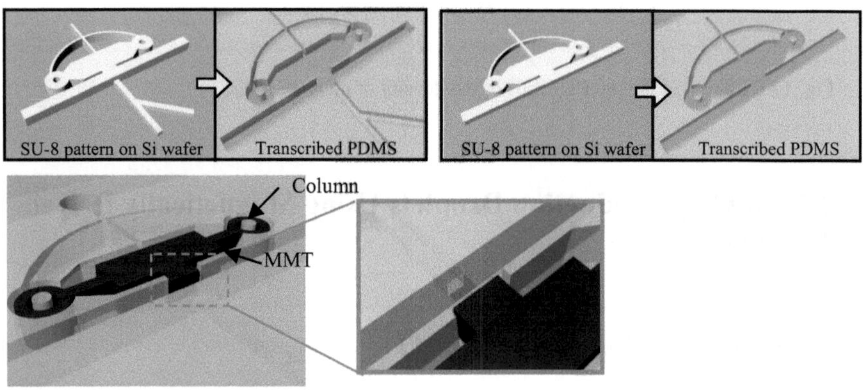

Fig. 12.7 Fabrication of 3-D microchannel and installation of MMT

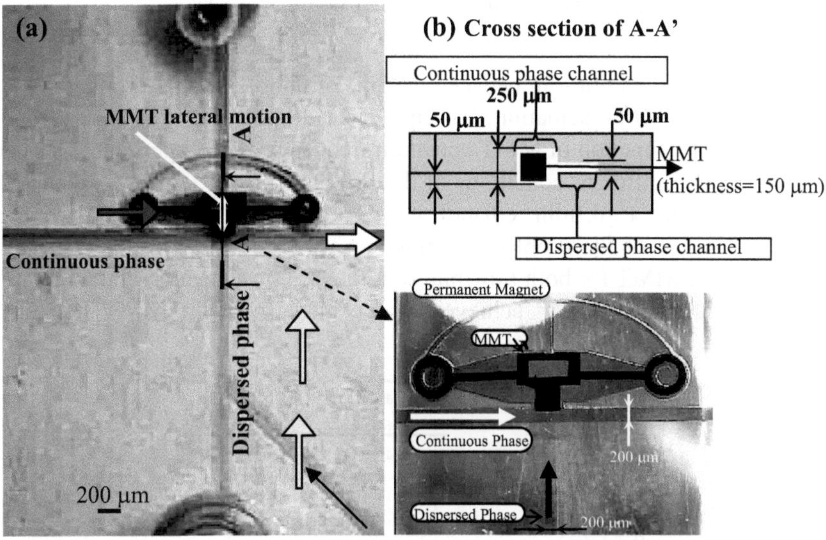

Fig. 12.8 (a)Top View of the MMT installed biochip (the flow is dyed with methylene blue) and (b) Cross section of A-A'

Figure 12.9 shows the photos of the droplet dispensing experiment. The liquid for continuous phase was ethanol and the dispersed phase was olive oil. The shape of the dispensing droplet was ellipse shape due to the limitation of the height of the microchannel of 200 μm, therefore the size of the droplets was evaluated by using the equivalent diameter. Figure 12.9 shows the representative condition (oil: 1 μm/L, ethanol: 5 μm/L) of droplet dispensing and Figure 12.9(a) shows the droplet dispensing without the actuation of MMT and Figure 12.9(b) shows the droplet dispensing with the MMT actuation (0.6 Hz). As you can see in the figures, the size of the droplet was effectively reduced with the actuation of MMT. Therefore it can predict that the size of the droplet and the frequency of the MMT has a certain relationship.

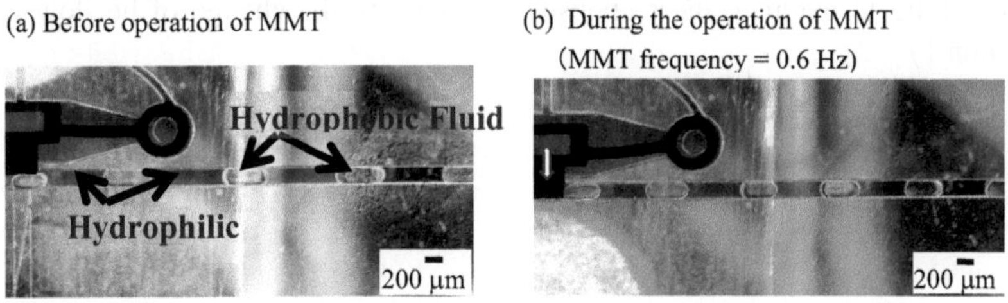

Fig. 12.9 Operation of droplet dispensing by MMT

12.3.2 Evaluation of Droplet Size

To evaluate the relationship between the size of the dispensing droplet and the frequency of the MMT actuation, the equivalent droplet size was measured by CCD images as a function of MMT actuation frequency. The experiment was carried out with two different conditions of the flow rate (1.oil: 1 μL/min, ethanol 5 μL/min and 2. oil: 1 μL/min, ethanol: 4 μL/min). As you can see in the Figure 12.10, the size of the droplet is monotonously reduced with increase of the frequency of the MMT for both flow rate conditions. Consequently, about the 15% reduction of droplet size was achieved by the increase of the MMT frequency from 0 to 5 Hz.

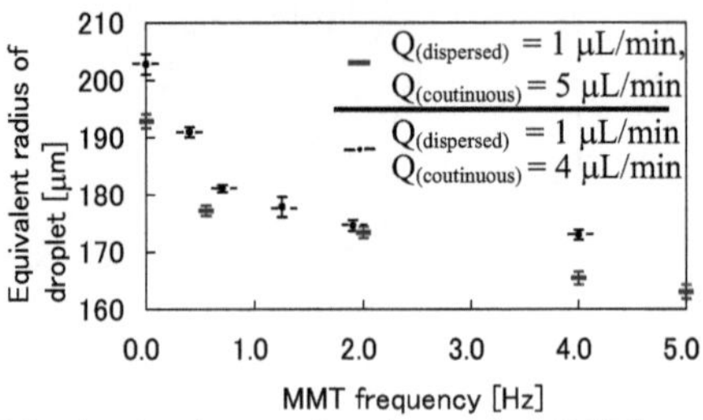

Fig. 12.10 Profiles of the droplet size as a function of the MMT frequency

It is important to control the size of the droplet to the size of the particle or cell which is enclosed in the droplet within a limited time. Therefore we have evaluated the response time of dispensing of the droplets by two different methods which are 1. by the change of the flow rate of the continuous and dispersed phase. 2. by the change of the frequency of MMT. To compare the time-lag under the same condition, first of all, we fixed the diameter of the droplets before and after the actuation at about 190 μm and 170 μm respectively. For example, the flow rate for oil of 1 μL/min and for ethanol of 5 μL/min produce the droplet whose size is about 190 μm, and hence we have set the flow rate condition as the initial setting. Also, the flow rate for oil of 1 μL/min and for the ethanol of 8 μL/min produced the droplet whose size is about 170 μm, therefore we have set the condition as second setting to evaluate the response time only by the change of he flow rate (condition 1).

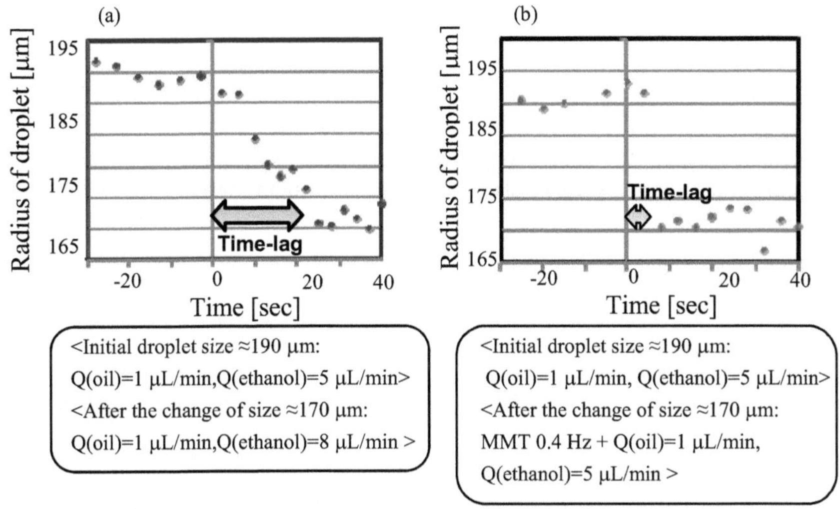

Fig. 12.11 Response of changing the size of the droplet by changing the flow rate

On the other hand, the droplet size becomes about 170 μm when the actuation frequency of MMT was set at 0.4 Hz with the initial setting of the flow rate (the flow rate for oil of 1 μL/min and for ethanol of 5 μL/min). Hence we have set the MMT frequency of 0.4 Hz as the second setting to evaluate the response time by the change of MMT actuation frequency (condition 2). The flow in the microchannel was controlled by two microsyringe pumps (Kd-Scientific model 230) for oil and ethanol, and which was used for both experimental conditions of 1 (by the change of flow rate) and 2 (by the change of the frequency of MMT). In summary, condition 1; the flow rate was set at 1 μm/min, 5 μm/min for oil and ethanol respectively before changing the size of the droplet, and the flow rate of ethanol was changed to 8 μm/min to change the size of the droplets, whilst condition 2; the flow rate was fixed to 1 μm/min, 5 μm/min for oil and ethanol respectively, and the MMT actuation frequency was set at 0.4 Hz to change the size of the droplet. Figure 12.11(a) shows the response time of the change of the droplet size by the change of the flow rate. It was about 20 sec. On the other hand the response time of the change of the droplet size by the change of the MMT frequency (Figure 12.11(b)) was about 7 sec. Therefore it was confirmed that the response time by the change of the MMT frequency is one third of that by the change of the flow rate for the condition.

Acknowledgments This work was financially supported by the Research and Development Program for New Bio-industry Initiatives and the Ministry of Education, Culture, Sports, Science and Technology Grants-in-Aid for Scientific Research (17040017 & 19016004).

References

1. Song JM, Griffin GD, Vo-Dinh T (2003) Application of an integrated microchip system with capillary array electrophoresis to optimization of enzymatic reactions. Analytica Chimica Acta, 487:75–82.
2. Kaigala GV, Hoang VN, Stickel A., Lauzon J, Manage D, Pilarskib LM and Backhouse CJ (2008) An inexpensive and portable microchip-based platform for integrated RT–PCR and capillary electrophoresis, Analyst, 133:331-338,.
3. Melamed MR, Lindmo T and Mendelsohn ML (1991) Flow Cytometry and Sorting, second edition, Wiley-Liss, New York, USA.
4. Fuhr G, Hagedorn R et al (1991) Linear motion of dielectric particles and living cells in microfabricated structures induced by traveling electric fields, Proc. of IEEE Micro Electro Mechanical systems, 259-264.
5. Arai F, Ichikawa A, Ogawa M, Fukuda T, Horio K and Itoigawa K (2001) High-speed separation systems of randomly suspended single living cells by laser trap and dielectrophoresis, Electrophoresis 22:283-288.
6. Shirasaki Y et al (2002) A Novel Biomolecule Sorter Using Thermosensitive Hydrogel in Micro Flow System. Proc. of the Micro Total Analysis Systems (μ-TAS2002), 925-927.
7. Ashkin A and Dziedzic JM (1987) Optical trapping and manipulation of viruses and bacteria", Science, 235:1517.
8. Pamme N (2006) Magnetism and microfluidics, Lab on a Chip 6:24-38.
9. Crick FHC and Hughes AFW (1949) The physical properties of cytoplasm-A study by means of the magnetic particle method, Experimental Cell Research 1:37-80.
10. Abbott JJ, Nagy Z, Beyeler F and Nelson BJ (2007) Robotics in the Small, IEEE Robotics & Automation Magazine, 92-103.
11. Yamanishi Y, Lin YC and Arai F (2007) Magnetically Modified PDMD Microtools for Micro Particle Manipulation, Proceedings of the 2007 IEEE/RSJ International Conference on Intelligent Robotics and Systems, 753-758.
12. Yamanishi Y, Lin YC and Arai F (2007) Magnetically Modified PDMS Devices for Active Microfluidic Control, μ-TAS2007, 883-885.
13. Yamanishi Y, Sakuma S and Arai F (2007) Magnetically Modified Soft Micro Actuator for Oocyte Manipulation, IEEE International Symposium on Micromechatronics and Human Science (MHS), 442-447.
14. Yamanishi Y, Sakuma S and Arai F (2008) High-accuracy Polymer-based Magnetically Driven Microtool Production and Application, Journal of Robotics and Mechatronics, 20(2):273-279.
15. Yamanishi Y, Sakuma S, Onda K, Arai F (2008) Biocompatible Polymeric Magnetically Driven Microtool for Particle Sorting, Journal of Micro - Nano Mechatronics, 4(1):49-57.

Chapter 13
Shape Memory Piezoelectric Actuator and Various Memories in Ferroelectric Materials

T. Morita[1], Y. Kadota[1], T. Ohashi[1] and T. Ozaki[1]

Abstract In general, the ferroelectric properties, such as a piezoelectricity, a permittivity and an electro-optic function, are controlled by external electrical field. To keep these properties, a contentious voltage-supply has been indispensable. On the contrary, this study proposes to realize memory effects on the ferroelectric materials. As one of the examples, the shape memory effect was demonstrated using the imprint electrical field. Control of the imprint was performed under the severe conditions of a very high electrical field and high temperature. After applying a pulse shaped voltage, the piezoelectric shape was kept under zero electric field. With the opposite pulse voltage, it was confirmed that the shape returned to the initial one. In addition to this shape memory effect, the optical transmittance memory effect and the magnetic force memory were also realized.

13.1 Introduction

Ferroelectric materials have multi-functional properties, such as a piezoelectricity, non-volatile charge and electro-optic functions. For a memory device, the ferroelectric random access memory (FeRAM) has been intensively studied [1-3]. This nonvolatile memory utilizes the remanant polarization in the ferroelectric materials; the principle of operation is based on charge detection. Another application of ferroelectric materials is for piezoelectric devices [4-6]. A large permittivity and electro-mechanical coupling factor are advantages for use of these materials as piezoelectric micro actuator and sensors. In general, however, it is thought that the strain that is obtainable for a piezoelectric actuator is so small, and the use of a large input voltage is indispensable.

For the operation of conventional piezoelectric actuators, the driving voltage is considered so as not to exceed the coercive electrical field. With a perfectly re-

[1] T. Morita, Y. Kadota, T. Ohashi and T. Ozaki

Graduate School of Frontier Sciences, The University of Tokyo
morita@k.u-tokyo.ac.jp

versed polarization, displacement of the piezoelectric results in the same position, because the piezoelectric strain versus electrical field is symmetric (butterfly hysteresis curve), so that reversal of the piezoelectric polarization is meaningless for actuator application.

In this study, we propose a shape memory piezoelectric actuator. Operation of the actuator is based on the polarization reversal; therefore, this approach is totally different from that used for traditional piezoelectric actuators. In the case where a piezoelectric butterfly curve is symmetric with respect to the piezoelectric strain-axis, the actuator will not have a shape memory effect. However, the butterfly curve can be asymmetric; for example, in the form of the piezoelectric thin films, asymmetric butterfly piezoelectric curves were reported to be due to an electrical imprint field. Figure 13.1 shows the principle of the piezoelectric shape memory effect by control of the imprint field.

Various functions in ferroelectric materials, not only piezoelectric strain, but also permittivity, refractive index, optical transmission and so on, have also butterfly characteristics against electrical field as shown in Fig.13.1. By using the imprint electrical field control technology, we can obtain the new memory effects in these properties. In addition, the shape memory piezoelectric actuator can generate another memory effect by combing other functional materials such as magnetostrictive material. In this report, after demonstrating the shape memory piezoelectric effect, the other examples in optical transmission memory effect using PLZT and in the magnetic force memory effect using a magnetostrictive-shape memory piezoelectric actuator composite are introduced.

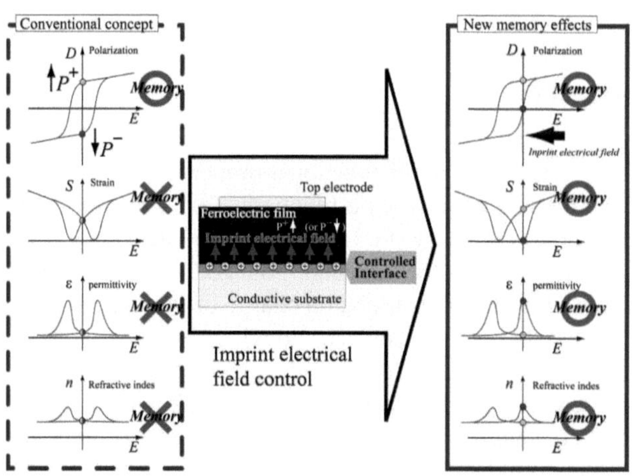

Fig. 13.1 Principle of the new memory effect using ferroelectric materials with imprint electrical field

13.2 Shape Memory Piezoelectric Actuator with Imprint Electrical Field

13.2.1 Driving Principle and Advantages of the Shape Memory Piezoelectric Actuator

There are some important advantages of a shape memory piezoelectric actuator. When a conventional piezoelectric actuator is utilized as a mechanical relay switch, the operation voltage must be continuously applied to maintain "off" or "on" conditions. Under these conditions, the power consumption is not zero, because of the leakage current and power supply to the amplifier that consumes electric power continuously.

On the contrary, the shape memory piezoelectric actuator does not require an electrical field to maintain an "on" and "off" condition, because it has two stable positions that are maintained without the use of an electrical field. If the switch mode must be changed from "on" to "off", the polarization is reversed with a pulsed voltage. After this operation no electrical fields is required, as shown in Fig.13.2. Therefore, after reversal of polarization the electrical source can be disconnected and the energy consumption becomes zero. Another advantage is a low voltage requirement. This claim could be considered as unusual, because the operation principle of the shape memory piezoelectric actuator is based on reversal of the polarization, which requires a large voltage. However, the voltage shape can be driven using a pulsed shape. A pulse-shaped voltage can be generated by combining a small voltage source with capacitors and transformer. After the accumulation of charge to the capacitor, the charge can then be used for the operation of the shape memory piezoelectric actuator thorough the transformer as a pulse-shaped voltage.

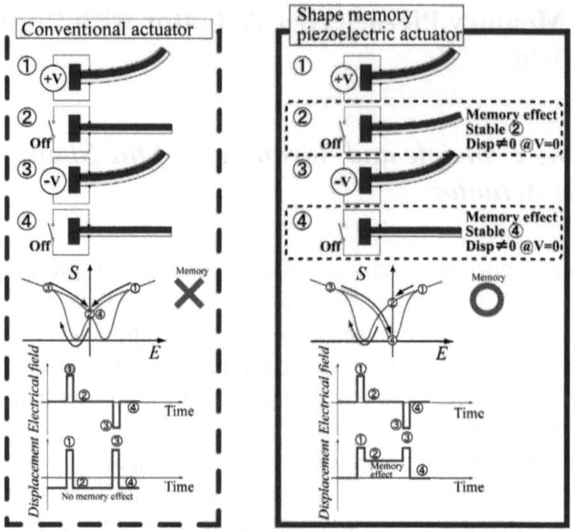

Fig. 13.2 Principle of the shape memory effect compared with conventional piezoelectric driving method

13.2.2 Experiments and Results

To confirm the operation principle for the shape memory piezoelectric actuator, a PZT unimorph type actuator was used as shown in Fig.13.3. The actuator was 0.6 mm thick, 37 mm long (within 28 mm part was PZT) and 13.4 mm wide. The actuator was supplied as a bimorph actuator (Nihon Ceratec Co., Ltd, LPD3713), with 0.2 mm thick stainless steel sandwiched by two 0.2 mm PZT layers. A signal wire for one PZT layer was disconnected, and one PZT layer was operated as a unimorph actuator for simplified operation.

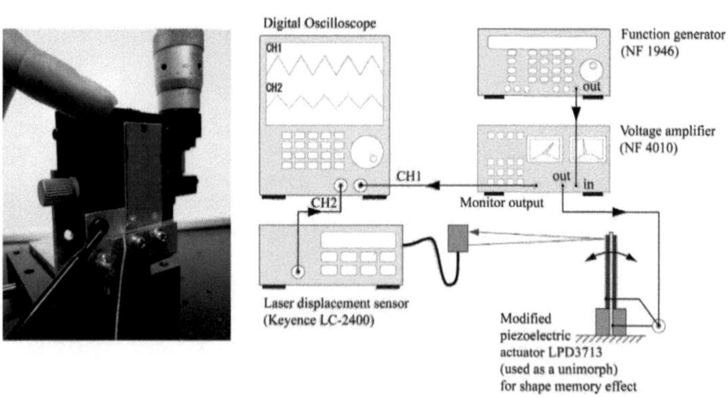

Fig. 13.3 Unimorph shape memory piezoelectric actuator and experimental setup

One end of the unimorph actuator was clamped and the operation voltage was supplied from a signal generator (NF Co., Ltd, WF1946) through an amplifier (NF Co., Ltd, 4010). The piezoelectric displacement was measured with a laser displacement sensor (Keyence Co., Ltd, LC2450). To apply the imprint electrical field, a 700 V DC voltage was applied at a temperature of 150 °C in an electrical oven (Yamato Co., Ltd, DKN302). The electrical field was applied from the top electrode to the bottom electrode (stainless plate). After four hours treatment, the imprint electrical field was obtained as shown in Fig.13.4.

A pulse operation was carried out using this unimorph shape memory piezoelectric actuator. With application of the pulse shaped voltage, the piezoelectric displacement was changed to another stable position, as shown in Fig.13.5. After return to zero electrical field, the piezoelectric actuator exhibited two stable positions, according to the polarization direction as shown in Fig.13.4. The gap between one stable position and the other corresponds to the gap displacement with zero electrical field on the shifted piezoelectric butterfly curve shown in Fig.13.4.

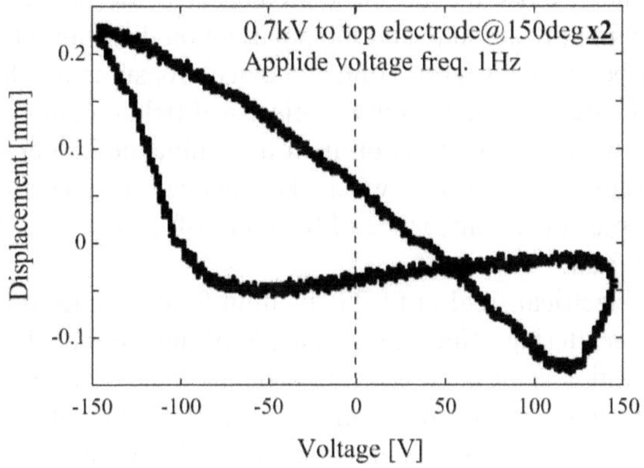

Fig. 13.4 Asymmetric butterfly piezoelectric strain

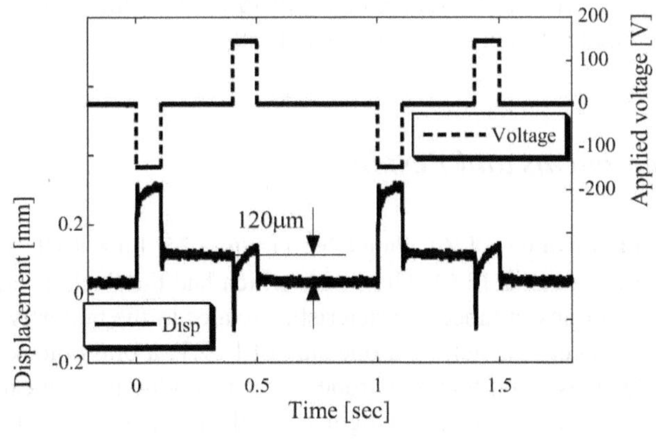

Fig. 13.5 Pulse operation of the shape memory piezoelectric actuator

In this study, a shape memory piezoelectric actuator was proposed and its operation was demonstrated. The shape memory piezoelectric actuator is based on the imprint electrical field, and this imprint electrical field was controlled by a high electrical field treatment at temperature.

13.3 Optical Transmittance Memory Effect

13.3.1 Background and Principle

PLZT is a transparent ferroelectric material that has various nonlinear electro-optic effects, such as variable birefringence and variable light scattering [7-9] which was applied for optical scanners, optical switches and optical shutters[10-11]. Conventionally, those devices need a continued external electrical field so as to remain a certain optical value and the polarization direction of ferroelectric materials does not be reversed. Those optical devices consume much electrical power and need complicated drivers to apply an electrical field continuously.

In this study, the memory effect on light transmittance is proposed using PLZT as an optical shutter or an optical switch. This optical memory realizes a pulse operation, and enables the maintenance of the on or off state in the absence of an external electrical field.

Applying an electrical field to PLZT, its light transmittance changes because of valuable a light scattering effect. The principle of this effect is based on refractive index discontinuities at domain and grain boundaries. Variation of ferroelectric domain size, density of strain-relieving 71° and 109° domains and proportion of rhombohedral to tetragonal grains and domains as functions of polarization induce the valuable light scattering effect. The relationship between the light transmittance and the intensity of the electrical field describes a butterfly-shaped curve, as shown in Fig.13.1. The light transmittance of PLZT becomes maximum when the electrical field matches a coercive electrical field.

13.3.2 Experiments and Results

An optical shutter using a PLZT plate (Zr : Ti = 65 : 35, La = 8.19 %, 5 x 1.5 x 0.5 mm^3) was fabricated (Fig.13.6). The PLZT which had the Pockels effect was used because the light transmittance characteristics relates to the butterfly-shaped curve against external electrical field. On one side of PLZT, a brass block (5 x 1.5 x 30 mm^3) was glued as a bottom electrode with a conductive adhesive (DOTITE 705A), and on the other side, a rectangular top electrode was pasted with the same adhesive. Their side surfaces (5x 0.5 mm^2) were polished to mirror finish using a precision polishing machine (Musashino Denshi MA-150) so that the laser beam can penetrate, whose spot size was 0.48mm.

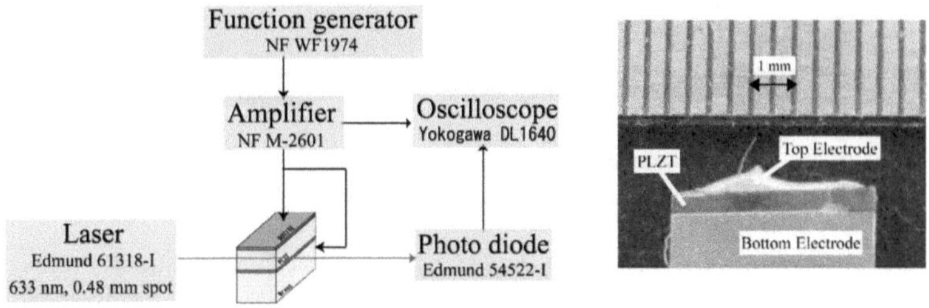

Fig. 13.6 Photo of the PLZT sample and experimental setup

In order to induce an imprint electrical field, a 2.5 kV/mm electrical field was applied in the thickness direction for 10 h at 120 °C in silicone oil. The electrical field was supplied from an amplifier (NF HVA4321) and the silicone oil was heated by an electric hot plate (IKA RH digital KT/C). The electrical field supply was stopped when the temperature of the silicone oil get enough low.

Before and after the electrical imprint field treatment, the change of the light transmittance of PLZT was measured. The laser (633 nm, 0.8 mW, 0.48 mm spot Edmund 61318-I) went through PLZT, and the electrical voltage was supplied from a function generator (NF WF1974) through a high-speed amplifier (NF 4010, M-2601). A photo diode (Edmund 54522-I) was set at about 300 mm distance from PLZT to detect the intensity of the penetrated laser. With a triangular electrical field (1.4 kVpp, 0.1 Hz), the relationship between the light transmittance and the electrical field was measured. After the electrical imprint field treatment, the relationship between the light transmittance became asymmetric and obtained the memory effect.

Using this sample, the light transmittance of PLZT was controlled with the pulse voltages, as shown in Fig.13.7. The intensity of the pulsed voltage was the same intensity of the triangular electrical field for measuring the butterfly-shaped curve. Without an external electrical field, the light transmittance maintained each stable value and was controlled depending on the polarization direction. The light transmittance memory effect was successfully demonstrated. The amount of the memory effect which was the difference between the minor light transmittance and the major light transmittance corresponded to the amount presumed from the asymmetric butterfly-shaped curve.

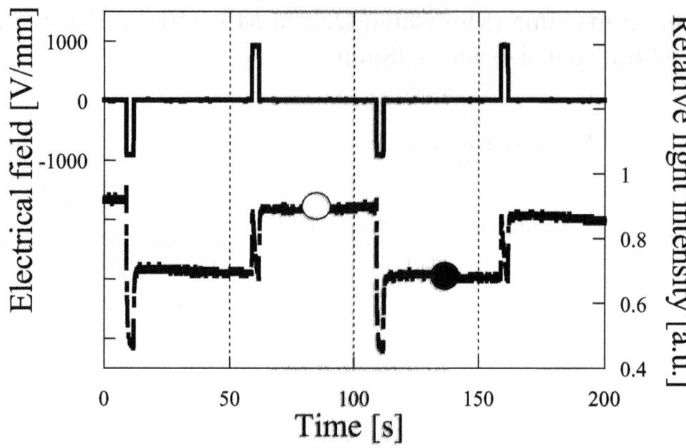

Fig. 13.7 The light transmittance controlled with the pulsed voltages

13.4 Magnetic Force Memory Effect with a Magnetostrictive-Shape Memory Piezoelectric Actuator Composite.

13.4.1 Principle of the Magnetic Force Effect

Magnetic actuators, such as stepping motors, voice coil motors, and solenoids are widely utilized in practical applications. However, there are difficulties in miniaturizing magnetic actuators, due to their complicated coil structure. Furthermore, the magnetic coils are operated with current flow, which results in Joule heating problems, and quick response is restricted due to their large inductive impedance. It is well known that the Joule heating causes most of the energy loss in magnetic actuators.

In recent years, magnetostrictive materials exhibiting giant magnetostriction (over 1000 ppm) have been produced, such as Terfenol-D ($Tb_xDy_{1-x}Fe_2$), and various applications have been investigated [12-13]. A voltage-controllable mechanism was proposed, using this material with a piezoelectric material and a permanent magnet [14-15]. In these studies, the shape change induced by the piezoelectric actuator was utilized to induce a permeability change in a magnetostrictive material. A continuous voltage supply to the piezoelectric actuator was required in order to maintain a desirable magnetic flux density.

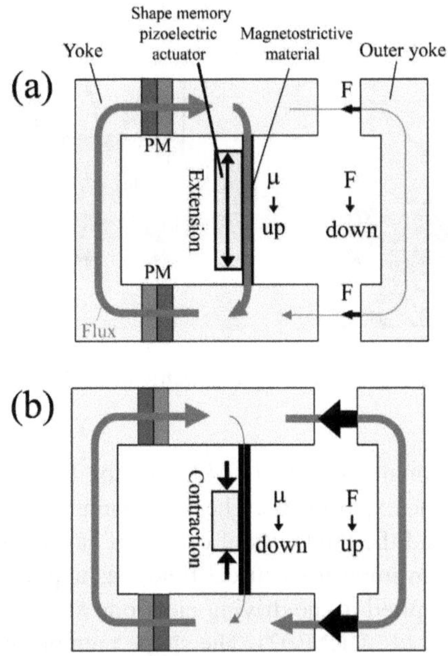

Fig. 13.8 (a) Extension and (b) contraction conditions for the composite model

In our study, a shape memory piezoelectric actuator and a magnetostrictive material were attached to form a composite structure. The permeability of magnetostrictive materials can be controlled by strain as shown in Figs.13.8(a) and (b). The magnetostrictive material can be controlled by voltage applied to the shape memory piezoelectric actuator. The magnetic flux from the permanent magnet flows on two flux paths: the magnetostrictive material and an outer yoke. When a positive voltage is applied, the magnetostrictive material expands, driven by the shape memory piezoelectric actuator as shown in Fig.13.8(a). In this condition, the permeability of the magnetostrictive material becomes large due to its strain. Therefore, a large portion of the magnetic flux from the permanent magnet moves to the magnetostrictive material. As a result, the magnetic force attracting the outer yoke is diminished. To the contrary, when a negative voltage is applied to the actuator, the magnetostrictive material contracts and the magnetic force increases (Fig.13.8(b)). Using this principle, the above two conditions occur when positive and negative voltage pulses are applied to the actuator. Either of two different magnetic forces can be maintained without consuming any electrical power.

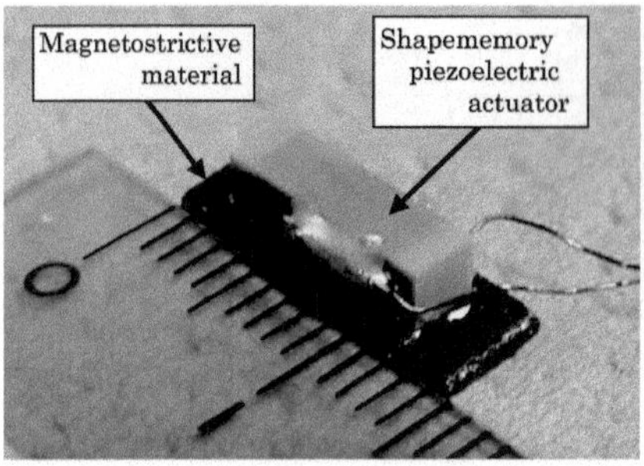

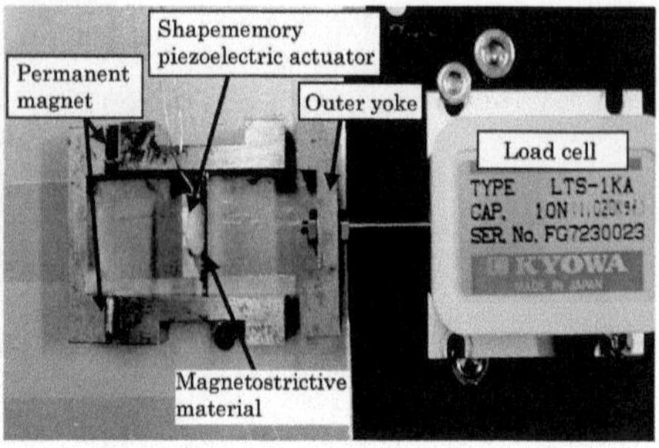

Fig. 13.9 A magnetostrictive-shape memory piezoelectric actuator composite and experimental setup

A multi-layer lead zirconate titanate (PZT) actuator (Nihon Ceratec Co., Ltd., PAC133J) was used in the experiments. The dimensions of the piezoelectric actuator were 3×3×10 mm, and the thickness of each PZT layer was 0.065 mm. To apply the memory effect, an imprint electrical field was applied to the actuator with a 350 V DC potential applied to the driving electrode at 180°C for 3 h in an electric oven (Yamato Co., Ltd., DKN302). The shape memory piezoelectric actuator and the Terfenol-D magnetostrictive material (Etrema Products Inc., $Tb_{0.3}Dy_{0.7}Fe_{1.92}$, 1×5×15 mm) were attached using an epoxy-based adhesive to form a composite structure. Figure 13.9 shows the magnetic circuit, which was composed of a permanent magnet (Nd-B-Fe, 0.24 T), silicon steel (Yoke), and the composite device. The magnetic force was measured using a load cell (Kyowa, LTS-1kA). The operating voltage was applied from a function generator (NF Co., Ltd., WF1946) through a voltage amplifier (NF Co., Ltd., 4010). A strain gauge was attached to the shape memory piezoelectric actuator to measure piezoelectric strain in the longitudinal direction.

The magnetic force memory was controlled using a pulsed voltage as shown in Fig.13.10. The magnetic flux density was dependent upon the change in permeability of the magnetostrictive material. The memory value of the magnetic force was 1 mN, which corresponded to that of the asymmetric butterfly curve at 0 V (not shown here). From these experiments, the expected memory effect was confirmed, and the magnetic force memory maintained a stable value at 0 V.

A composite of a shape memory piezoelectric actuator and a magnetostrictive material was proposed to achieve a magnetic permeability memory effect. The memory effect realized a magnetic force using this system. A device was constructed and successfully operated under a pulsed voltage. In contrast to conventional magnetic devices, this device enables pulsed voltage operation, avoiding the typical Joule heating problems. The simple coil-less structure is an advantage, allowing for easier miniaturization. Optimization of the magnetic circuit is ongoing, and one goal is to maintain a larger magnetic force memory for practical applications.

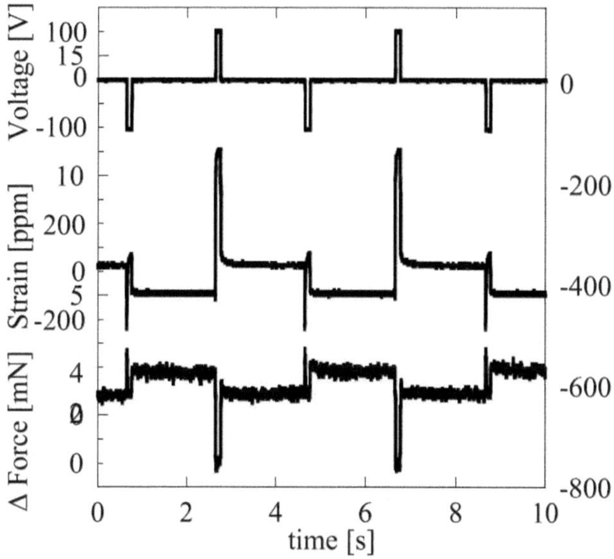

Fig. 13.10 Change in the piezoelectric strain and the magnetic force driven by a pulsed voltage

13.5 Conclusions

It was demonstrated that the various memory effects can be obtained by controlling the imprint electrical field in ferroelectric materials. In this report, we showed only the shape memory effect, the optical transmittance memory effect and the magnetic force memory. However, the imprint electrical field is not limited to only these memories. At least, we have already verified to provide the permittivity memory effect [16] and refractive index memory effect [17].

Control of the imprint was performed under the severe conditions of a very high electrical field and high temperature. Similar to a FeRAM imprint, this shape memory effect is thought to come from the induced charge at the boundary face between the ferroelectric material and an electrode. The mobility of the induced charge has to be taken into account; therefore, for practical applications, reproducibility is important.

Acknowledgments This research was partially supported by the Ministry of Education, Culture, Sports, Science, and Technology through a Grant-in-Aid for Scientific Research on Priority Areas, No. 438, "Next Generation Actuators Leading Breakthroughs", and also by the Murata Science Foundation, the Futaba Electrons Memorial Foundation, Japan Chemical Innovation Institute and Support Center for Advanced Telecommunications Technology Research (SCAT).

References

1. Kohlstedt H, Mustafa Y, Gerber A, Petraru A, Fitsilis M, Meyer R, Bottger U and Waser R (2005) Current status and challenges of ferroelectric memory devices. Microelectronic Engineering 80:296-304
2. Itoh K, Watanabe T, Kimura S and Sakata T (2003) Reviews and Prospects of high density RAM technology. Proc. of 2000 International Semiconductor Conference 13-22
3. Ramesh R, Aggarwal S and Auciello O (2001) Science and technology of ferroelectric films and heterostructures for non-volatile ferroelectric memories. Materials Science & Engineering R-Reports 32:191-236
4. Morita T (2003) Miniature piezoelectric motor. Sens. and Actu. 103: 291-300
5. Muralt P (2000) Ferroelectric thin films for micro-sensors and actuators: a review. J. of Micromechanics and Microengineering 10: 136-146
6. Setter N (2002) Piezoelectric Materials in Devices. EPFL Ceramics Laboratory, Lausanne
7. Land E C (1974) Variable birefringence, light-scattering, and surface-deformation effects in PLZT ceramics. Ferroelectrics 7:45-51
8. Smith D W, Land E C (1972) Scattering-mode ferroelectric-photoconductor image storage and display devices. Appl. Phys. Lett. 20:169-171
9. Land E C, Smith D W (1973) Reflective-mode ferroelectric-photoconductor image storage and display devices. Appl. Phys. Lett. 23:57-59
10. Kosaka T, Fukunaga D, Uchiyama K, Shiosaki T (2007) Epitaxial growth of PLZT thin films and their electro-optic properties. Proc. ISAF2007 843-845
11. Hikita K, Tanaka Y (1989) Dependence of electro-optic properties of PLZT upon the chemical-compositions. Ferroelectrics 94:73-80.
12. Bayrashev A, Robbins P W, and Ziaie B (2004) Low frequency wireless powering of microsystems using piezoelectric-magnetostrictive laminate composites. Sens. and Actu. 114:244–249
13. Ryu J, Carazo V A, Uchino K, and Kim E H (2001) Magnetoelectric properties in piezoelectric and magnetostrictive laminate composites. Jpn. J. Appl. Phys. 40:4948-4951
14. Ueno T and Higuchi T (2006) Novel composite of magnetostrictive material and piezoelectric actuator for coil-free magnetic force control. Sens. and Actu. 129:251-255
15. Ueno T, Keat S C, and Higuchi T (2007) Linear step motor based on magnetic force control using composite of magnetostrictive and piezoelectric materials IEEE Transactions on Magnetics 43:11-14
16. Kadota Y, Hosaka H and Morita T (2008) Utilization of the permittivity memory effect for position detection of a shape memory piezoelectric actuator. Jpn. J. Appl. Phys. 47:217-219
17. Ohashi T, Hosaka H and Morita T (2008) Refractive index memory effect of ferroelectric materials induced by electrical imprint field. Jpn. J. Appl. Phys. 47:3985-3987

Chapter 14
Design and Fabrication of Micro Pump for Functional Fluid Power Actuation System

Yutaka TANAKA[1] **and Shinichi YOKOTA**[2]

Abstract A prototype model of micro pump module for micro fluid power actuation systems is fabricated and experimentally investigated. The functional fluid; ECF (Electro-Conjugate Fluid) is used to be pumping function for the micro pump module of the fluid power actuation system. The fluid power source using the ECF can be embedded in the main body of the micro fluid power actuation systems and can be effective as driving source of the actuator. Some kind of a novel needle-type of micro electrode in the pump module is fabricated by a micro electric discharge machining. The shape and gap of the electrodes on the delivery pressure of the pump due to the ECF jet are experimentally investigated. The pressure due to the ECF jet strongly depends on the number and shape of the needles and the gap of the electrodes.

14.1 Introduction

At present the research on micro-nano technology has been carried out as the seeds in a wide range of research fields. Especially, MEMS technology has a high potential key technology to realize a miniaturization, low power consumption, high density and high integration in many fields such as medical, sensor, information and communication, biotechnology and aerospace industry. Recently, according to advance of semiconductor technology, story of the micro-machine is not a

[1] Yutaka TANAKA
Faculty of Engineering and Design, Hosei University, Fujimi, Chiyodaku, Tokyo 102-8160, Japan

E-mail: y_tanaka@hosei.ac.jp

[2] Shinichi YOKOTA
Precision & Intelligent Laboratory, Tokyo Institute of Technology, Nagatsuta, Midori-ku, Yokohama 226-8503, Japan

E-mail: syokota@pi.titech.ac.jp

dream because of development of micro processing and fabrication technologies. In 1989 gear trains and motors on a wafer have been firstly reported by using the semiconductor micro fabrication technology in U.S.A [1]. Semi-conductor processing technology applied to these fabrications has touched off since the latter half of 1980's, and the research specialized in microscopic environment has begun to be paid to attention rapidly all over the world. There are many micro integrated elements such as micro mirrors and acceleration sensors applied to the semiconductor manufacturing process for the practical use. These elements are made on the flat surface of the silicon wafer with fine fabrication and high integration.

Recent increasing trend of strong demands to develop smaller and functionally more reliable components and works of various applications (not only in semiconductor industry) can be observed in every advanced country of this world. Moreover, it is the most important technical issue to develop novel concept of devices with the best use of characteristics on not only a miniaturization of conventional products but also emphasis of future in microscopic environment.

Various actuation principles, for example piezoelectric, electrostatic and thermal expansion have already been reported for the micro actuation system [2]. However, fluid power actuation system has a great potential to realize large output power density comparing to the conventional actuation system under the microscopic environment [3]. In conventional fluid power actuation system, however, there are many mechanical parts in the system, for example valves for actuator control and pumps for power source.

Our final goal is to develop a new micro fluid power actuation system with accumulating, integrating, miniaturizing and high performance on one single-chip as shown in Fig.14.1. The system consists of four modules, a signal and sensor module, a pump module, a valve module, and an actuator module. This is all-in-one design and fabrication unit consisting of the actuator, the valve, the pump and the sensor modules and functions.

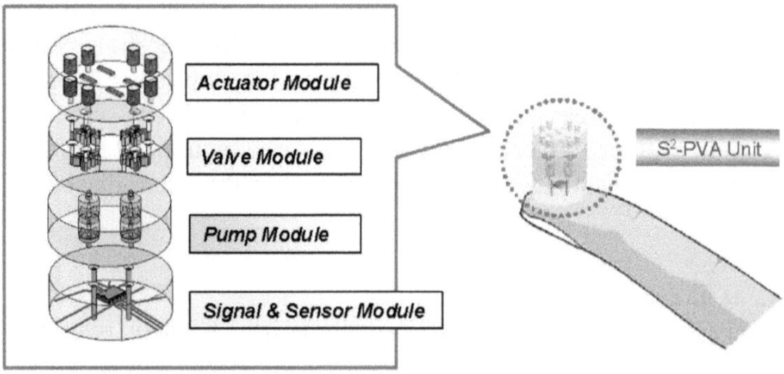

Fig. 14.1 Concept of micro fluid power actuation system

In this chapter in order to realize micro fluid power actuation systems, a high performance micro pump module using functional fluids is proposed, fabricated, and experimentally investigated. A prototype model of micro pump module for micro fluid power actuation systems is fabricated and experimentally investigated. The functional fluid; ECF (Electro-Conjugate Fluid) is used to be pumping function for the micro pump module of the fluid power actuation system. The fluid power source using the ECF can be embedded in the main body of the micro fluid power actuation systems and can be effective as driving source of the actuator. Some kind of a novel needle-type of micro electrodes in the micro pump module is fabricated by a micro electric discharge machining. The shape and gap of the electrodes on the delivery pressure of the pump due to the ECF jet are experimentally investigated.

14.2 Electro-Conjugate Fluid

The electro-conjugate fluids (ECF) recently developed by Otsubo and Edamura [4] are dielectric functional fluids having high insulation. The ECF occurs the strong jet flow that is impressed high electrical voltage of direct from a positive electrode to a negative electrode inserted into the fluid. The phenomenon is well known as one of the electrodynamics effects. The ECF can directly convert electric energy into kinetic energy of the fluid without mechanical moving parts. It is able to be a much attractive approach for applications in novel micro pump devices [5, 6, 7]. In our previous study a small planar pump using the ECF as driving source of the actuator has been fabricated [8].

It is well known that the fluid velocity of the jet generated by means of the ECF is almost in proportion to the applied voltage between the positive and negative electrodes. The pump utilizing the principle referred to above, which is of the simple structure because of being devoid of movable portions, is endowed with excellent features such as compactness, exceedingly high integration, extremely high density with the micro pump. Such excellence in performance makes it possible to suppress generation of noise and vibration. Thus it can safely be affirmed that the pump is adequate to be used for micro-driving. The Dibutyl decanedioate (DBD), which has long been put to practical use as standard ECF, is made use of in this study. DBD is the electro-conjugate fluid with the density of 936.34 kg/m^3 (@20 degrees Celsius) and viscosity of 9.07×10^{-3} Pa·s (@20 degrees Celsius).

14.3 ECF Pump Module

The cylindrical type of micro ECF pump has been deigned and fabricated. Figure 14.2 illustrates a proposed cylindrical ECF pump for micro fluid power actuation system. The positive needle electrode with a number of array pins is located on one side of a narrow tube. The negative electrode with ring geometry is assigned to outlet port of the tube. The chamber in the tube is filled with the ECF. When a dc voltage is applied between the positive and negative electrodes, the ECF jets are locally generated from the tip of positive needle electrodes to the negative ring electrode by applied non-uniform electric field. Due to the ECF jet flow, the local flows are generated in the tube and the pumping function with output pressure and flow rate is obtained.

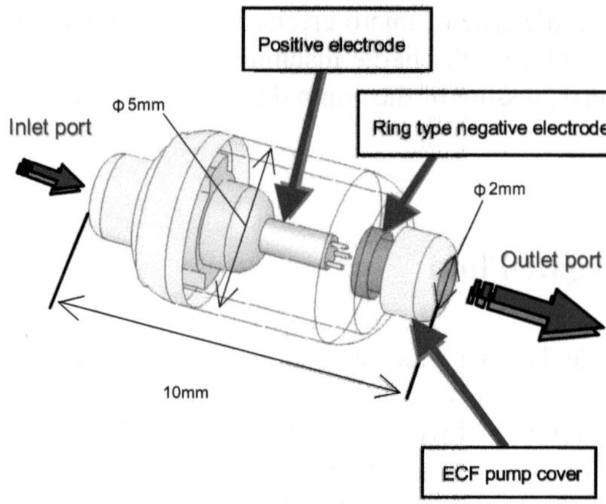

Fig. 14.2 Schematic of cylindrical ECF pump

The ECF jet intensity depends on the applied electric field, and the strength of the electric field is also related to the electrode configuration. The shape and gap between the electrodes are the most important factor to design and improve the pumping performance. In our proposed cylindrical ECF pump, the configuration of the electrodes and the gap distance between the positive and negative electrodes strongly depend on the ECF jet intensity. In our study we made several electrode patterns to investigate experimentally how is the strength of the ECF jet influenced by the type of electrodes and gaps.

Some kind of the novel needle electrode with the micro pins for the cylindrical ECF pump is fabricated by a micro Electrical Discharge Machine (micro EDM). The micro EDM has a manufacturing potential as the most precise and convenient way of such precise and complicated electrode.

The processing procedure of the positive electrode by the micro EDM is illustrated in Fig.14.3. At first a rod electrode is formed to a cuspidate needle electrode by a wire electro-discharge grinding (WEDG) method [9]. Next step the cuspidate electrode with negative polarity is fed into a plate electrode to make a sharp edged groove or hole. The groove or hole array pattern is repeatedly formed according to the configuration of the array pins of the needle type electrode. The polarity of a new thick rod electrode is then reversed, and the new rod electrode is fed into the plate electrode on the grooves or holes. As a result the machining electrode with sharp edged pins is fabricated according to the configuration of the structure. The configuration of the needle-type electrode depends on the configuration of the plate electrode with the grooves and holes.

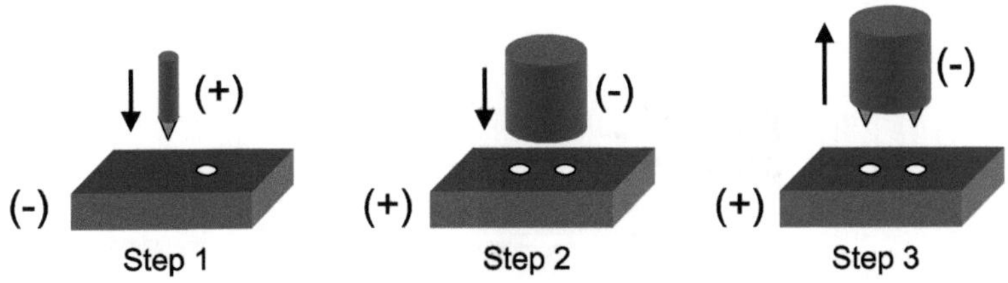

Fig. 14.3 Process of needle electrode fabricated by micro EDM

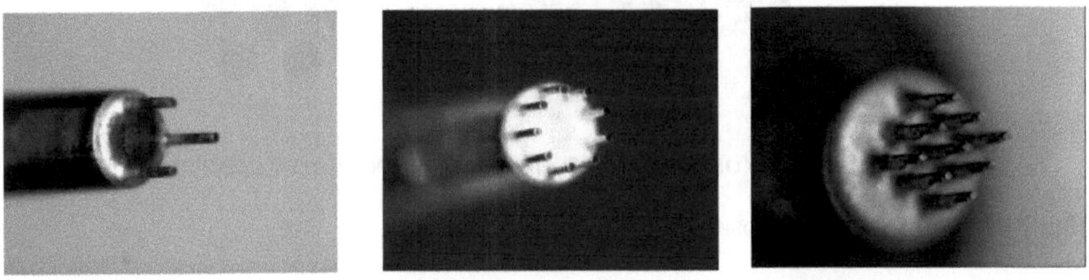

(a) Circular array of 3 pins (b) Circular array of 9 pins (c) Square array of 9 pins

Fig. 14.4 Typical examples of fabricated three types of positive needle electrodes

The typical examples of needle electrodes fabricated by the micro EDM are shown in Fig.14.4. Circular arrays with three pins (a) and nine pins (b) electrodes and square array with nine pins (c) needles are fabricated on the tip of an initial electrode with a diameter of 1 mm. Each pin has a sharp tip and a diameter of 100 μm. We consider that the strong ECF jet can be obtained by increasing the number of array pins on the needle type of the positive electrode and by putting the configuration of the array pins in a circular arrangement along the ring type of the negative electrode. As a material of electrode, we selected a cemented carbide for positive electrode which has high-strength and easily fabricated by EDM and a brass for negative electrode which has easily fabricated by a machining tool.

Figure 14.5 shows typical examples of a pair of positive and negative electrodes for the ECF pump. The positive needle electrode has three or six pins with an each diameter of 100 μm on the tip of the initial electrode with 1 mm diameter. Negative electrode has also three or six holes with an each inner diameter of 300 μm and the equal configuration according to the three or six pins of the positive needle electrodes.

The diameter and arrangement of the negative and positive electrodes on the delivery pressure of the pump due to the ECF jet is experimentally investigated. In supplementary experiments, the isomorphic pair of electrodes has much potential to supply the output pressure and flow rate by the ECF jet. The magnitude of the output pressure in the case of the six pins and holes type of electrode has a maximum potential to generate the strong ECF jet in our experiments. The flow pattern in the case of the multi-pins and holes type of electrodes must be numerically and experimentally investigated, since the output power strongly depends on the flow pattern of the ECF jet.

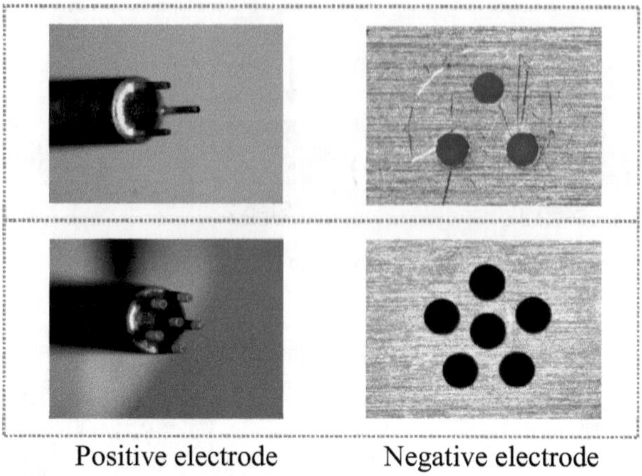

Fig. 14.5 Pair of fabricated positive and negative electrodes

14.4 Pumping Performance

14.4.1 Experimental Setup

The shape, arrangement and gap of the electrodes on the delivery pressure and flow rate of the pump due to the ECF jet are experimentally investigated. The pressure due to the ECF jet strongly depends on the number and shape of the needles and the gap distance of the electrodes. The experimental apparatus for measurement of pressure is illustrated in Fig.14.6. The needle type of the positive electrode is attached with a special fixture on a micrometer stage for feeding. The ring type of the negative electrode is installed in the inlet of a manometer tube. The gap distance between the positive and negative electrodes is adjusted by the micrometer. The output pressure and flow rate are measured by pressure and flow sensors according to the applied dc voltage.

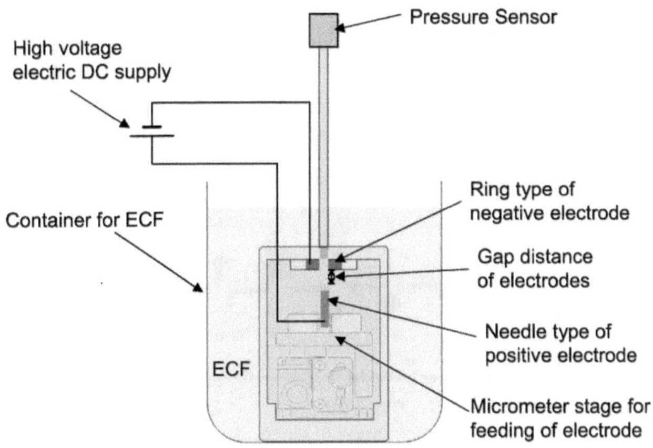

Fig. 14.6 Experimental apparatus for pumping performance

14.4.2 Effect of Gap Distance

The three types of positive electrode as shown in Fig.14.4 and an original electrode without manufacturing are used to measure the output pressure by changing the applied voltage and electrodes gap. The ring type of negative electrode has an inner diameter of 1.5 mm. Relationship between the gap distance and the output pressure is plotted in Fig.14.7. The applied voltage is kept at 5 kV. As the gap distance becomes shorter, the output pressure becomes larger, because the ECF jet depends on the intensity of the electric field. When the electrode gap is adjusted at less than 200 μm, the output pressure cannot be measured because of discharging between electrodes. In the case of the original electrode without manufacturing, the output pressure is measured to be unstable at the gap distance of less than 1.0 mm. Comparing to the original electrode without manufacturing, the needle type electrodes have much potential to supply the output pressure by the ECF jet. The magnitude of the output pressure in the case of the square array type is similar to the one in the case of the circular array type. On the other hand, the circular array type with three pins of electrode has a maximum potential to generate the strong ECF jet in our experiments. Opposed to our expectation about the change of number of the pins, the circular array with three pins electrode has a little larger potential to generate the ECF jet than the circular array with nine pins electrode. The flow pattern in the case of the circular array type electrodes must be numerically and experimentally investigated, since the output pressure strongly depends on the flow pattern of the ECF jet. It is necessary to search into the number of pins on the needle type of electrode depending on the pump performance.

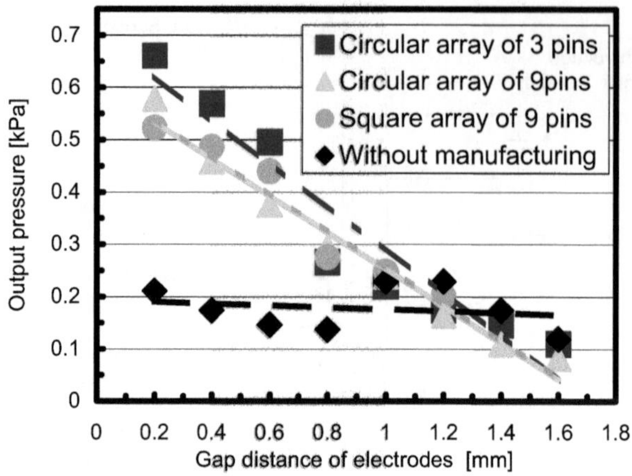

Fig. 14.7 Experimental results of relationship between distance of electrodes and output pressure

14.4.3 Effect of Inner Diameter

Relationship between the applied voltage and the output pressure with change of the inner diameter of the negative electrode D_1 is plotted in Fig.14.8. The case of six pins and six holes type of electrodes, shown in Fig.14.5, is also plotted in Fig.14.8. The diameter of the positive electrode and the gap distance between the negative and positive electrodes is kept at 100 μm and 200 μm, respectively. These conditions are selected by an optimal value of the previous experiments.

The output pressures have changes of quadratic curve according to the applied voltage as shown in Fig.14.8. The inner diameter of the negative electrode has a great effect to the magnitude of output pressure. As the inner diameter becomes smaller, the output pressure becomes much larger.

Comparing to the original electrode with one pin and hole electrode, the isomorphic pair of electrodes shown in Fig.14.5, have much potential to supply the output pressure by the ECF jet. The magnitude of the output pressure in the case of the six pins and holes type of electrode has a maximum potential to generate the strong ECF jet in our experiments.

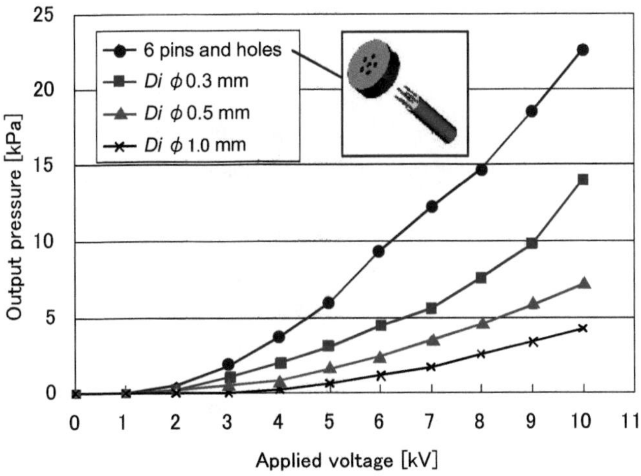

Fig. 14.8 Experimental results of relationship between applied voltage and output pressure

14.5 Cylindrical Type of ECF Pump

According to the experimental results of pumping performance, the prototype of the cylindrical micro pump module as shown in Fig.14.9 is designed and fabricated. A chassis of the pump is fabricated at one process by a 3D optical rapid prototyping method; a micro stereolithography system. Figure 14.10 shows a comparison between the output pressure of the previous developed planar ECF pump [8] and the developed cylindrical ECF pump. The magnitude of the output pressure in the case of the cylindrical ECF pump is much larger than one in the case of the planar ECF pump. The size of the cylindrical ECF pump is much smaller than the planar ECF pump, volume ratio of 1/12. At the maximum, the output pressure of 6.1 kPa (@5 kV) and volumetric flow rate of 172 mm^3/s (@5 kV) have been obtained for the cylindrical type of ECF pump module. The cylindrical ECF pump has a great potential to obtain the output power under the design and fabrication for the micro fluid power actuation system.

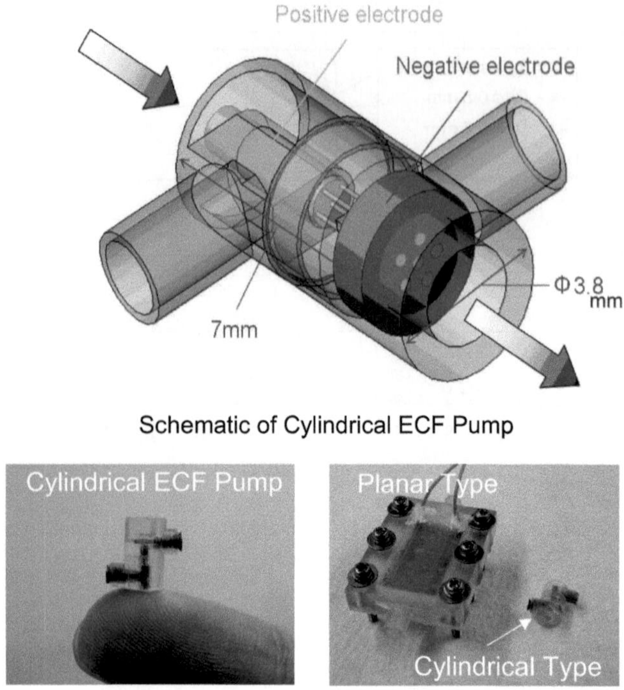

Fig. 14.9 Fabricated micro ECF pump modules

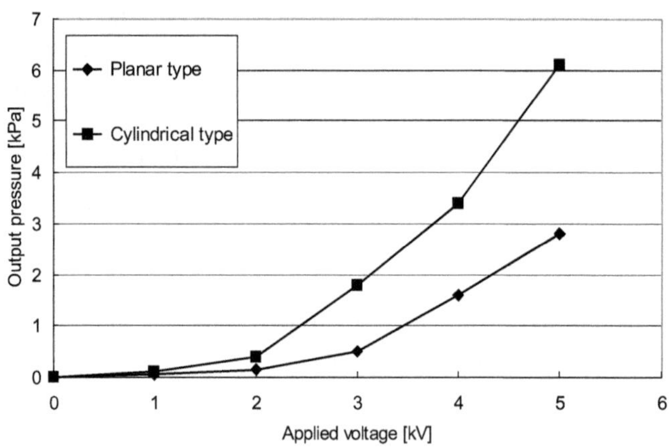

Fig. 14.10 Output pressure for planar type and cylindrical type of ECF pump modules

14.6 Conclusions

The prototype model of the micro pump module using ECF is fabricated and experimentally investigated. Some kind of the novel needle-type of micro-electrode in the pump module is fabricated by the micro electric discharge machining. The shape, arrangement and gap distance of the electrodes on the delivery pressure and flow rate of the pump due to the ECF jet are experimentally investigated. The pressure due to the ECF jet strongly depends on the number and configuration of the pins and the gap distance of the electrodes. The static characteristics of the output pressure for the micro pump module are experimentally investigated. The output pressure of the pumping function is made greater in a shape of a quadratic function complying with the applied voltage. The arragement and configuration of the pin and hole on the electrodes strongly depends on the output pressure and flow rate. According to the experimental results of pumping performance, the miniaturized and high-powered cylindrical micro pump module is fabricated and compared to the previous planar ECF pump. The cylindrical type of the micro ECF pump has a great potential to obtain the output fluid power under the design and fabrication of the micro fluid power actuation system. At the next step of our study, an optimal configuration of the electrodes is selected to realize a high performance of the pump module.

Acknowledgments This study has been supported in part by Grant-in-Aid for Scientific Research in Priority Areas "Next-generation actuators leading break-through" of the ministry of Education, Culture, Sports, Science and Technology of Japan (No. 17040023 and No.19016020).

References

1. Fan LS, Tai YC, Muller RS (1989) IC-Processed electrostatic micro motor. Sensors and Actuators 20:41-48
2. Park JH, Yoshida K, Yokota S (1999) Resonantly Driven Piezoelectric Micro Pump Fabrication of a Micro Pump Having High Power Density. Mechatronics 9 No.7: 687-702
3. Yoshida K, Yokota S (1993) Study on High-power Micro Actuator Using Fluid Power. Proc of 6th FLOMEKO:120-130
4. Ohtsubo Y, Edamura K (1997) Dieelectric Fluid Motors. Applied Physics Letters 71 (3): 318-320
5. Yokota S, Kondoh Y, Sadamoto A, Ohtsubo Y, Edamura K (2001) A Micro Motor Using Electro-conjugate Fluids (Proposition of Stator Electrode-Type (SE-type) Micro ECF motors). JSME International Journal 44 (C-3):756-762
6. Yokota S, Kawamura K, Takemura K, Edamura K (2005) High Integration Micro-motor Using Electro-Conjugate Fluids (ECF). Journal of Robotics and Mechatronics 17 (2):142-148
7. Woo-Suk Seo, Yoshida K, Yokota S, Edamura K (2007) A High Performance Planar Pump Using Electro-conjugate Fluid with Improved Electrode Patterns. Sensors and Actuators A: Physical 134:606-614

8. Tanaka Y, Ziegelheim J, Yokota S (2006) Design and Fabrication of Actuation System Using Functional Fluid. Proc. 19th International Conference on Hydraulics and Pneumatics: 218-228
9. Yamazaki M, Suzuki T, Mori N, Kunieda M (2004) EDM of Micro-rods by Self-drilled Holes. Journal of Materials and Processing Technology 149:134-138

Chapter 15
Intelligent Actuators for Mechatronics with Multi-Degrees of Freedom

Making Mechatronic Systems Simple, Smart and Reliable

Koichi SUZUMORI[1] and Shuichi WAKIMOTO[1]

Abstract This chapter details intelligent servo actuators for mechatronic systems with multi-degrees of freedom. The intelligent actuators are equipped with micro sensors and a micro controller, realizing local control functions of motion, force, and compliance, a communicating function, and a motion generating function.

Three examples of intelligent actuators are described: (1) an intelligent electrical motor and its application to a snake-like robot; (2) pneumatic intelligent cylinders and their application to active polyhedron linkage mechanisms; and (3) a new control method, called multiplex pneumatic transmission drive, for pneumatic systems.

15.1 Introduction

This chapter details new methods to design and fabricate intelligent servo actuators for mechatronic systems that have multi-degrees of freedom (multi-DOF) of motion. The intelligent actuators are equipped with micro sensors and a micro controller, realizing local control functions of motion, force, and compliance, a communicating function between host computers and other actuators, and a motion generating function. These result in reliable and simple distributed control systems and greatly reduce the number of electric wirings or pneumatic tubes connected to the actuators [1, 2, 3].

[1] Koichi SUZUMORI and Shuichi WAKIMOTO

Graduate School of Natural Science and Technology, Okayama University

The goal of this research is to develop intelligent actuators for small mechatronic systems that have multi-DOF of motion such as micro-robots.

An example multi-DOF mechatronic system is shown in Fig.15.1. This example was developed for a physical human-machine interface that makes the operators feel as if they were actually touching 3D continuous virtual objects on a PC. The prototype shown on the left consists of thirty linear actuators, forming a linkage mechanism of icosahedrons [4]. On the right is an image of the final goal of this research, which consists of a great number of intelligent miniature actuators and is expected to work as a pre/post processor for analysis of finite elements methods.

Fig. 15.1 Examples of multi-dof linkage mechanism (left: prototype, right: final goal)

The development of intelligent actuators is one of the keys to realize these mechatronic systems. These intelligent actuators are required to have motion and force sensing functions, control functions with local signal processing, and communication functions between the other actuators and the host computer, which reduce the number of electrical/pneumatic cables.

There are many other examples of multi-DOF mechatronic systems such as robots and micromachines. The goal of this research is to develop methods to design and fabricate intelligent actuators for the increasing requirements of these systems.

This chapter describes three examples of intelligent actuators and their applications that which have been developed in our laboratory: (1) development of an intelligent electrical motor with a programmable system-on-chip (PSoC) based micro processor and its application to a snake-like robot; (2) development of pneumatic intelligent cylinders in which micro-optical encoders, pressure sensors, micro valves and a micro CPU are built and their application to active polyhedron linkage mechanisms; and (3) a new control method, called multiplex pneumatic transmission drive, for pneumatic systems consisting of many pneumatic actuators.

15.2 Intelligent Electrical Motor Realizing Snake-Like Robot

An intelligent electrical motor has been designed and developed for a snake-like pipe inspection robot. The motor consists of a 3[W] DC motor, a reduction gear with 249 reduction ratio, an optical encoder, and a micro controller, resulting in the "intelligence" of a communication function, a trajectory generating function, and a local motion/force servo control function.

Figure 15.2 shows the developed actuator and snake-like robot consisting of twelve intelligent actuators [5, 6]. The robot moves like a snake and travels through pipes. Each actuator receives simple commands on motion from the host computer, generates motion reference, and realizes servo control on force and position. It negotiates complex pipelines consisting of T-shaped branches, elbows, vertical pipes, and different diameter pipes.

Fig. 15.2 PSoC based micro controller built in the intelligent electrical motor (left) and its application to a snake-like robot with twelve joints (right)

Figure 15.3 shows control architecture of the intelligent motor. The micro servo controller mounted on the motor consists of a micro processor (PSoC CY8C 27243) and a tiny motor driver chip. PSoC is a one-chip micro computer made by the Cypress Semiconductor Corp. On the processor, users can customize various types of I/O ports such as pulse counters, digital I/Os, A/D converters, D/A converters, communication ports, PWM module, and operational amplifier. These I/O ports are suitable for integrating intelligent actuators, making it possible to design and fabricate tiny controllers. For this motor design, two pulse counters for the encoder, a pulse width modulator for the motor driver, and an I2C communication port for the host computer are integrated into a PSoC processor, as shown in Fig.15.3.

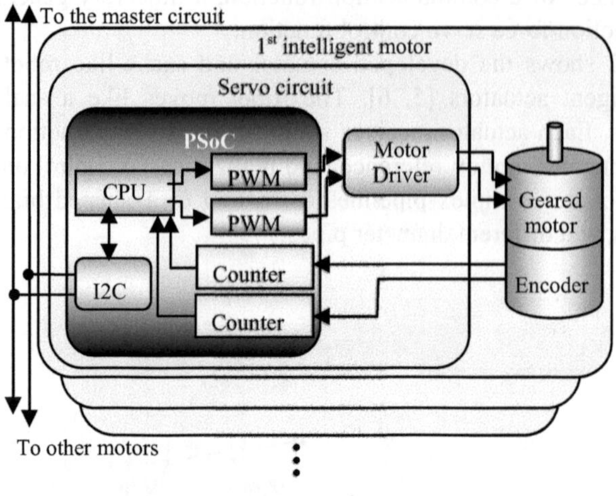

Fig. 15.3 Control architecture of the motor

Two communication protocols are used in this system: one is a RS232C between the host computer and the master circuit that is mounted on the end of the robot, and the other is an inter-integrated circuit (I2C) protocol between the PSoC and the master circuit. I2C communication needs only two lines of serial clock line (SCL) and serial data line (SDL) as shown in Fig.15.3. Connecting twelve motors serially through I2C and a power supply line makes the robot work, resulting in greatly reducing the number of the electrical cables between the motors.

After receiving the command to start from the host computer, the PSoC starts generating the trajectory of the sinusoidal motion and the position servo control. It detects the motor motion from the encoder and controls the duty ratio of the PWM signals to the motor on the basis of the position error.

During the locomotion of the robot, the PSoC is always adjusting the trajectory to adapt the robot motion to the shape of the pipe. If the motor does not keep its motion on the target trajectory, the PSoC find out that the robot joint is touching the pipe wall and so adjusts the control parameters of the amplitude and the center position of the sinusoidal motion. This makes the robot motion suitable even if the pipe diameter changes or the pipe bends.

Figure 15.4 shows examples of pipe negotiating experiments. As shown in the figure, the robot automatically negotiates several kinds of pipes such as diameter changing, elbow, T-branch and vertical pipes. The intelligent motors greatly reduce the number of the electrical cables between the actuators and the host computer and make the control system robust to electrical noises.

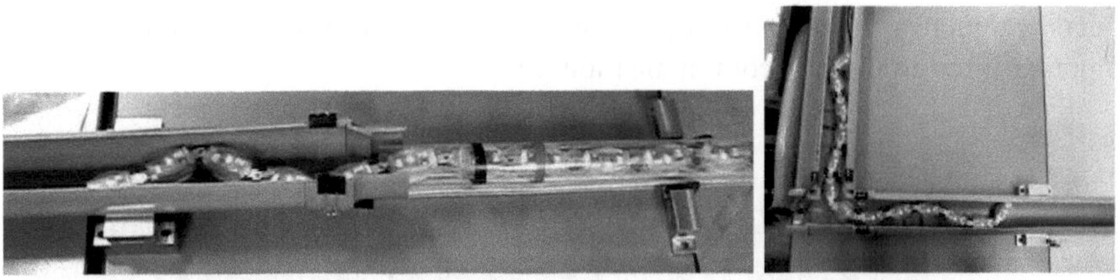

Fig. 15.4 Pipe negotiation experiments: in changing diameter pipe (left) and in T-branch (right)

15.3 Intelligent Pneumatic Cylinders for Active Linkage Mechanisms

Although pneumatic actuators generally have several advantages over other actuators such as their lightweight, simple structure, cheapness, compliance and high power/weight ratio, they require a compressor and pneumatic tubes and are not as controllable as electrical actuators. Improvement of these disadvantages of conventional pneumatic actuators will greatly expand the application fields of pneumatic actuators.

On the basis of this, we have developed intelligent pneumatic actuators. In this section, two examples of our research are described.

15.3.1 Intelligent Pneumatic Cylinder Realizing Active Polyhedron

As described in the introduction, the active polyhedron is a linkage mechanism working as a physical human-machine interface. In this section, intelligent pneumatic cylinders developed for this mechanism are explained [7, 8, 9].

The purpose of this research is to realize a new physical human-machine interaction that can exchange distributed information of force and deformation between people and machines using a distributed model to many force-presentation points. As over a hundred actuators are necessary for this system, development of cheap and light actuators is essential.

Figure 15.5 shows the intelligent pneumatic cylinder developed for this purpose. The cylinder is almost the same size as a conventional cylinder and is equipped with two micro-optical encoders to detect position of the piston rod as shown in Fig.15.5. In the optical encoder chip a micro LED and two pairs of micro optical lens and photo detector are fabricated. On the surface of the piston rod, 0.16mm stripe marker lines are fabricated by oxidizing the stainless steel rod surface using a YAG laser marker. The minimum/maximum lengths of the cylinder are 113 [mm] and 169 [mm], which means the moving stroke is 56 [mm]. On the cylinder, a PSoC processor is mounted that detects the pulse signals from the encoders, calculates the piston rod position, and sends the position data to the host computer through the I2C communication line.

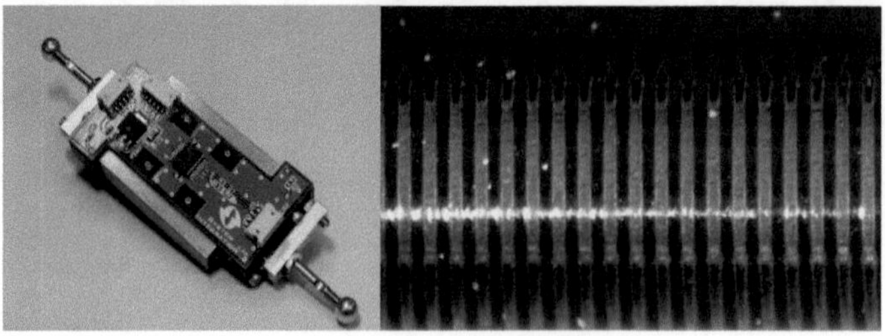

Fig. 15.5 Intelligent pneumatic cylinder and the strip on the guide rod

Figure 15.6 shows the experimental results of a position servo control experiment carried out as a performance evaluation test of the developed intelligent cylinder. It works well as a pneumatic actuator.

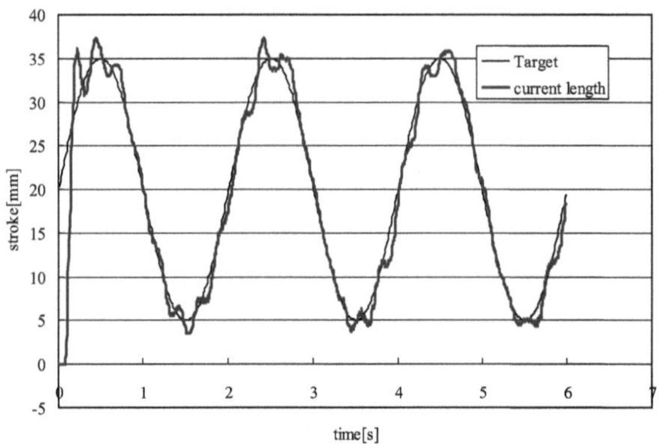

Fig. 15.6 Experimental results of position tracking

The developed 120 intelligent cylinders were assembled into the 80-faced polyhedron with 120 degrees of freedom shown in Fig.15.7. The polyhedron changes its shape to become a small ball 529 [mm] in diameter as shown in the left of Fig.15.7, a big ball 836 [mm] in diameter, and a flat plate as shown in the right of Fig.15.7.

Fig. 15.7 Active polyhedron consisting of 120 intelligent pneumatic cylinders

The polyhedron works as a physical human-machine interface. Figure 15.8 shows an experiment of handling a virtual object on a PC. The shape of the virtual 80-faced polyhedron is constructed with the length data of 120 cylinders on a PC, enabling operators to change the shape of the object in the PC in a similar way to handling clay. In addition, operators feel the reaction forces for altering the virtual object. Reaction forces can be calculated on the basis of the physical property model on the PC, and they are sent to each actuator controller to generate the force. Operators can handle the virtual objects as if they were actually touching them.

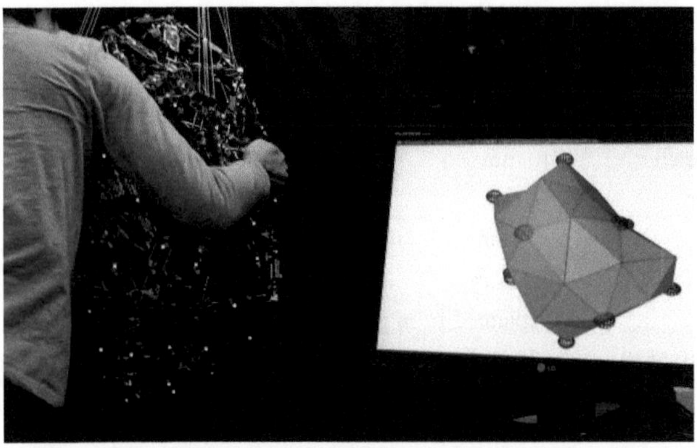

Fig. 15.8 Active polyhedron working as a physical human-machine interface

15.3.2 Intelligent Pneumatic Cylinder Realizing Intelligent Chair

Another new intelligent pneumatic cylinder has been developed that consists of five elements in a single device combining software and hardware architecture design [10, 11, 12]. An optical encoder chip to read constructed laser strips rod and a pressure sensor to detect the chamber pressure are used with two unit valves for the cylinder movements. These hardware parts are controlled using algorithms on PSoC chip board. Figure 15.9 and Table 15.1 show the structure and the specifications of the designed intelligent cylinder.

Figure 15.10 shows an example of the experimental results. Force is controlled by manipulating the pressure differences in the two chambers of the cylinder. PI control is applied to change the PWM duty cycle for the desired output force. As shown in Fig.15.10, the response of the output force follows the step reference force.

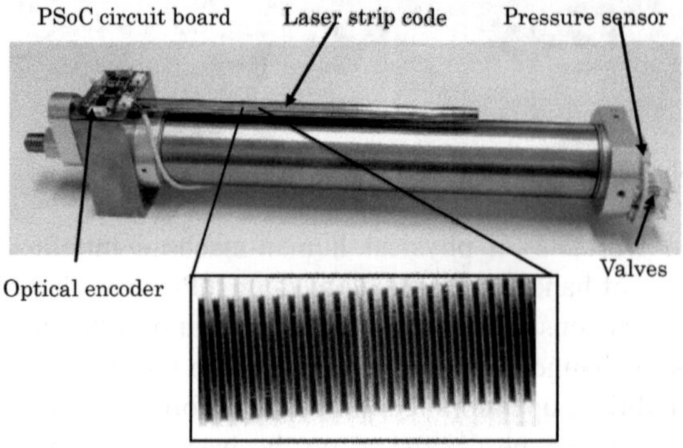

Fig. 15.9 Intelligent cylinder parts and zoomed laser strip code on the guide rod

Table 15.1 Specifications of Intelligent Cylinder

Available Elements	Encoder, Strip rod, Pressure sensor, PSoC board, Valves
Length	388 mm
Cylinder diameter	40 mm
Rod diameter	16 mm
Rod stroke	200 mm
Laser resolution	0.169 mm
Laser strip	Separated from piston rod, together with guide rod
Max force at 0.6MPa	120 N

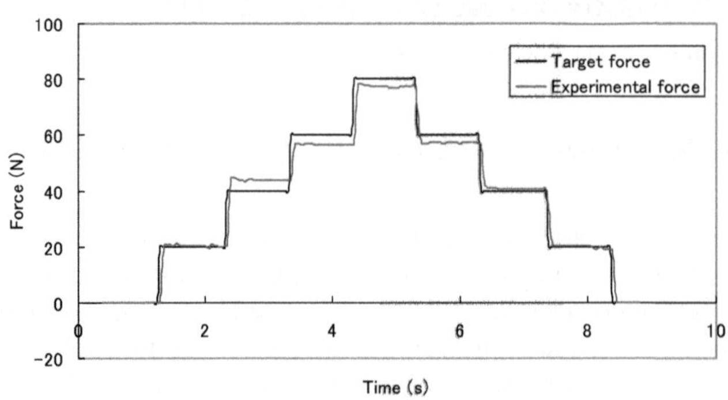

Fig. 15.10 Force Control Experiment

The developed intelligent pneumatic cylinder is applied to an intelligent chair tool (ICT) application, which consists of 20 actuators in the vertical plane and 16 actuators in the horizontal plane. Figure 15.11 shows the overall ICT structure with a combination of 36 links of intelligent pneumatic cylinders to receive physical information and give feedback responses on stiffness and damping from the user dynamics, making it a useful tool for creating future chairs.

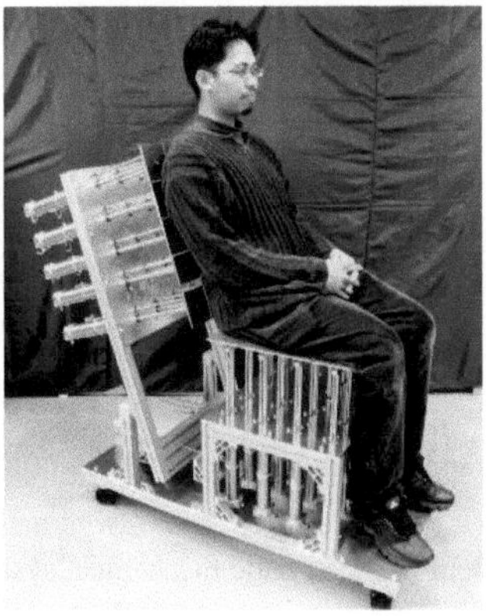

Fig. 15.11 Proposed Intelligent Chair Tool (ICT)

15.4 A New Pneumatic Control System Using Multiplex Pneumatic Transmission

A new pneumatic valve and its operating system have been developed to control pneumatic actuators using air vibration in air supply lines [13, 14]. This valve and operating system realize a new control method for pneumatic systems consisting of many pneumatic actuators only through air supply lines.

The driving mechanism of this valve and operating system is based on a novel idea: superimposing pneumatic waves into the air supply line drives any selected valve(s) connected to the air supply line. Each valve is designed to have a different natural frequency to be driven and is activated when pneumatic vibration at the natural frequency is applied to the air supply line.

Figure 15.12 shows the outline of the pneumatic mechanism to which the proposed control system and the valves are applied. As shown in the figure, an oscillator built in the air supply line vibrates the supply air. Each actuator is equipped with the proposed pneumatic valve(s), which has a natural frequency for activation. Superimposing the pneumatic vibrations at the frequency corresponding to the natural frequency of its valve causes mechanical resonance of the oscillating bodies in the valve to drive the actuator.

Since the superimposed pneumatic waves are transmitted in the air supply line as control signals and also as valve drive energy, electrical cables, which need to be connected to each valve in conventional pneumatic drive systems, are not needed in this system. This results in very simple pneumatic systems with multi-DOF.

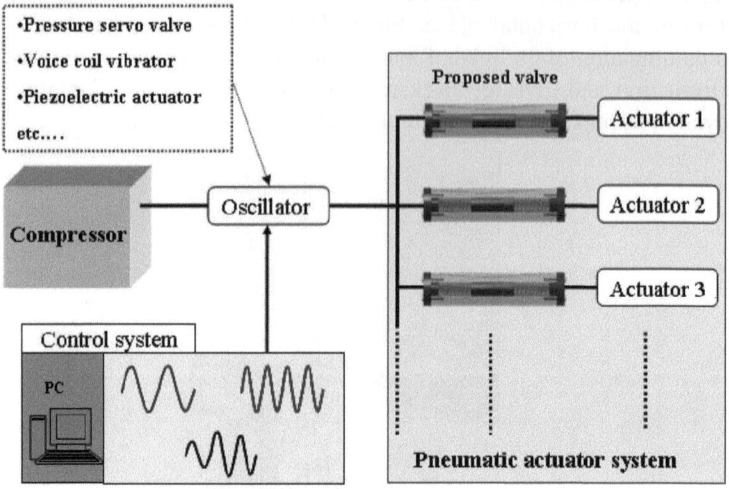

Fig. 15.2 Proposed control system for pneumatic mechanical systems with multi-DOF

Figure 15.13 shows a structure and working principle of the proposed valve. The valve has two oscillating bodies, m_1 and m_2, which have through-holes for air flow, and are supported elastically by two springs, k_1 and k_2, respectively. This mechanical system acts as a spring-mass vibration system. Without the pneumatic oscillation at the natural frequency, two bodies contact by the spring force and no air-flows through the valve, while applying air oscillation of the natural frequency causes the bodies to oscillate separately to cause air to flow through the valve.

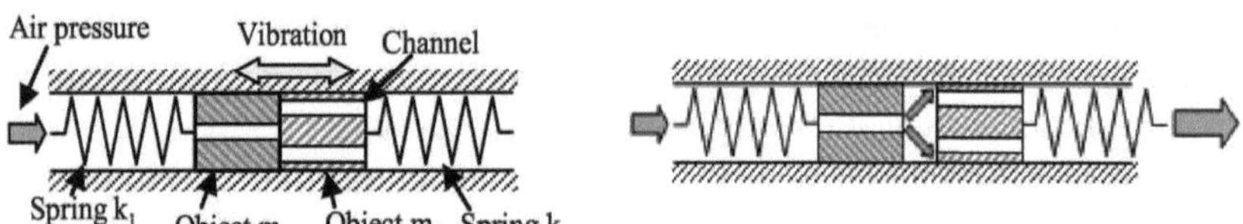

(a) Non-resonance frequency (b) Resonance frequency

Fig. 15.13 Basic structure of the pneumatic on/off valve that is activated when the pneumatic vibration at the natural frequency is applied to the air supply line

Currently, we can successfully control five cylinders independently using this system. We are now trying to increase the number of the controlled cylinders.

15.5 Conclusions

Three examples of intelligent actuators and their applications, which have been developed in our laboratory, were described in this chapter: (1) An intelligent electrical motor with a PSoC based micro processor has been designed, developed, and applied to a snake-like robot, which negotiates variously shaped pipes automatically. (2) Pneumatic intelligent cylinders in which micro-optical encoders, pressure sensors, micro valves and a micro CPU are built have been designed, developed, and applied to active polyhedron linkage mechanisms and a virtual intelligent chair. (3) A new control method for pneumatic systems consisting of many pneumatic actuators has been developed.

While most of the intelligent actuators detailed in this chapter are still at the development stage, some of them have been commercialized, for example, developed intelligent pneumatic cylinders have been modified and are on the marketplace as cylinders with motion detectors function.

The authors believe that intelligent actuators have great potential for future mechatronics.

Acknowledgments This research was supported by a Grant-in-Aid for Scientific Research on Priority Area (No. 438) "Intelligent Actuators for Multi-Degrees-of-Freedom Mechatronics (16078209)" from the Ministry of Education, Culture, Sports, Science and Technology of Japan. The authors would like to Prof. Takefumi Kanda, Okayama University, for his useful technical advice, Mr. Kazutoshi Kono from Koganei Co. Ltd., for manufacturing prototype cylinders, and our students in our Laboratory for their assistance.

References

1. Suzumori K (2007) Intelligent Actuators for Mechatronics with Multi-Degrees-of-Freedom. The 4th Public Symposium on Next-Generation Actuators Leading Breakthroughs:73-76
2. Suzumori K (2008) New Pneumatic Actuators Producing Breakthrough in Mechatronics. The 7th JFPS International Symposium on Fluid Power:197-202
3. Suzumori K, Kanda T, Kosaka K et al (2006) Intelligent Servo Actuators for Multidegrees of Freedom Mechatronics. ACTUATOR 2006:128-131
4. Ochi J, Suzumori K, Tanaka J et al (2005) Development of Active Links for Physical Man-Machine Interaction. Journal of Robotics and Mechatronics 17, 3:293-301
5. Wakimoto S, Suzumori K, Takata M et al (2003) In-Pipe Inspection Micro Robot Adaptable to Changes in Pipe Diameter. Journal of Robotics and Mechatronics 15, 6:609-615
6. Kuwada A, Wakimoto S, Suzumori K et al (2008) Automatic Pipe Negotiation Control for snake-like robot. 2008 IEEE/ASME International Conference on Advanced Intelligent Mechatronics:558-563
7. Suzumori K, Tanaka J and Kanda T (2005) Development of an Intelligent Pneumatic Cylinder and Its Application to Pneumatic Servo Mechanism. AIM 2005:479-484
8. Ogawa H, Kosaka K, Suzumori K et al (2006) Force-Presentation Method for Active Polyhedron for Realizing Physical Human-Machine Interaction. 2006 IEEE International Conference on Robotics and Automation :3941-3947
9. Suzumori K (2009) Development of Active 80-faced Polyhedron for Haptic Physical Human-Machine Interface. IEEE/RSJ International Conference on Intelligent Robots and Systems
10. Ahmad 'Athif Mohd Faudzi, Suzumori K and Wakimoto S (2009) Development of an Intelligent Pneumatic Cylinder for Distributed Physical Human-Machine Interaction. Advanced Robotics 23, 23, 1-2:203-225

11. Ahmad 'Athif Mohd Faudzi, Suzumori K and Wakimoto S (2009) Development of an Intelligent Pneumatic Cylinder for Distributed Physical Human-Machine Interaction. Advanced Robotics 23, 23, 1-2:203-225
12. Ahmad 'Athif Mohd Faudzi, Suzumori K and Wakimoto S (2008) Distributed Physical Human Machine Interaction Using Intelligent Pneumatic Cylinders. 2008 International Symposium on Micro-NanoMechatronics and Human Science:249-254
13. Nishioka Y, Suzumori K, Kanda T et al (2008) Pneumatic Valve Operated by Multiplex Pneumatic Transmission. Journal of Advanced Mechanical Design, Systems, and Manufacturing, 2, 2:222-229
14. Nishioka Y, Suzumori K, Kanda T et al (2008) A New Pneumatic Control System Using Multiplex Pneumatic Transmission. The 7th JFPS International Symposium on Fluid Power:439-442

Chapter 16
Actuation of Long Flexible Cables Using Ciliary Vibration Drive

Masashi KONYO[1] and Satoshi TADOKORO[1]

Abstract Long flexible cables have difficulty in handling their movement with just pulling or pushing. We proposed an actuation mechanism for long flexible cables to get active mobility using a ciliary vibration drive. The ciliary vibration drive generates driving force on a cable by vibrating inclined thin sting or wire cilia. Ciliary bending and recovery movement during vibration makes cilia tips stick and slip rapidly and generates distributed driving force on the cable. We made observation and modeling of physical phenomena of cilia movement to design the optimal ciliary vibration drive. We also determined optimal parameters such as material, a diameter, density and an inclination angle of cilia, and interval of vibration motors on a trial basis. We also developed an active scope camera which was installed the proposed mechanism. A prototype of the active scope camera showed good performance in practical rescue activities. A prototype of a scope camera 8 m long crawls at a maximum speed of 47 mm/s, climbs slopes of 20 deg, surmounts obstacles 200mm high, follows walls, and turns on floors. Experiments at Collapsed House Simulation Facility demonstrate its practical advantage in rubble pile.

16.1 Introduction

Long flexible cables have difficulty in handling their movement with just pulling or pushing. For example, scope cameras and fiberscopes, which are widely used in urban search and rescue and for pipe inspection, sometimes get stuck due to their flexibility. Figure 16.1 illustrates the problem in inserting them manually into narrow gaps. They cannot climb slopes up and surmount obstacles when the pushing force of the operator cannot be delivered for them. This is the limitation of a passive drive. Robotics is expected to drastically improve their abilities by active actuation.

[1] Masashi KONYO and Satoshi TADOKORO

Graduate School of Information Sciences, Tohoku University

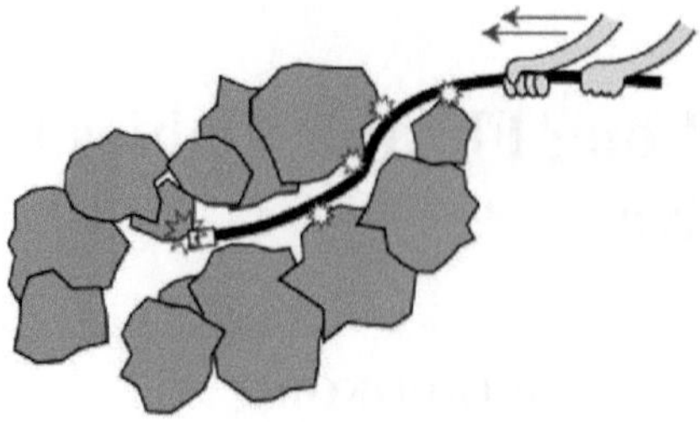

Fig. 16.1 Problems of insertion of a scope camera

In this study, a ciliary vibration drive mechanism is applied to add them active mobility. A ciliary vibration drive is one of a smart and flexible mechanism suitable for the active cable system. A number of inclined cilia can generate driving forces by vibration. Some micro-robots [1-3] and industrial part feeders have used such actuation mechanisms. Our challenges are applying distributed ciliary vibration drive mechanisms for a long flexible cable and controlling its distributed driving forces to get arbitrary shapes and motions of the cables.

Following research topics have been studied:

- Developments of ciliary vibration mechanism fit for a long flexible cable [4, 5]
- Numerical analyses of the driving mechanism of ciliary vibrations [5, 7]
- Modeling of a flexible cables and controlling cable's motion by distributed driving force [9]
- Development of the active scope camera system [4-7]
- Control method of a turn direction of the active scope camera [6, 8]
- Evaluation of the active scope camera for rescue activities [6]

In this chapter, the mechanism of ciliary vibration drive and its optimal design are described. In addition, an application of the ciliary drive mechanism applied for the active scope camera system is introduced.

16.2 Ciliary Vibration Drive Mechanism

The ciliary vibration drive generates driving on a cable by vibrating inclined thin sting or wire cilia. Ciliary bending and recovery movement during vibration makes cilia tips stick and slip rapidly and generates distributed driving force on the cable. Cilia covering the cable surface of the cable (Figure 16.2) vibrate and generate driving over the entire flexible cable regardless of location and contact angle. This design is applicable to flexible cables such as scope cameras.

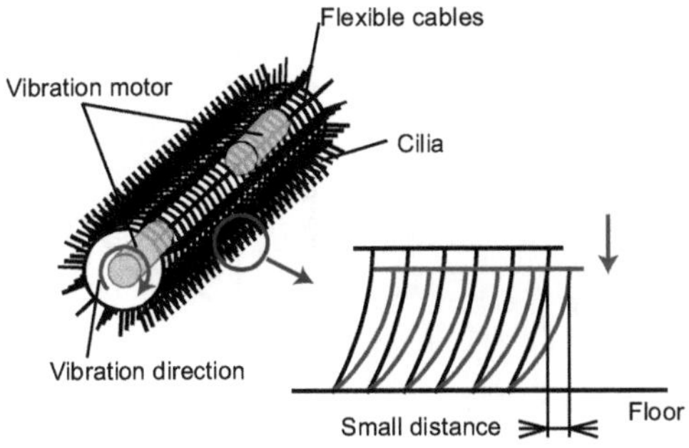

Fig. 16.2 Ciliary vibration drive

This architecture has the following advantages and disadvantages:

- Distributed cilia do not interfere with cable flexibility.
- Component can be small.
- Shape of mechanisms can be arbitrarily modified.
- Driving is distributed over the cable.
- Driving is small.
- Contact conditions sometimes adversely affect driving.

16.3 Observations and Modeling of Ciliary Vibration Drive

To study ciliary vibration mechanism, we used a test piece for observation experiments and a target for modeling [7]. Figure 16.3 shows the test piece.

Aluminum rods forming the core were similar in diameter and line density to the cable of a video scope (Olympus IPREX MX IV7630X2), which we used in research. Its rigidity helped ensure stable contact with the floor, and the small damping factor enabled motor vibration to be transferred to all cilia. The 400 mm length reduced pitching not found in an actual scope camera. The cilia is made of nylon and the length of cilia is 7.5 mm.

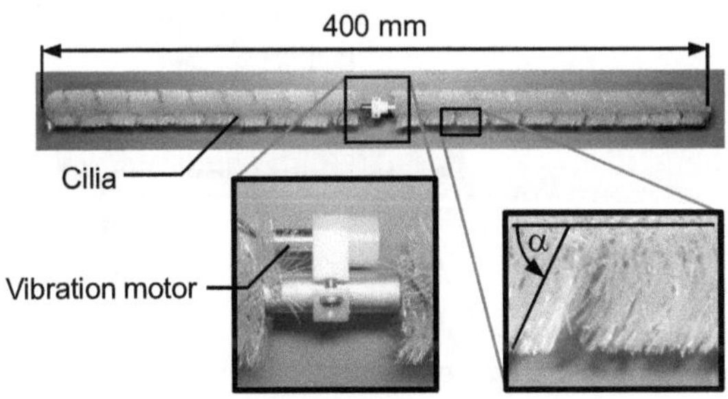

Fig. 16.3 Test piece of ciliary vibration drive

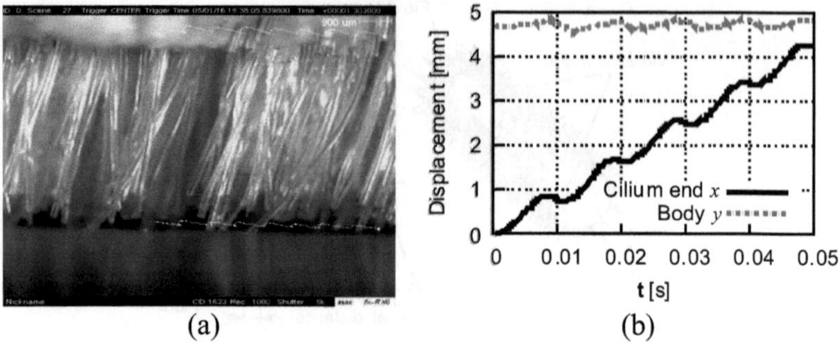

Fig. 16.4 Cilia movement in a capture image and trajectories

In the experiments, the test piece was set on a plate and its movement in stable driving was observed using high-speed camera at a frame rate of 1000 fps. Cilia movement was tracked by image pattern matching. Figure 16. 4(a) and (b) show an example of the capture image and cilia trajectories which represents body movement y on the vertical axis and the cilium tip movement x on the horizontal axis. Our observations confirmed that cilia moved down diagonally to the front when cilia descended and the cilium tip stopped against the ground (t = 0.01 [s]). The entire cilia moved up keeping the same location in the x direction when ascending and cilium tips slipped on the ground (t = 0.015 [s]). Figure 16.5 shows total iteration of the activation mechanism. Differences in movement between cilia and tips of cilia appear to be related to driving. Cilia tips rapidly stick and slip on a floor through vibration. Driving force was generated by sticking to the floor efficiently.

For modeling the ciliary vibration mechanism, parameters required to construct a dynamic model involve eight factors: 1) cilia bending, 2) cilia rotation, 3) cilia number, 4) friction between cilia ends and the run surface, 5) difference in forward and backward friction, 6) cilia sticking based on surface shape, 7) motor vibration, and 8) angle of cable body. While cilia bending and rotation are a basic kinetic parameter, bending is very small, i.e., the length-flexure is 0.8 %. The coefficients of static and dynamic friction are important in determining cilia tip stick-slip.

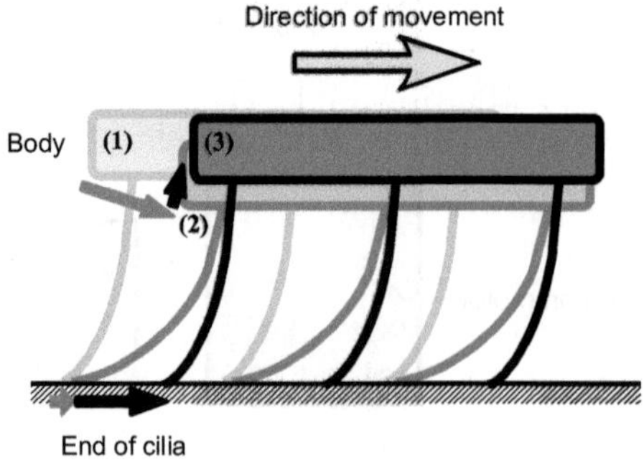

Fig. 16.5 Observed driving mechanism

Although it is difficult for a numerical model to construct an ideal model including all of these factors, we focused on (1) characteristic vibration of the cilia and (2) stick-slip contact between cilia tips and the floor. To determine the effects of these phenomena, we constructed (i) pseudo linear spring model and (ii) stick-slip friction model. Observation through relative velocity - displacement phase plane analysis indicated that stick-slip phenomena were closely related to efficient driving ciliary vibration driving. Using these models, we determined the optimal design for driving vibration frequency and the inclination angle of cilia. See literature [7] for more information.

16.4 Optimal Design for Active Scope Camera

16.4.1 Inclination Angle of Cilia

Inclination angle of cilia has a major effect on the driving force. The test pieces as shown in Figure 16.3 were used for performance experiments. Five conditions of the inclination angle of cilia $\alpha = \{5, 15, 30, 40, 50\}$ degrees with a variation of 5 degrees were tested. Elapsed time for moving 800 mm on a concrete plane surface was measured. The vibration motor was 16 mm long and weighed 1.21 g. The applied voltage was 3.0 V, and the rotation frequency was 8,000 rpm. The generated centrifugal force was 0.75 N and its direction was perpendicular to the center rod.

The experimental relationship between the inclination angle of cilia α and the driving velocity indicates a peak at $\alpha = 15$ degrees as shown in Figure 16.6. It was revealed by observation of high-speed camera that cilia always slip and never stick when α is large, and horizontal component of drive force becomes small under small α. Therefore, the optimal inclination angle of cilia was determined to be 15 degrees for the active scope camera.

16.4.2 Diameter and Density of Cilium

Optimal mechanical parameters of a cilium were experimentally obtained using the test piece. Four types of nylon cilia shown in Table 16.1 were experimented. Diameters of type A, C and D were almost the same, and that of B was larger. Densities of type A and B were low, and those of C and D were high. The angle of inclination of cilia was 15 degrees with a variation $\pm$ 5 degrees for all the types. The other conditions were the same as the last section.

Figure 16.7 shows relation between types of cilia and driving velocity. Type B indicated the best performance. Comparison of A and B with C and D showed that the low-density configuration had higher traveling performance. Comparison between A and B showed that the larger diameter was better, which might mean that the higher rigidity is better. Therefore, type B was selected for fabrication of the active scope camera.

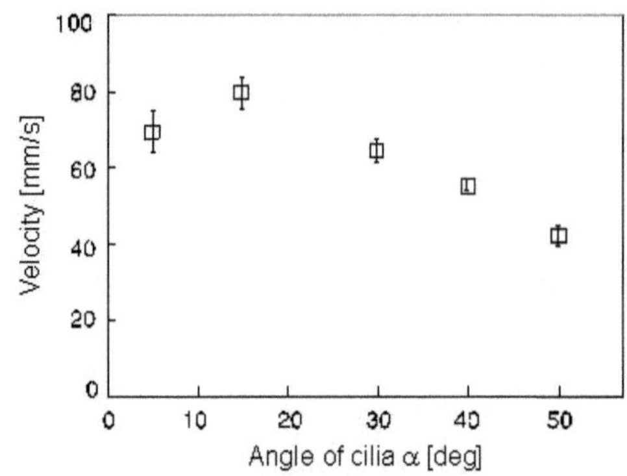

Fig. 16.6 Effects of cilia angle on driving velocity

Table 16.1 Cilia type

Type	Diameter of cilium [mm]	Density [number/mm^2]
A	0.10	6.2
B	0.13	6.2
C	0.093	24
D	0.094	34

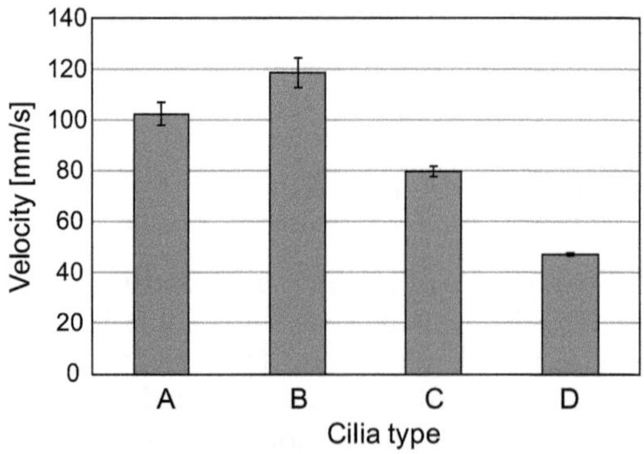

Fig. 16.7 Effects of cilia diameter and density on driving velocity

16.4.3 Interval of Vibration Motors

Flexibility and damping characteristics of the cable of scope cameras absorb vibration. Driving force far from a vibration motor is less because the amplitude is smaller. Attenuation of cilia vibration according to the longitudinal position on the cable is measured in order to optimize the interval of vibration motors installed.

Only one motor vibrated, and the amplitudes of eight points of 400 mm long at intervals of 20 mm were measured. Vibration of cilia was recorded by a high-speed camera. The frequency of rotation was about 9000 rpm, and the centrifugal force generated was about 0.95 N. The experimental result showed that the amplitude near the motor was 130 μm and it gradually decreased as the distance from the motor is larger, and the amplitude became 1/3 at the point 180 mm from the motor.

This concluded that the motor interval should be less than 360 mm to generate enough driving force. On the other hand, if the density of motors is too high, interference between adjacent motor motions spoils the output force, and the part where the motor is installed becomes nodular, which decreases the flexibility and sometimes catches obstacles. Hence, the optimal distance was determined to be 300 mm.

16.5 Prototype of Active Scope Camera

The optimal design parameters, which were obtained by the fundamental experiments in the last section, were applied to an industrial video scope 8 m long, which has been used in actual rescue activities.

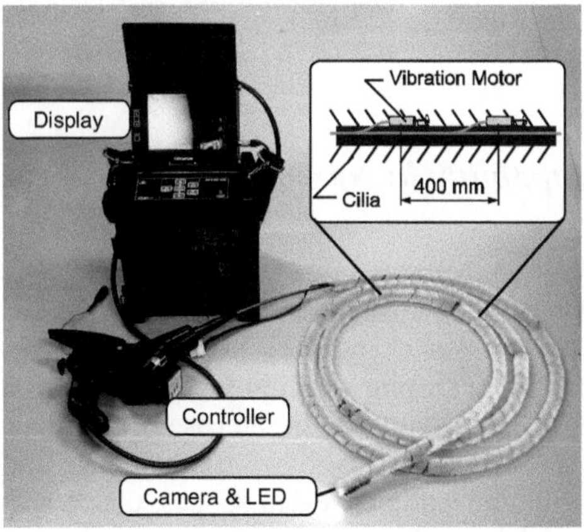

Fig. 16.8 Prototype of Active scope camera

The system mechanism of the produced active scope camera is shown in Figure 16.8. Twenty-four vibration motors were installed on the cable of scope camera with an interval of 300 mm on the basis of the fundamental experiments as mentioned in 16.4.3. The motors were attached to the cable using sealed cases, and the parts were covered by heat shrinkable tube for waterproofing. The cable surface where the motor cases exist also has short cilia in order to remove unevenness to avoid hooking.

The core of the active scope camera is an industrial video scope, Olympus IPLEX MX IV7680X2 8 m long, which is used by first responders. It has a CCD camera and a bending head operated by a manual controller. The image taken by the camera is shown in a LCD display on the control unit. Portability of this active scope camera is high because one person can carry it, the system is driven by batteries, and the total weight is 7.1 kg. The diameter of this active scope camera is 24 mm, which satisfies the required specifications explained in the introduction.

16.6 Fundamental Performance of Active Scope Camera

16.6.1 Motion Speed

The speed was measured on concrete floor, which is typical in urban search and rescue. The elapsed time for moving 1 m was measured. The length of contact between the active scope camera and the floor was 7 m. The applied voltage was 3.6 V per each motor, and the frequency of rotation was about 9000 rpm with generating centrifugal force 0.95 N per motor.

The resultant maximum speed was 46.7 mm/sec, which was about a half of the experimental result of the test piece. This is probably because the flexibility and damping of the cable spoiled the generated force as well as the interference between the motors.

16.6.2 Turning Capability in Narrow Gaps

Turning performance of the active scope camera in narrow gaps was evaluated by a wooden driving course with a U-shape as shown in Figure 16.9. The running direction of the scope camera was changed by bending the head with contacting the sidewall in the course. When turning width in Figure 16.9 was more than 160 mm, it was able to turn 180 degrees around. In the case of less than 160 mm, a part of the motor case got stuck because of higher rigidity than the other parts. If we compare this result with conventional video scopes, the turning performance was considerably improved. It was very difficult for the conventional video scope without this mechanism to turn around at an angle of 180 degrees.

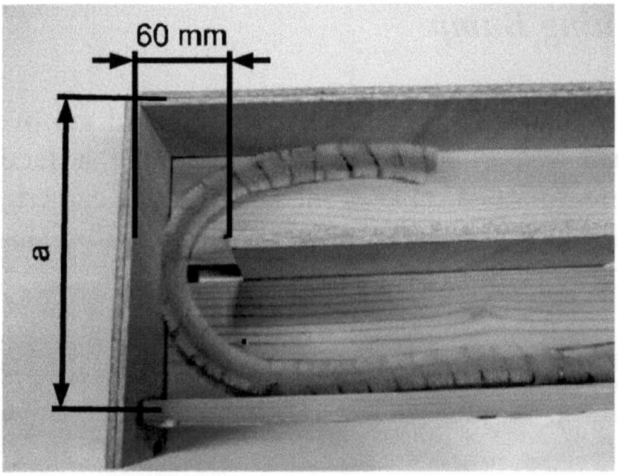

Fig. 16.9 Turning motion in narrow gaps

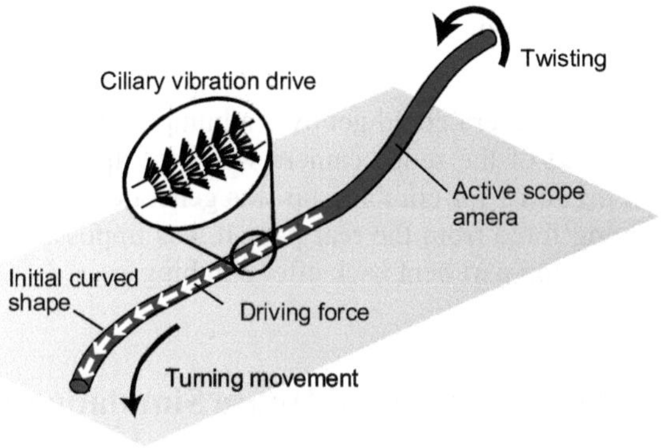

Fig. 16.10 Turning in open space by twisting the root

16.6.3 Turning Capability in Open Space

In open flat space, there is no wall to guide. Therefore, the strategy in narrow gaps, where the moving head guides the direction to follow walls, cannot be applied.

The active scope camera turns by itself when the top part is gradually curved, because slant driving force is applied. Because stiffness of the cable in the twisting (roll) direction is high, the top part makes roll motion when the root is twisted manually. Then it changes the direction of the gradual curve of the top part, and turns. Therefore, control of the twisting angle enables the scope camera change the direction in open space as shown in Figure 16.10.

The turning capability in open space without the side wall was evaluated by the minimum turning radius. The minimum turning radius was about 1 m when the ground was a flat linoleum-covered floor. Hence, the active scope camera can change the direction of motion even if there is no side wall to follow.

16.6.4 Surmounting Bump

The height of bump that the active scope camera could get over was measured. The bump was made of a wood block, and the running surface was a linoleum-covered floor. The rear part of cable of the active scope camera was inserted in a polyvinyl pipe as a cover in order to prevent the cord from skidding and to efficiently apply driving force from the rear part.

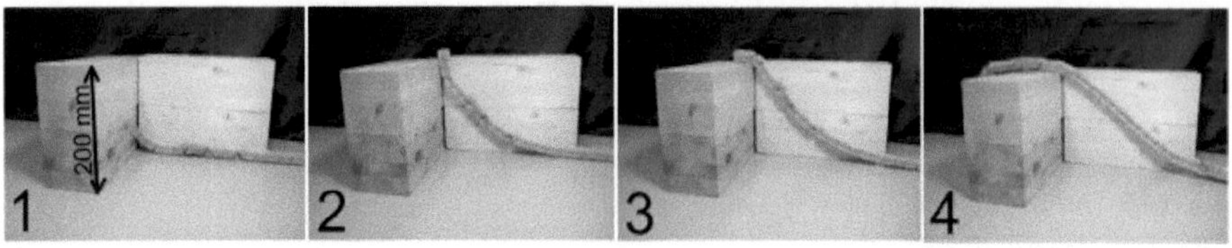

Fig. 16.11 Surmounting bump

The active scope camera could get over a bump 200 mm high as shown in Figure 16.11. The head of the scope camera was bent upwards to contact the wood block, and driving force for climbing up was generated. The contact pressure was applied by driving force from the rear part. It was impossible to surmount higher bumps because the top part bent back after climbing.

16.7 Search Test in Collapsed House Simulation Facility

Synthetic evaluation of the motion in realistic conditions was performed in Collapsed House Simulation Facility of International Rescue System Institute, Kobe Laboratory.

Wooden blocks, coat hangers and dishes were scattered on plywood floor in a collapsed house. Dummy victims (two adult arm and an infant) were buried. A polyvinyl tube 2 m long was inserted into the rubble pile as shown in Fig. 16.12 to guide the active scope camera. Two operators controlled the scope for urban search exercise. One inserted the cable, and the other watched the video display and controlled the head bending. The operators could not see inside of the house, but they have known the approximate positions of the victims. The active scope camera was controlled to search the victims using only the image information from the scope camera.

Figure 16.12 shows three routes experimented. The dummies exist at the end of the routes. Thick solid lines indicate the moving route of the active scope camera. These results demonstrated that the active scope camera 8 m long was able to search victims deep in rubble piles.

Fig. 16.12 Running experiments in Collapsed House Simulation Facility of International Rescue System Institute, Kobe Laboratory

16.8 Conclusions

This chapter described the mechanism of ciliary vibration drive and the application for the active scope camera system for urban search and rescue.
Optimal mechanical design parameters such as material, a diameter, density and an inclination angle of cilia, and specifications and density of vibration motors were obtained on the basis of fundamental experiments of test pieces and prototypes.

In order for this system developed is actually used by first responders, intensive practical testing at their exercises is necessary. Supports for operators' skill by controlling the distributed driving forces is also important. In addition, modeling based on the physical contact and dynamic actuation of the cables will contribute to optimize the driving mechanism and enhance drive performance effectively.

Acknowledgments This research was supported in part by `Special Project for Earthquake Disaster Mitigation in Urban Areas (DDT Project),' and supported in part by a Grant-in-Aid for Scientific Research on Priority Areas `Next-Generation Actuators Leading Breakthroughs.'

References

1. Hatsuzawa T, Hayase M and Toshiaki O (2003) A linear actuator based on cilia vibration. Sensors Actuators: A. Physical, vol.105, no.2:183-189
2. Fukuda T, Mitsumoto N, Arai F and Matsuura H (1993) A study on Micro Robot (1st Report, Design, Experiment and Mathematical Model of Micro Mobile Robot). Transactions of the Japan Society of Mechanical Engineers, vol.59 no.562:1787-1794
3. Ioi K (1999) A Mobile Micro-Robot Using Centrifugal Forces. IEEE/ASME International Conference on Advanced Intelligent Mechatronics :736-741
4. Isaki K, Niitsuma A, Konyo M, Takemura F, Tadokoro S (2006) Development of an Active Flexible Cable Driven by Ciliary Vibration Mechanism. Proc. 10th International Conference on New Actuators :219-222
5. Isaki K, Niitsuma A, Konyo M, Takemura F and Tadokoro S (2006) Development of an Active Flexible Cable by Ciliary Vibration Drive for Scope Camera. IEEE/RSJ International Conference on Intelligent Robots and Systems :3946-3951
6. Hatazaki K, Konyo M, Isaki K, Tadokoro S and Takemura F (2007) Active Scope Camera for Urban Search and Rescue. IEEE/RSJ International Conference on Intelligent Robots and Systems :2596-2602
7. Konyo M, Isaki K, Hatazaki K, Tadokoro S and Takemura F (2008) Ciliary Vibration Drive Mechanism for Active Scope Cameras. Journal of Robotics and Mechatronics, Vol.20, No.3 :490-499
8. Sawata K, Konyo M, Saga S, Tadokoro S and Osuka K (2009) Sliding Motion Control of Active Flexible Cable Using Simple Shape Information. Proc. the 2009 IEEE International Conference on Robotics and Automation :3736-3742, Kobe

Chapter 17
Micro Cilium Actuators in Group

Nobuyuki IWATSUKI[1] and Koichi MORIKAWA[1]

Abstract Aiming to realize novel micro mobile robots or micro conveyers, artificial micro cilium actuators in group should be developed. A pipemorph actuator which is composed of a thin needle as an internal electrode, piezoelectric thin film cylindrically surrounding the internal electrode, and external electrodes attached on the orthogonal four surfaces of the piezoelectric film is adopted as an artificial cilium and then the effective processes based on the hydrothermal method the and electrostatic flocking to fabricate the micro pipemorph actuators in group are proposed and experimentally examined. The cilium actuators in group which are fabricated by using partial nickel plating of titanium wire can convey small paper sheets or plastic sheets at the maximum speed of 1.53μm/min.

17.1 Introduction

Recently high performance micro machines such as micro mobile machines which can move and work in human body or micro conveyers which can transport a cell or a DNA are strongly required. In order to obtain excellent mobility for these micro machines, it may be effective to arrange a lot of micro actuators and to cooperatively drive them. In the natural field, a paramecium drives its cilia to generate progressive wave and moves in water. If we can fabricate artificial micro cilium actuators in group, we will obtain high performance micro mobile robot as shown in Fig.17.1.

Thus the authors have aimed to develop artificial micro cilium actuators in group[1] which are composed of 'a piezoelectric pipemorph actuator'[2]. This actuator was composed of a thin metallic needle as an internal electrode, piezoelectric ceramics layer surrounding the needle and multiple external electrodes on the

[1] Nobuyuki IWATSUKI and Koichi MORIKAWA

Department of Mechanical Sciences and Engineering,

Graduate School of Science and Engineering

Tokyo Institute of Technology

piezoelectric thin film. By applying alternating voltage between the internal and external electrodes at resonant frequencies of the actuator, two-dimensional displacement of the tip could be expected.

Here, the fabrication process to arrange lots of small pipemorph actuators on a plate or shell is proposed and experimentally examined. The results of several experiments to drive the actuators themselves and to convey a small object are also shown.

17.2 Fabrication of Micro Cilium Actuators in Group

17.2.1 Fundamental Fabrication Process

Figure 17.2 shows the proposed fundamental process to fabricate micro cilium actuators in group[1],[5]. At first the partial Ni plating is done at the tip of thin needles of 50-100 microns in diameter made of pure Ti, which will be internal electrodes of pipemorph actuators, as shown in Fig.17.2(a), Since it is very difficult to join Ti materials by soldering , this partial Ni plating will enable the soldering. Next, PZT ceramics thin film will be fabricated on the Ti needles by using the hydrothermal method[3] as shown in Fig.17.2(b). Since PZT ceramics extract layer is extracted only on the surface of Ti, Ti needles are thus used. In this case, the partial Ni plating will play a role as masking material against the hydrothermal method. Then Ti needles coated with PZT ceramics thin films will be arranged perpendicular to a base plate made of Ni by using electrostatic flocking method[4]. At the same time, the needles will be joined at the partial Ni plating part by soldering as shown in Fig.17.2(c). Finally, two pairs of external electrodes will be printed on the PZT ceramics layer by the vacuum deposition as shown in Fig.17.2(d).

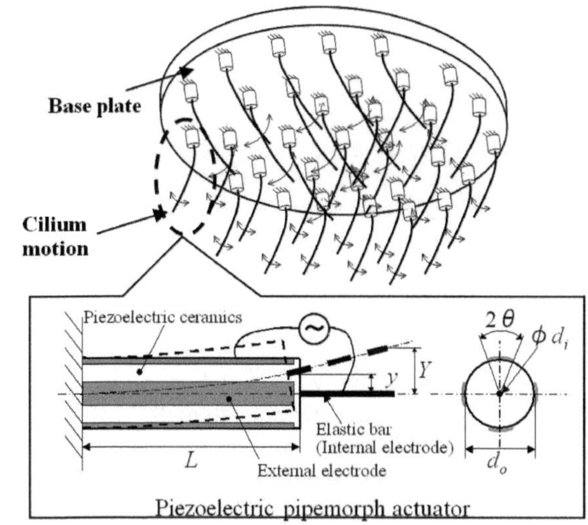

Fig. 17.1 Microrobot driven by artificial micro cilium

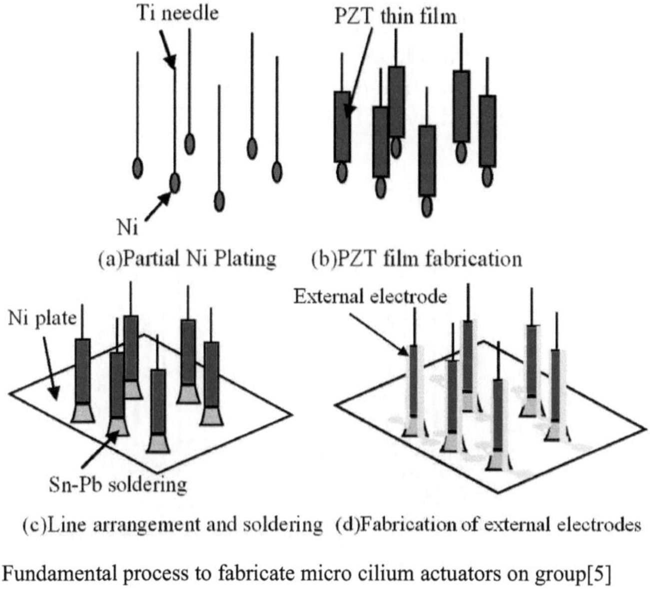

Fig. 17.2 Fundamental process to fabricate micro cilium actuators on group[5]

17.2.2 Partial Ni Plating

A Ti wire of 50-100 microns in diameter was wrapped around a rectangular shaped jig made of the Teflon and was connected with the anode, and the nickel plate was also connected with cathode in the Sulfamate solution bath as shown in

Fig.17.3. The half of the Ti wire was soaked in the Sulfamate solution bath which was stirred with a hot stirrer device and a temperature controller[6].

Figure 17.4 shows a SEM photograph of the Ni plated Ti wire obtained after 10 min plating with 5A/dm^2 in current density. The Ni plating with high sticking strength of 7 microns in film thickness was observed.

17.2.3 Fabrication of PZT Thin Film with the Hydrothermal Method

The obtained Ti wire which was partially Ni-plated on the jig was put in a Teflon case with ionic solution for the hydrothermal method[7],[8] and was set in the autoclave so as to make Ti wires always soak in the ionic solution as shown in Fig.17.5. Figure 17.6 shows a scanning electron microscopic photograph of PZT ceramics thin film fabricated on a Ti wire of 100 microns in diameter. The boundary of the Ni-plated and non-plated parts is shown in the upper photograph in Fig.17.6[9]. The lower photograph shows the non-plated parts. Cubic crystals of PZT ceramics of about 6 microns in side len²gth on non-plated part can be observed while few crystals of PZT are observed on Ni-plated part. It was thus confirmed that Ni-plating played a role as the masking material for the hydrothermal method to fabricate the PZT thin film on Ti wires. Then by cutting the wires, a lot of Ti internal electrodes coated by PZT ceramics thin film and with Ni plated tip were obtained.

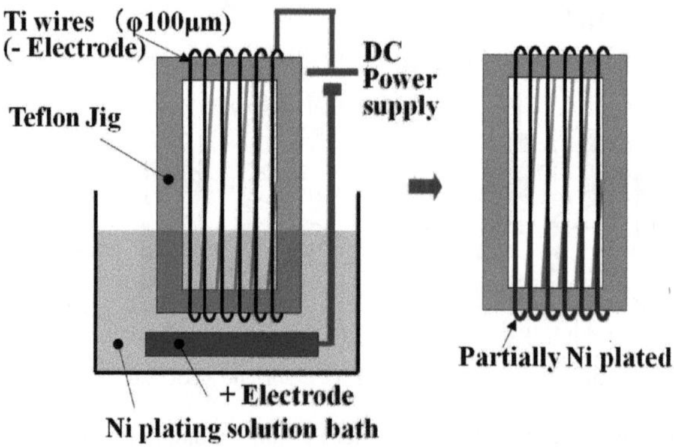

Fig. 17.3 Partial Ni plating

Fig. 17.4 Fabricated Ni film on Ti wire

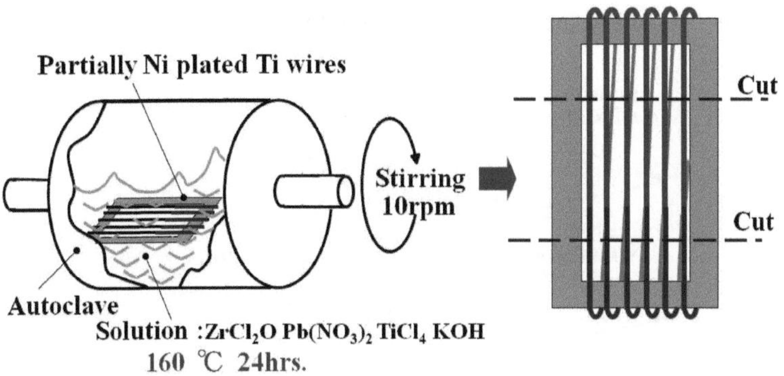

Fig. 17.5 The hydrothermal method to fabricate PZT ceramics thin film

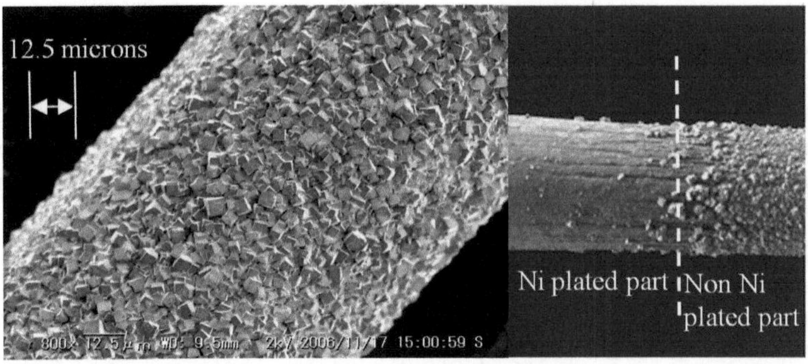

Fig. 17.6 SEM photograph of PZT ceramics thin film on Ti wire with 100 microns in diameter

17.2.4 Electrostatic Flocking and Jointing by Soldering

A device to arrange and joint the internal electrode with partial Ni plating and PZT ceramics thin film to a base plate is illustrated in Fig.17.7. A pair of rectangular flat plate electrodes are set up horizontally and in parallel. A lot of tthin needles coated by PZT ceramics thin film were put on the lower electrode. By applying high electric field of about 2.5kV/cm between the pair of electrode plates, the needles will fly up to the upper electrode along electrical flux lines and they will contact vertically to the upper electrode plate. Flux and melted solder are spread on the upper electrode plate beforehand and the electrode plate is heated. After arranging the needles in the same direction on the upper electrode plate, the electrode plate is cooled and the needles can be jointed.

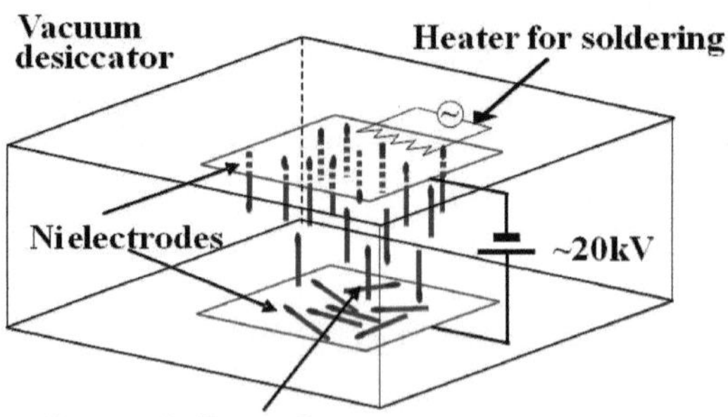

Fig. 17.7 Device for electrostatic flocking and jointing with soldering

Since the electrostatic flocking has been originally developed to arrange short polymer fibers in the same direction and to joint them with adhesive for decoration use, we thus first examined this method for various metallic needles of 5-10 mm in length of various materials such as Ti needles, Sn-plated copper needles, Ti needles coated by PZT ceramics thin film. The experimental results are summarized in Table 17.1. Ti needles of 100 microns in diameter, Sn-plated copper needles of 150 microns in diameter and Ti needles coated by PZT ceramics thin film of 50 and 100 microns in diameter could fly up to the upper electrode plate all together by applying high electric field of about 2.5kV/cm. Since Ti needles of 200μm in diameter could not fly up, it was thus confirmed that too heavy metallic needles were not suitable for the electrostatic flocking. And Ti needles coated by PZT ceramics thin film of 100μm in diameter could fly up to the upper electrode plate.

Then, electrostatic flocking with the partially Ni plated Ti needles coated by PZT ceramics thin film with 100 microns in diameter and 5-10 mm in length were carried out. A few needles with PZT thin film also kept standing in Sn-Pb solder on the upper electrode as shown in Fig.17.8. However, they were not strongly jointed then they came off by applying tensile force[9]. Therefore we tried to improve the electrostatic flocking with soldering especially on the temperature and process time for soldering and on washing of material wires with diluted sulphuric acid just after the hydrothermal method. It was revealed that the oxidation of solder and Ni plating strongly affected decrease of wet property of Pb-Sn and made jointing strength lower. Through the above improvement, the jointing strength increased up to 0.24GPa. After patterning solder spots with cream solder, electrostatic flocking with soldering was tried to joint micro cilium actuators at the spots[10]. Figure 17.9 shows a photograph of the jointed actuators.

Table 17.1 Results of fundamental electrostatic flocking experiments

Material	Dia. μm	Result
Copper with enamel	120	A few wire only stood up on the lower electrode at 2.5kV/cm
Titanium	100	Several wires flew up to the upper electrode at 2.5kV/cm
Titanium	200	No wire flew up even at 5.0kV/cm
Tin plated copper wire	150	All wires flew up to the upper electrode at 2.5kV/cm
Titanium with PZT thin film	50	All wires flew up to the upper electrode at 2.5kV/cm
Titanium with PZT thin film	100	All wires flew up to the upper electrode at 2.5kV/cm

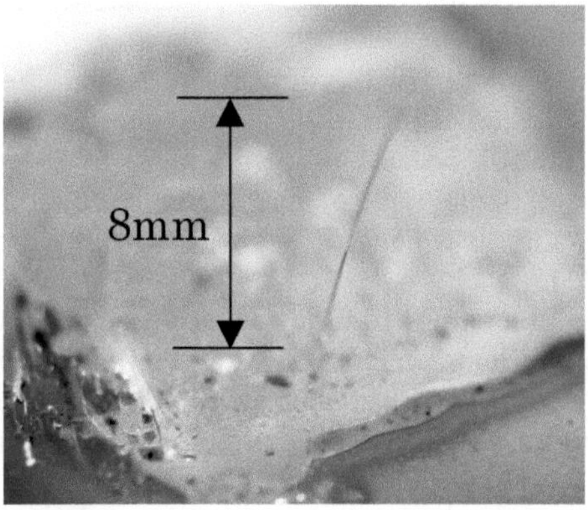

Fig. 17.8 Jointed micro cilium actuator

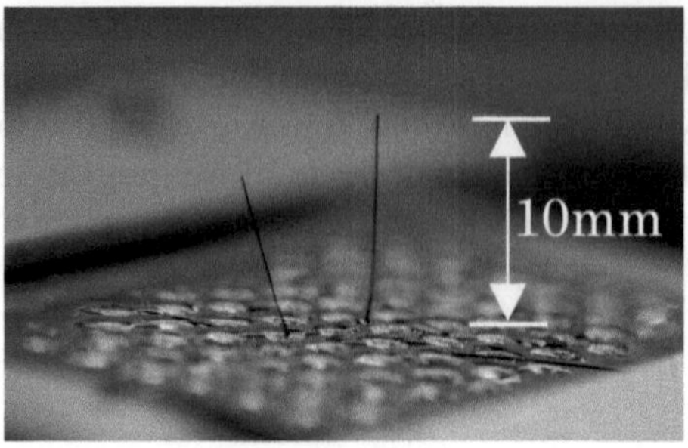

Fig. 17.9 Micro cilium actuators jointed at solder spots

17.2.5 Jointing Electrodes with Ni Plating

Since it was difficult to joint internal electrodes to a base plate by soldering in the electrostatic flocking, a new method to do so directly with Ni plating was proposed. In this case the internal electrodes were arranged in parallel to a base plate so as to increase joint area.

The proposed method is illustrated in Fig.17.10 and is explained as follows:

1. A copper tape is attached on a rectangular jig 1 made of Teflon. A Ti thin wire is then wrapped around the jig 1. They are covered with a jig 2 made of copper tape.
2. The copper tape part is soaked in the Ni sulfamate solution bath and is then plated with Ni.
3. The top and bottom of the Ti wires are cut and are detached with copper tape.
4. Ni plated part is flaked off from the copper tape. In this case it is easy to flake the copper tape off because of the flexibility of the copper tape.

Figure 17.11 shows a photograph of the internal electrode array jointed to a base plate with Ni plating in which six Ti wires of 100 microns in diameter are spaced with 0.5mm[11].

After fabricating PZT ceramics film on the electrode array with the hydrothermal method, the micro cilium actuator array can then be obtained. Moreover, by accumulating many micro cilium actuator arrays as multilayer structure, a two-dimensional micro cilium actuator group can be fabricated. The fabrication process is shown in Fig.17.12[12] and is explained as follows:

1. PZT ceramics thin film is fabricated with the hydrothermal method on the Ti electrode array jointed with the Ni plating. And internal electrodes are wired with a common earth.
2. A plastic layer is fabricated on the Ni base plate so as to insulate between internal electrodes and external electrode.
3. One side of the PZT coated internal electrodes is masked with a sheet. And the external copper electrodes and a power source pad are fabricated on PZT coated internal electrodes and insulation layer, respectively, with the vacuum

deposition. In this case, each pair of internal and external electrodes is fabricated.

4. By repeating the processes (1) – (3) in the direction perpendicular to the power source pad, the micro cilium actuators in group can then be fabricated[13].

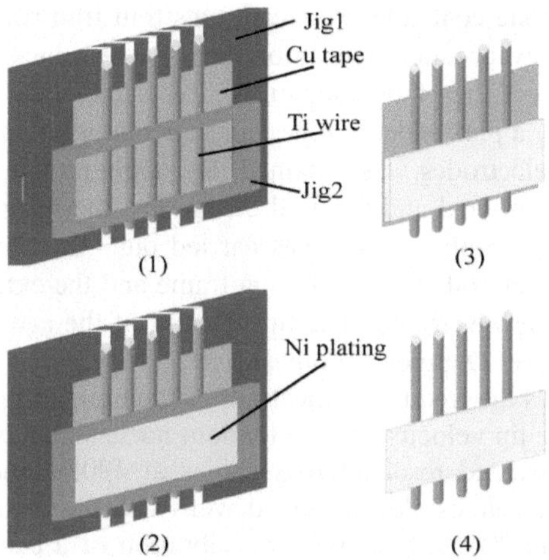

Fig. 17.10 Jointing electrodes with Ni plating

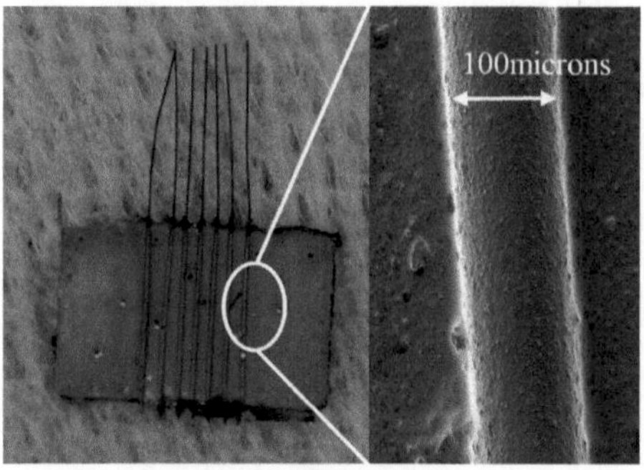

Fig. 17.11 Internal electrode array jointed with Ni plating

17.3 Driving Experiments of Micro Cilium Actuators in Group

17.3.1 Piezo Drive of a Micro Cilium Actuator

An internal Ti electrode coated by PZT ceramics thin film was set just under the ceiling of a vacuum evaporation deposition device with adhesive, and an external electrode of copper was then fabricated on the exposure side of the PZT ceramics thin film. Resultantly a piezoelectric pipe morph actuator having only a pair of internal and external electrodes was obtained. By applying the alternating current voltage between the internal and external electrodes, the resonant driving test of the piezoelectric pipemorph actuator was carried out. As seen in Fig.17.13, one edge of the internal electrode was fixed on a frame and the external electrode was put in a pair of copper electrodes. The tip velocity of the actuator was measured with a microscopic laser Doppler vibrometer.

Figure 17.14 shows the frequency spectrum of the transfer function between the input voltage and the tip velocity when a random noise signal was given as the input voltage. There existed resonant frequencies at 490Hz and 2970Hz, respectively. Resonant frequencies almost agreed well with the theoretical natural frequency of the 1st and 2nd mode of bending vibration of a cantilever made of Ti round bar. It was thus confirmed that the fabricated micro cilium actuator was driven by its piezoelectricity.

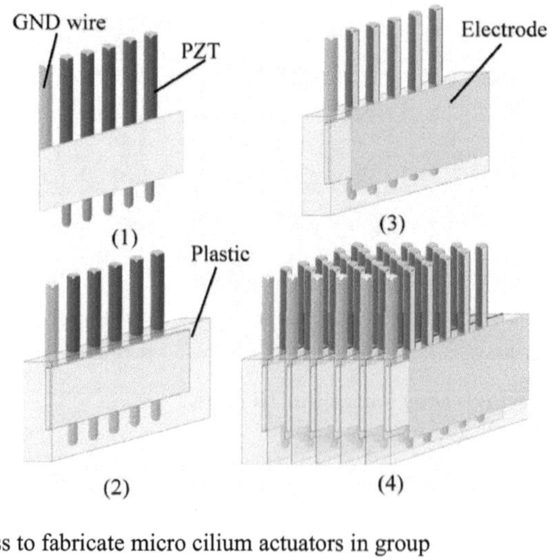

Fig. 17.12 Process to fabricate micro cilium actuators in group

Fig. 17.13 Driving experiment of a micro cilium actuator

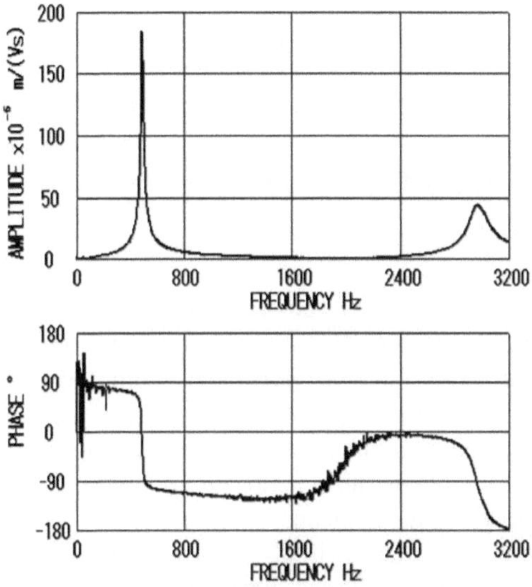

Fig. 17.14 Frequency response of tip velocity of a micro cilium actuator

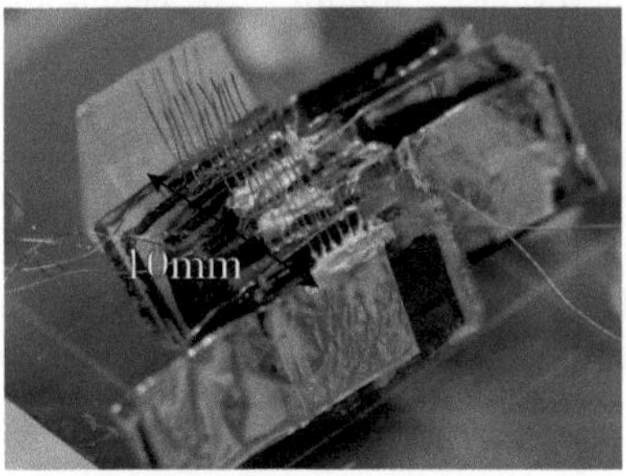

Fig. 17.15 A prototype of micro cilium actuators in group

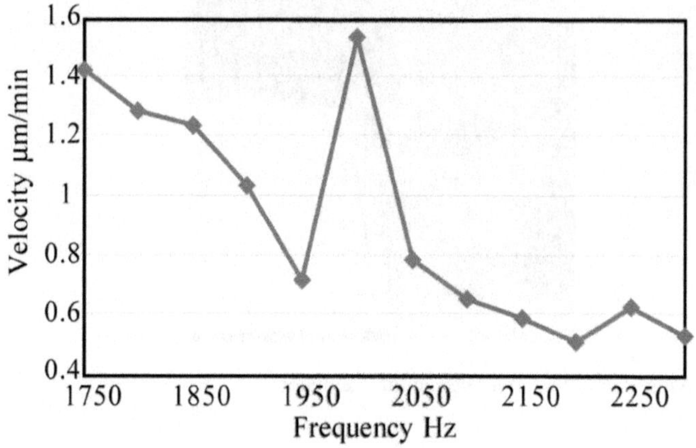

Fig. 17.16 Frequency characteristic of conveying velocity

17.3.2 Sheet Conveying Experiment of Micro Cilium Actuators in Group

Figure 17.15 shows the micro cilium actuators in group, which is fabricated with the method mentioned in section 17.2.5 by accumulating 6 micro cilium actuator arrays with a slant angle of 45 degree. Each micro cilium actuator array has 6 micro cilium actuators of 6mm in length. Since there exists the difference of frictional force in the reciprocation of vibrating cilium actuator tip because of its slant angle against a contacting plane, the actuator group is expected to generate driving force as a slanted fiber drive mechanism.

Sheet conveyance experiments with the micro cilium actuators in group carried out. Various kinds of small rectangular sheets made of paper or plastics were put on the tips of micro cilium actuators, and sinusoidal voltage with 5Vp-p in amplitude was then applied between internal and external electrodes. Moving velocity was measured with a microscope. Figure 17.16 shows the relation between the frequency of the applying voltage and conveying velocity for the square paper sheet with 6mm in side length. The conveying velocity takes the maximum value at about 2000Hz. This frequency agrees almost well with the natural frequency of the micro cilium actuator array. The results including conveying velocity, power and energy efficiency of the conveyance experiments for various kinds of sheets are summarized in Table 17.2. The micro conveyer of micro cilium actuators in group could convey a small paper sheet of 2mg in mass at maximum speed of 1.5 microns/min and also could convey a polystyrene plate of 51mg in mass at 0.375microns/min[13],[14].

Table 17.2 Results of micro conveying experiments

Sheet	Paper	OHP sheet	Polystyrene plate
Side length	6mm	6mm	6mm
Mass	2mg	5.26mg	51mg
Coefficient of static friction	0.4~0.7	0.3	—
Coefficient of moving friction	—	0.3	0.17~0.19
Conveying speed	1.53μm/min	0.12μm/min	0.375μm/min
Output power	5.000×10^{-13}W	1.031×10^{-13}W	3.124×10^{-12}W
Energy efficiency	6.000×10^{-4} %	1.219×10^{-4} %	3.786×10^{-3} %

17.4 Conclusions

Aiming to develop micro cilium actuators in group, its fabrication process using the nickel plating, the hydrothermal method and the electrostatic flocking method was proposed and its driving performances were experimentally examined. The results obtained are summarized as follows:

1. By using the hydrothermal method with the movable jig in autoclave, PZT piezoelectric ceramics thin film with cubic crystals of 6 microns in side lengths could be fabricated on Ti thin wire of 100 microns in diameter.
2. Ti electrodes coated with PZT ceramics thin film with Ni plated tip could be arranged vertically to a base plate made of Ni and be jointed to the base plate by Ps-Sn soldering while electrostatic flocking.
3. (3)A new method to joint Ti internal electrodes to a base plate directly with Ni plating was proposed and a micro cilium actuator array of 6mm in length and of 100 mm in diameter and two-dimensional micro cilium actuators in group were fabricated.
4. (4)A micro cilium actuator with 100 microns in diameter was fabricated and examined its frequency response of admittance. Since the 1st and 2nd resonant frequencies which agreed almost well with the natural frequencies of bending vibration of Ti round bar were confirmed, it was then validated that the fabricated actuator had piezoelectricity.
5. Micro cilium actuators in group, which was fabricated by combining the cilium actuator array as multi layers could convey small paper sheets or plastic sheets with 2mg mass at the maximum speed of 1.53microns/min

Acknowledgments This research has been carried out under the support of the Grand-in-aid for Scientific Research on Priority Areas, No. 438, "Next-Generation Actuators Leading Breakthroughs". The authors thank to the related all for their supports. The authors also wish to express their gratitude to Prof. Minoru Kurosawa, Tokyo Institute of Technology, Nihon Tokushu Kogyo Co. Ltd., and Prof. Tadashi Hattori and Dr. Daiji Noda, University of Hyogo for their kind advices on the hydrothermal method, electrostatic flocking and Ni plating, respectively.

References

1. Iwatsuki N (2005) Development of Micro Cilium Actuators in Group. Proc 1st Sympo on the Grand-in-aid for Scientific Research on Priority Areas, No. 438, "Next-Generation Actuators Leading Breakthroughs": 59-62(in Japanese)
2. Iwatsuki N, Hayashi I and Hayashi T (1989) Multi-DOF Actuator made of PMN-Epoxy Composite. Proc JSPE Autumn Conf: 1021-1022(in Japanese)
3. Shimoura K et al (1991) Preparation of Lead Zirconate Titanate Thin Film by Hydrothermal Method. Jpn J Physics: 2174-2177
4. Noguchi H, Nishida N and Nakagawa T (1989) Electrostatic Flocking of Metal Fibers. Proc JSPE Spring Conf: 1029-1030(in Japanese)
5. Iwatsuki N, Morita K and Morikawa K (2006) Development of Micro Ciliary Actuators in Group Proc10th Intl Conf on New Actuators: 752-755
6. Kimura T, Fukibara H and Hattori T (2006) Manufacture of Ni Microprobe Using X-Ray Lithography and Plating Method -Behavior of Ni-plated Materials at a Constant Temperature Under Stress-. IEEE Trans on Sens and Micromachines 126(5): 195-199
7. Morita T et al (1998) A Cylindrical Micro Ultrasonic Motor Using PZT Thin Film Deposited by Single Process Hydrothermal Method. IEEE Trans on Ultrasonics, Ferroelectrics and Frequency Control 45(5): 1178-1187
8. Kanda T, Morita T and Kurosawa M (2000) A Flat Type Touch Probe Sensor Using PZT Thin Film Vibrator. Sensors and Actuators 83: 67-75
9. Iwatsuki N et al (2007) Fabrication of a Micro Cilium Actuator and Its Driving Experiments. Proc 11th Intl Conf on Mechatronics Tech: 1-6
10. Maeda M, Iwatsuki N and Morikawa K (200) Electrostatic Flocking of Micro Cilium Actuators in Group. Proc JSPE Spring Conf: 637-638 (in Japanese)
11. Iwatsuki N, Nishida Y and Morikawa K (2008) Fabrication of Micro Cilium Actuators in Group and Driving Experiments. Proc 11th Intl Conf on New Actuators: 342-345
12. Iwatsuki N, Morikawa K and Nishida Y (2008) Fabrication of Micro Cilium Actuators in Group with Nickel Plating and Sheet Conveyance Experiments. Proc 12th Intl Conf on Mechatronics Tech: 1-6
13. Iwatsuki N, Nishida Y and Morikawa K (2008) Fabrication of Micro Cilium Actuators in Group and Driving Experiments. Proc 11th Intl Conf on New Actuators: 342-345
14. Nishida Y, Iwatsuki N and Morikawa K (2009) Fabrication of Micro Cilium Actuators in Group - Fabrication of Cilium Actuator Array and Group and Their Driving Experiment - , J Jpn Soc AEM, 17(1): 2-7 (in Japanese)

Chapter 18
Novel Thin-Type High-Speed Ultrasonic Motors and Gyro-Moment Motors

Manabu AOYAGI[1], Takehiro TAKANO[2], Hideki TAMURA[3] and Yoshiro TOMIKAWA[3]

Abstract The goals of the present research are to develop new applications of an ultrasonic motor and to develop a new actuator based on the gyro-moment. The present paper considers three types of ultrasonic motor, which have stator vibrators constructed from PZT plate bonded on a metal shim, multilayered PZT, or $LiNbO_3$ single crystal plate. We have succeeded in developing ultrasonic motors that have features such as a thickness of less than 1 mm, lead-less construction, low driving voltage, and single-phase driving with reversible motion. These motors were confirmed to provide high-speed operation (approximately 10,000 rpm), have a small input power requirement of less than 1 W, and have a low driving voltage of less than 5 Vrms. In addition, several types of gyro-moment motors were developed and were confirmed to provide adequate performance when operated by various driving sources, including magnets, PZT plates, and finger tapping.

18.1 Introduction

Ultrasonic motors can produce high torque at low speed, thereby eliminating the need for gears and enabling direct drive. This is one of the major advantages of the ultrasonic motor (USM). However, in applications such as disk drives or micro

[1] Manabu AOYAGI

Graduate School of Engineering, Muroran Institute of Technology

[2] Takehiro TAKANO

Depertment of Information and Communication Engineering, Tohoku Institute of Technology

[3] Hideki TAMURA and Yoshiro TOMIKAWA

Graduate School of Science and Engineering, Yamagata University cooling fans, which involve a small load and constant high-speed rotation, high torque is not necessary.

In recent years, personal digital assistants, laptop computers, and other portable electric devices have become smaller, necessitating more compact and higher performance spindle motors. In general, a thin electromagnetic d.c. motor (DCM) has been widely used in such applications. Although miniaturization of the spindle motor has been actively investigated, thin DCMs (with a thickness of less than 1 mm) having the necessary performance are difficult to realize. In contrast, there is flexibility in choosing the shape of the USM, and USMs having a thickness of less than 1 mm can be achieved. Small USMs are able to yield greater mechanical output power than an electromagnetic motor of similar size.

The shape of the USM can be selected or designed based on the application. If USMs were capable of spinning faster and providing continuous operation over long periods of time, their application would become much more widespread. The present study addresses new challenges in developing new applications of USMs. The present paper considers design concepts and presents experimental results for various types of high-speed ultrasonic spindle motors that are proposed herein.

18.2 Concept and Investigated Items

As listed in Table 18.1, several types of thin USM have been considered, and research into thin ultrasonic spindle motors has been based on the following design considerations.

1. Small volume, simple structure, and simple drive: the USM and associated drive circuit used in portable electric equipment are required to have minimal volume. Therefore, a single-phase-drive motor, which requires only a simple drive circuit, is suitable for such applications. A number of single-phase-drive motors that use longitudinal or flexural vibration of a thin-plate type stator vibrator are proposed.
2. Low drive voltage (multilayer ceramics): Portable electric equipment re-quires low-drive-voltage operation. The use of multilayer structure piezoceramics can reduce the drive voltage. A single-phase-drive motor using a diagonally symmetrical multilayered ceramic plate is examined.
3. Lead-free ($LiNbO_3$ rectangular plate USM): Ordinary piezoelectric ceramics (PZT), which are widely used in various applications, contain lead. However, a lead-free PZT will be required in the near future. As a solution to this problem, stator vibrators using $LiNbO_3$, which is a piezoelectric single crystal, are investigated.
4. New operating principle (Gyro-moment motors): It is not necessary to be particular about a friction drive as an ultrasonic motor. A new operating principle using a gyro-moment generated by vibrations is proposed.

The design concepts and types of spindle USM considered in the present study are listed in Table 18.1 and are described in the following.

Table 18.1 Design concepts and motors considered in the present study

Concept and Item	1) Simple structure and drive	2) Low drive voltage	3) Lead free	4) New operating principle
Motor types				
A: PZT plate with shim	X	-	-	-
B: Multilayer ceramics	X	X	-	-
C: Single crystal ($LiNbO_3$)	X	-	X	-
D:Gyro-moment motor	x	x	x	X

18.3 Motor Type A: Annular Plate-Type USM Using the Non-Axisymmetric ((1,1)) Vibration Mode

A vibrating-type ultrasonic motor using a plate vibrator is suitable for obtaining high-speed rotation and can be easily constructed to have a thickness of less than 1 mm. Therefore, we herein consider ultrasonic motors that use a thin annular plate stator vibrator that vibrates in the non-axisymmetric ((1,1)) mode, in which a single vibrating element is arranged. This type of ultrasonic motor has a simple structure and is easier to position around a rotating shaft than a rectangular plate-type ultrasonic spindle motor having similar output power [1].

18.3.1 Operating Principle

The stator vibrator using an annular plate was designed by the finite element method analysis (ANSYS), as shown in Fig.18.1(a). The total thickness of this stator vibrator is 0.6 mm. Its vibration modes, including the movement of a vibrating-piece arranged in the plate, are shown in Fig.18.1(b). The stator vibrator is composed of piezoelectric ceramics in the form of an annular plate to which thin metal plates are bonded to the upper and lower surfaces, and the electrodes are formed by two piezoceramic sections. Poling in the two sections is in the thickness direction, but in opposite directions in each section. Therefore, by applying a driving electric voltage of the same phase to each electrode, the non-axisymmetric ((1,1)) mode shown in Fig.18.1(b) can be generated. At the same time, the vibrating-piece has a flexural displacement with phase difference from the non-axisymmetric ((1,1)) mode, as shown in the figure, so that an elliptic motion is generated at the tip of the vibrating-pieces. Such a motion causes the rotor to rotate by a friction force.

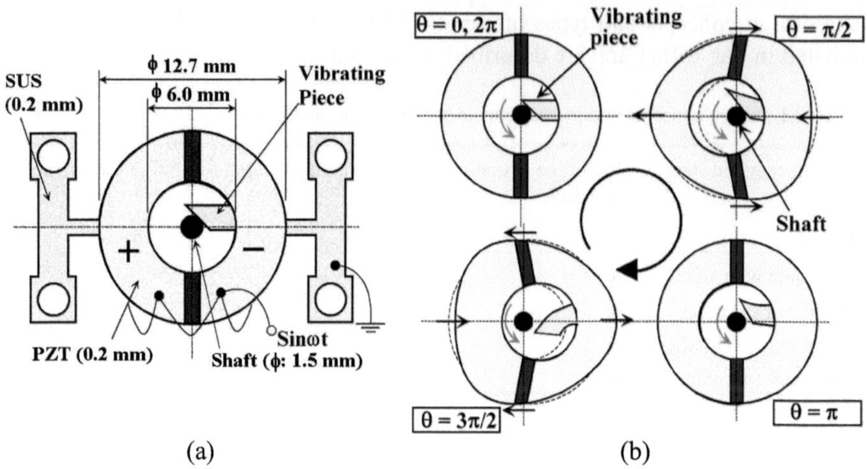

Fig. 18.1 (a) Construction of a stator vibrator and (b) vibration mode and operating principle

18.3.2 Measurement Results

This stator vibrator had a resonance frequency of 172 kHz and an input admittance of 163 mS. The stator vibrator is positioned on the measurement apparatus shown in Fig.18.2 and is supported by four screws at ends of T-type bars on both sides of the vibrator. The measured rotational speed and input power with respect to applied voltage of the prototype motors are shown in Fig.18.3. A rotational speed of 6,300 rpm and an input power of 0.91 W were obtained under a driving voltage of 30 Vpp and a driving frequency of 170.9 kHz.

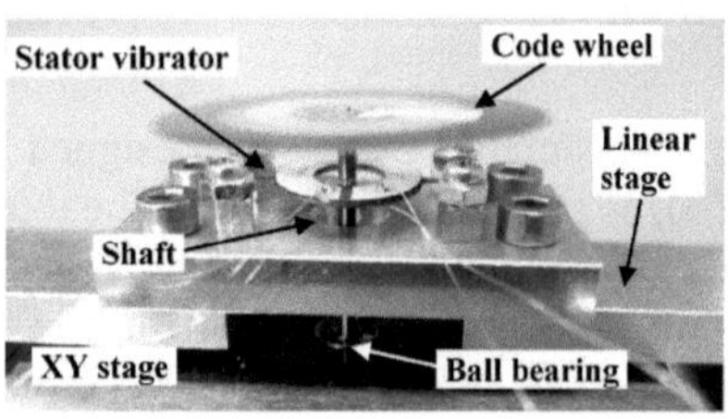

Fig. 18.2 Measurement apparatus of the USM using an annular plate-type vibrator

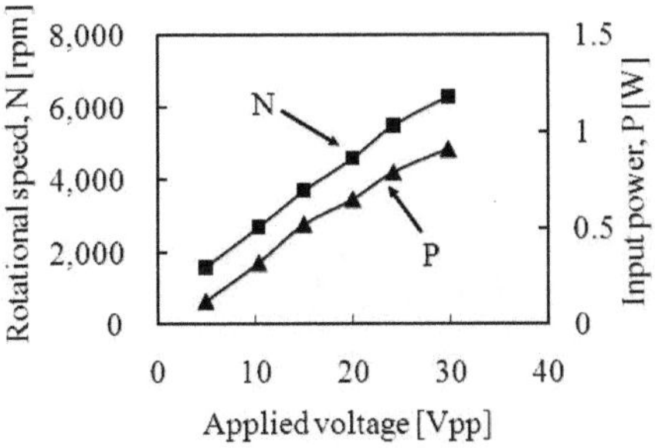

Fig. 18.3 Measured rotational speed and input power with respect to applied voltage

18.4 Motor Type B: Ultrasonic Motor Using a Multilayer Piezoceramic Vibrator

18.4.1 Mode-Coupled Vibrator and Vibration Modes

An ultrasonic motor based on the coupling of longitudinal and flexural vibrations using a diagonally symmetric form piezoelectric vibrator can be used to realize a single-phase-drive-type motor [2]. Figure 18.4 shows the vibrator used in the present study, where the vibrator is constructed of eight layers of thin ceramic plate and has a thickness of 1.0 mm. Additional external asymmetry is given to the vibrator by cutting the short-side edges. Figures 18.5(a) and 18.5(b) show two mode patterns of vibrations simulated by FEM. That is, the first longitudinal mode (L1-mode) and second flexural mode (F2-mode) are coupled. As shown in Fig.18.5, the lower-mode and upper-mode vibrations are dominated by the longitudinal and flexural vibration components, respectively.

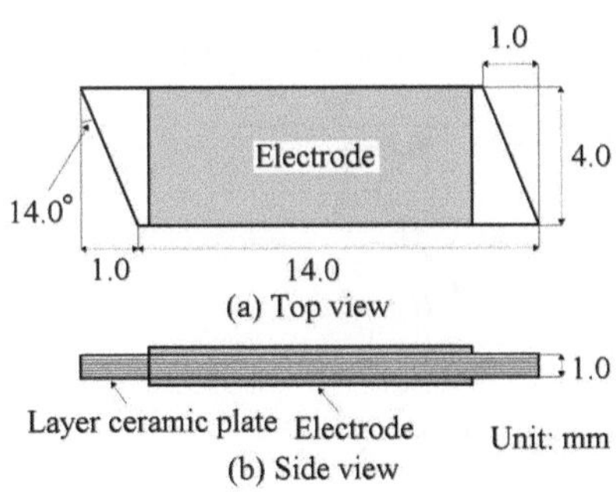

Fig. 18.4 Diagonally symmetrical multilayer piezoceramic vibrator

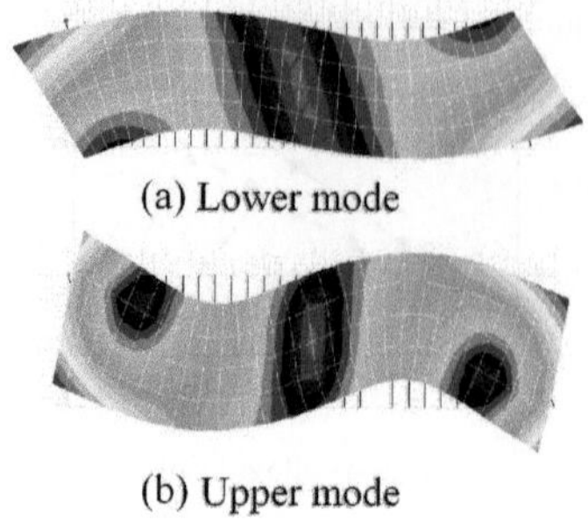

Fig. 18.5 Mode-coupled vibration modes simulated by FEM

18.4.2 Operating Principle and Motor Construction

We considered a motor in which the driving force is delivered from point D in Fig.18.6. When the vibrator is contacted to a rotary shaft, the surface of the shaft is pushed in the radial and circumferential directions by the vertical and horizontal displacements u_V and u_H, respectively, at the contact point. The shaft can then be rotated. When the vibrator is stretched, the direction of the horizontal displacement is opposite to each other in the lower mode and the upper mode. Thus, if we switch the driving frequency from the lower mode to the upper mode, the rotational direction of the motor will be inverted.

A photograph of the vibrator is shown in Fig.18.7. Friction material (alumina) is bonded to the contact surface of the vibrator. Figure 18.8 shows the test motor construction. The vibrator is gently supported by the supporting jig at the four vibration nodes of the F2 mode. The supporting jig and vibrator are pressed against a rotor having a diameter of 1 mm. This setup is appropriate for a thin-type USM.

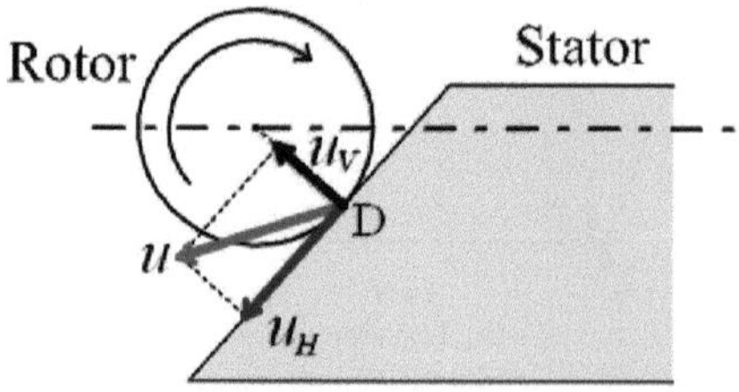

Fig. 18.6 Operating principle of the motor

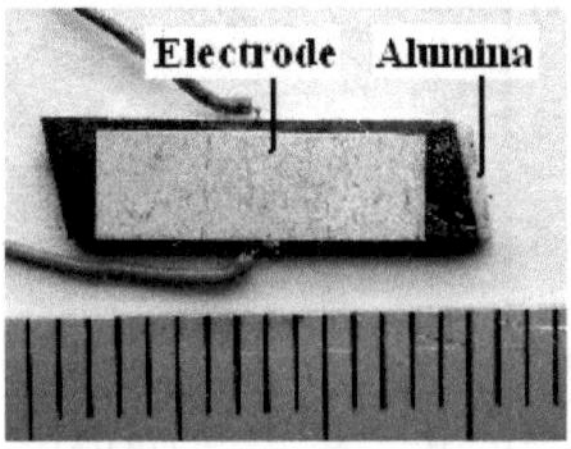

Fig. 18.7 Photograph of the stator vibrator

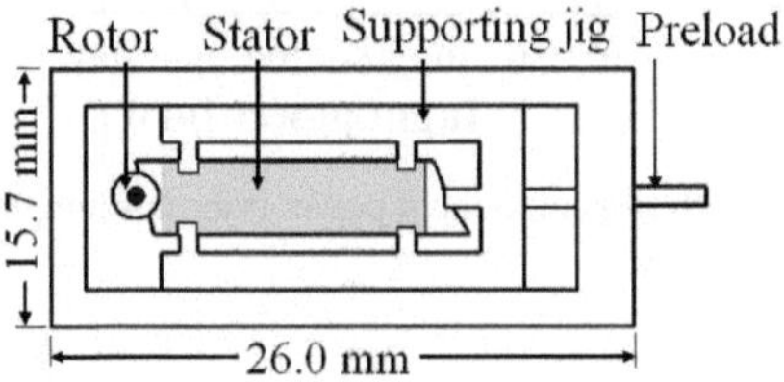

Fig. 18.8 Test motor construction

18.4.3 Operating Characteristics of the Test Motor

Figure 18.9 shows the rotational speed of the ultrasonic motor using the lower mode as a function of input power, where the rotor diameter is 1.0 mm. The motor rotates stably even at high rotational speed. The rotational speed and the input power were over 8,000 rpm and 300 mW, respectively. The results indicate that it is possible to construct a simple miniature motor that is suitable for high rotational speed at low input power and voltage. We expect this motor to be applied to driving actuators of mobile apparatuses.

Figure 18.10 shows the load characteristics of the ultrasonic motor using a different vibrator (length: 30 mm, width: 8 mm, thickness: 2 mm) and a rotor having a diameter of 6 mm. The motor has a maximum efficiency of 46% at a torque of 3 mN-m, an input power 300 mW, and an input voltage of 3 Vrms.

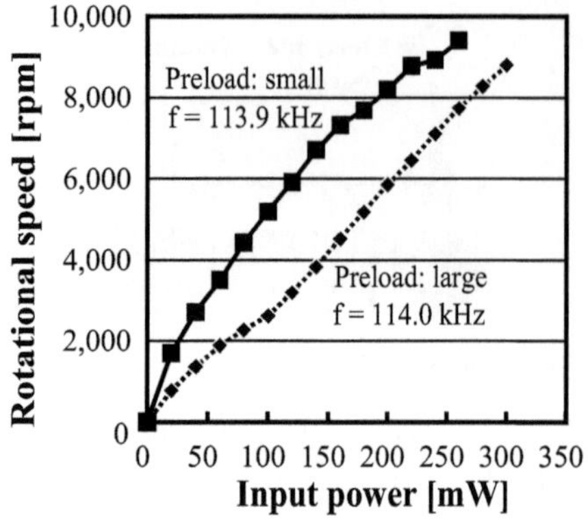

Fig. 18.9 Rotational speed with respect to input power (vibrator length: 15 mm, rotor diameter: 1.0 mm)

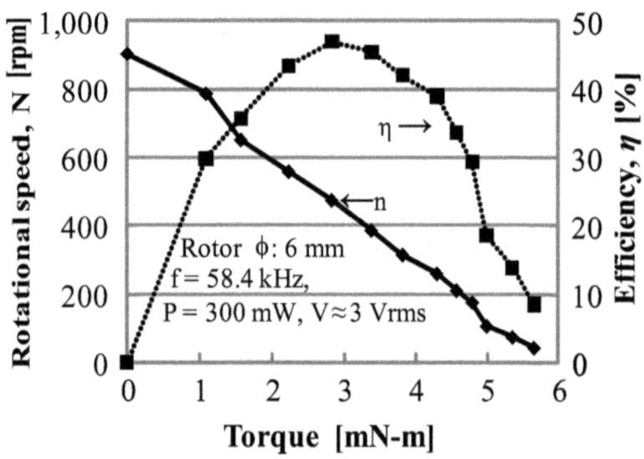

Fig. 18.10 Load characteristics of the mo-tor (vibrator length: 30 mm, rotor diameter: 6.0 mm)

18.5 Motor Type C: Piezoelectric Single Crystal Ultrasonic Motor

Lead-free piezoelectric materials are required in environmentally sensitive applications, and we have proposed several USMs that use a bulk wave vibrator made of $LiNbO_3$ piezoelectric single crystal.[3,4,5] The piezoelectric generative force is smaller in $LiNbO_3$ than in piezoceramics containing lead. Nevertheless, for a miniature resonant-type USM, $LiNbO_3$ has the advantages of low dielectric and mechanical losses at high driving frequencies and large vibrational velocity. [6] In addition, the crystal anisotropy of $LiNbO_3$ can easily realize mode-coupling in rectangular shapes. Therefore, we can obtain a single-phase driven USM using a simple thin structure.

18.5.1 Principles and Fundamentals of the Design

The mechanism of the USM is based on the elliptical motion on the contact surface of the stator vibrator by a two-phase drive, and elliptical vibration usually cannot be obtained by single-phase driving around a single resonant mode. However, a single-phase driven USM is realized when the rectilinear motion of the contact point lies at an oblique angle to the surface, as shown in Fig.18.11. Resonant modes with oblique vibration are obtained by combining a longitudinal mode and a flexural mode. These modes are then referred to as coupling modes.

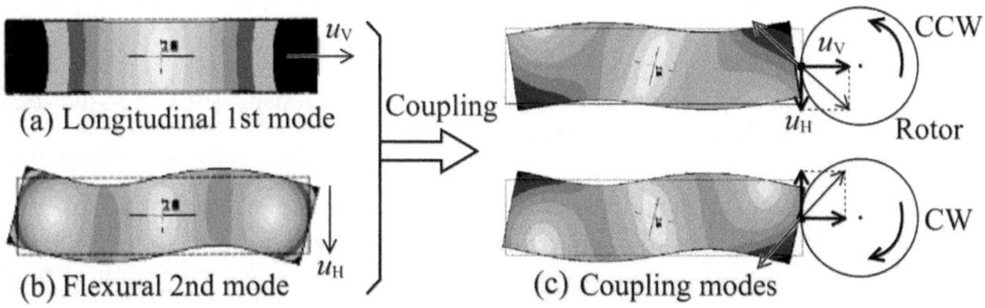

Fig. 18.11 Principle of operation for the single-phase driven USM using coupling modes. The direction of rotation can be switched by selecting one of the modes

An X-rotated Y-plate of $LiNbO_3$ can strongly excite in-plane bulk vibrations by the piezoelectric transverse effect at the X-rotation angles, θ, ranging from 120° to 160°, and the longitudinal and flexural modes appear independently in the rectangular vibrator. On the other hand, by applying the second rotation Φ in the y'-axis, as shown in Fig.18.12(a), the longitudinal and flexural modes are combined to convert two coupling modes, which are called the upper and lower modes, that correspond to the resonant frequency. This occurs because the second rotation changes the elastic compliances, especially s^E_{35}, into non-zero values, as shown in Fig.18.12(b). Since the propagation wave in the length direction is bent by a function of the anisotropic s^E_{35}, the components of the longitudinal and flexural waves are converted alternately each other and are coupled. Consequently, coupling modes can be realized in vibrators having a symmetrical outer shape, and these vibrators can be driven by a uniform single-phase electric field in the thickness direction.

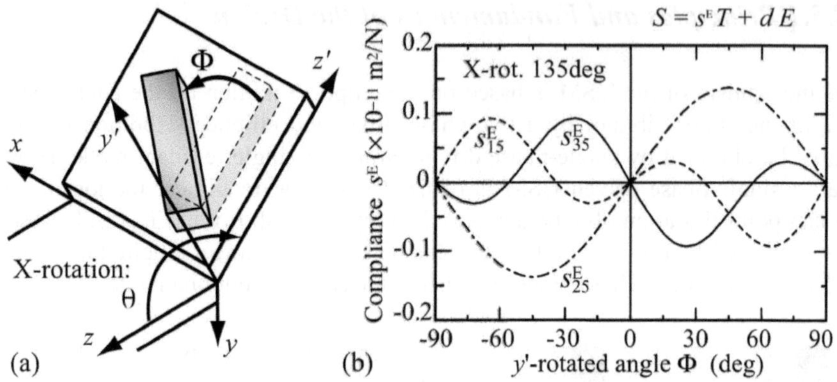

Fig. 18.12 (a) Coordinate system of the X-rotated Y-cut $LiNbO_3$ plate. (b) The elastic compliance correlates with the second rotation angle

The vibrator dimensions and crystal rotation angles were determined by FEM. For example, vibrator was designed so that the displacement ratio u_H/u_V approached 1 in both the upper and lower modes.[7,8] A test structure and its fundamental characteristics are shown in Fig.18.13(a). However, the electromechanical coupling factor k of the basic structure varies widely for the upper and lower modes. Therefore, we modified the electrode pattern to equalize the coupling factors. The electrode, as shown in Fig.18.13(b), is adapted to the internal stress, and the modified vibrator provides a coequal coupling factor.

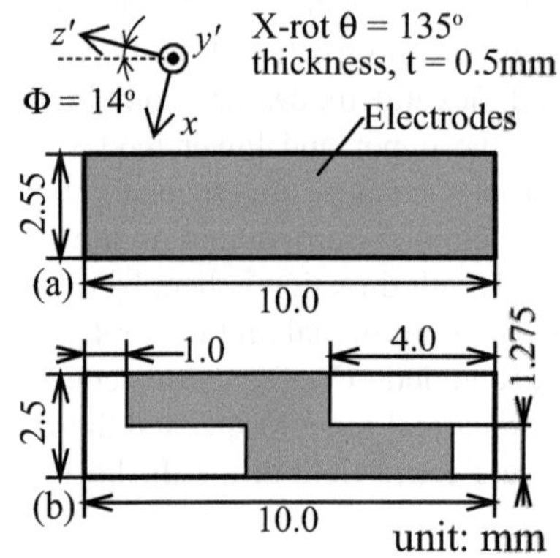

Mode	lower	upper	lower	upper
Q	1,125	612	1,218	1,688
f_0 (kHz)	281.17	286.40	280.107	298.84
R (Ω)	335.6	348.8	259.7	534.8
L (mH)	213.6	118.6	179.7	480.8
C (pF)	1.50	2.60	1.80	0.59
C_d (pF)	30.2	30.3	18.1	18.1
k (%)	11.6	40.9	30.3	30.9

Fig. 18.13 Dimensions of the test structure and its equivalent circuit constants

18.5.2 Experimental Results for an Ultrasonic Motor

A vibrator with a zirconium ball as an antifriction material is fixed by the conductive structure, which is used for electric power feeding, at the center points of both surfaces, as shown in Fig.18.14(b). The stator vibrator is preloaded to the rotor shaft, the diameter of which is 5 mm. The measurement system is shown in Fig.18.14(c).

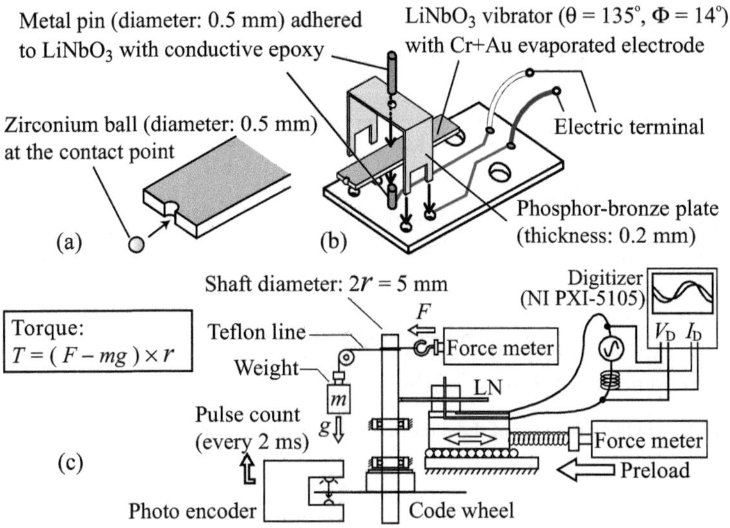

Fig. 18.14 Experimental setup. (a) Contact point with the antifriction material, (b) construction of the stator vibrator, and (c) the measurement system

The rotation characteristics with respect to the driving frequency sweeping are shown in Fig.18.15. The direction of rotation is switched by the frequency. The interval of frequency for the switching is wider in the modified electrode-type motor than in the basic motor, and the wider interval provides stability and ease of mode selection.

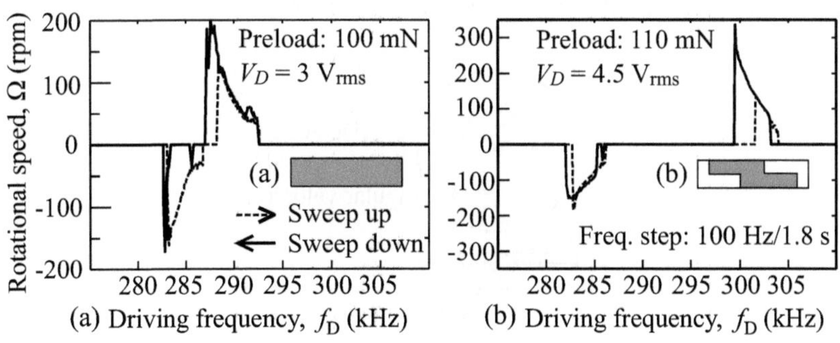

Fig. 18.15 Rotation characteristics correlate with sweeping drive frequency

The load characteristics of the modified electrode-type motor are shown in Fig.18.16, and the maximum efficiency is approximately 4 to 6%. The difference in the characteristics by the rotation direction and several problems must be improved by the structural modifications. However, it is found that a potentiality of the single crystal ultrasonic motor which consumes low power exists.

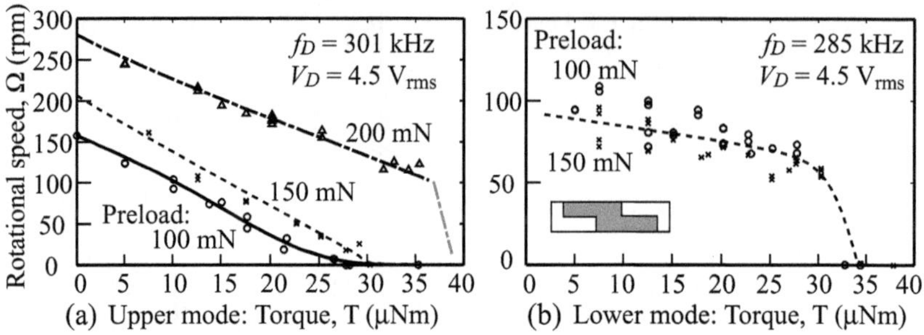

Fig. 18.16 Load characteristics of the modified electrode-type motor

18.6 Motor Type D: Gyro-Moment Motors

We also investigated a new motor, the operation principle of which depends on a vibratory gyro. Figure 18.17 shows the disk-type vibratory gyro, and its basic principle is shown by the equivalent circuit in Fig.18.18. By inputting the moments M_1 and M_2 and considering the phase difference between them, the moment M_3 is generated and rotates the rotor. This motor is referred to herein as the gyro-moment motor (GMM). [9, 10] A number of GMMs are shown in Figs. 18.19 and 18.20. The CW and CCW rotation of the rotor can be determined easily by setting the phase difference between the moments M_1 and M_2.

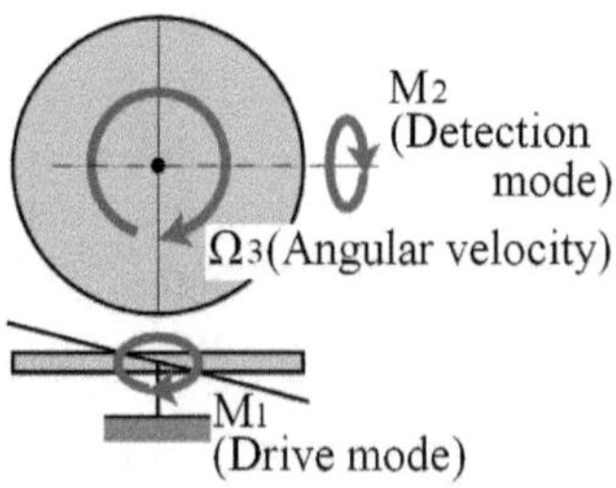

Fig. 18.17 Operation principle of the disk-type gyro-sensor

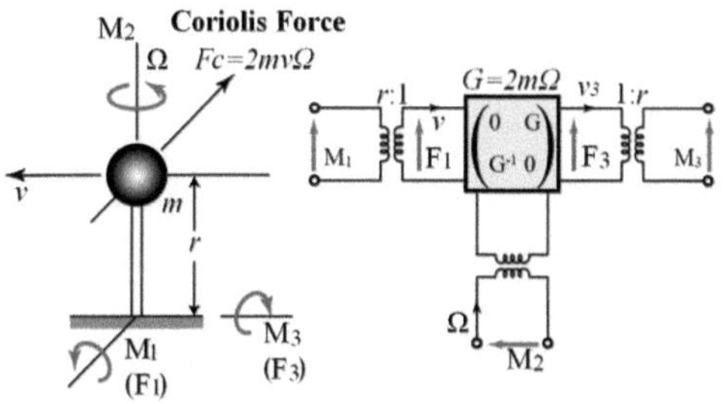

Fig. 18.18 Gyro-sensor and its equivalent circuit

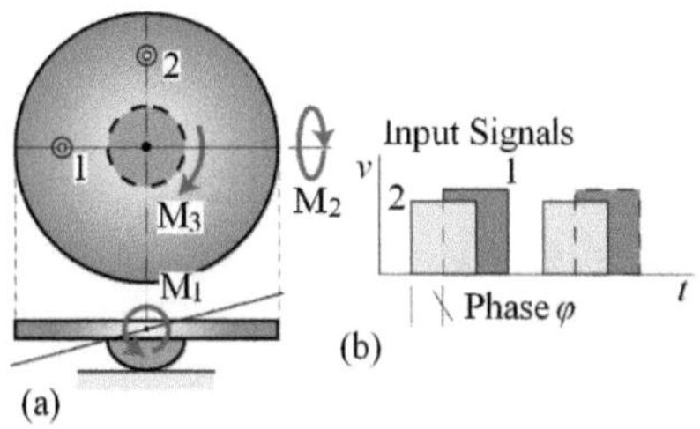

Fig. 18.19 Disk-type gyro-moment motor based on the gyro-sensor shown in Fig.17.18

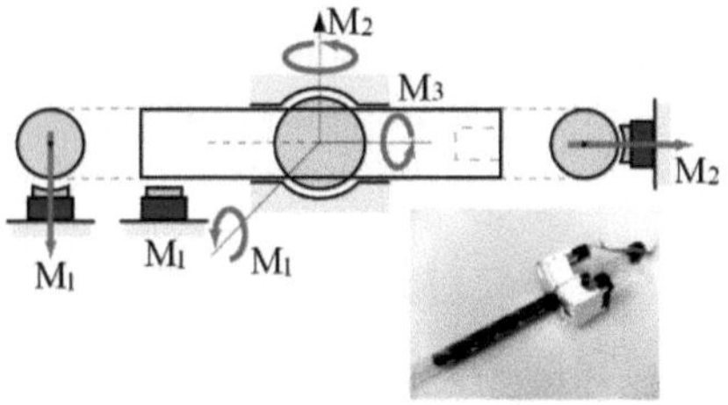

Fig. 18.20 Rod-type gyro-moment motor

18.6.1 Driving Method of the GMM

The driving method of the electromagnetic-type GMM is shown in Fig.18.21, which shows the directions of rotor rotation for the driving methods. In Fig.18.21, EM1~EM4 denote electromagnetic transducers.

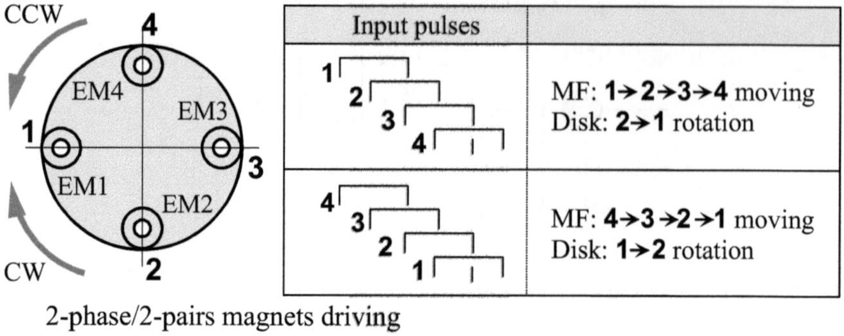

Fig. 18.21 Direction of rotor rotation for driving

18.6.2 Some Prototype GMMs

We produced a number of prototype GMMs:

1. Disk-type GMM: An example of the disk-type GMMs is shown in Fig.18.22. The measured characteristics were as follows:

Driving pulse signal: two-phase-four-pole pulse

Duty ratio:	50%	Driving pulse voltage:	5.0 (V)
Pulse frequency:	30.0 (Hz)	Revolution speed:	1,020 (rpm)

Fig. 18.22 Prototype disk-type motor

2. Slender bar-type GMM (Screw motor): In Fig.18.20, a drill bar (ϕ: 5.4 mm, length: 97 mm) was used as a bar rotor, which was inserted in a stator constructed from an acrylic pipe (external and inner diameters: 8 mm and 6 mm, respectively), which was filled with water. The measured characteristics of the slender bar-type GMM shown in Fig.18.20 are as follows:

Driving pulse signal: two-phase two-pole pulse,

Duty ratio:	50%,	Driving current:	74 (mA),
Pulse frequency :	15.0 (Hz),	Rotational speed:	90 (rpm)

18.6.3 Application to Toys

The GMM considered herein may also be applied in toys, such as those described below.

1. Finger-operated disk-type GMM-toy: Fig.18.23
2. Double-rotor-type GMM-top with CW and CCW rotations: Fig.18.24
3. GMM-top with self rotation and spiral rotation in the opposite directions: Fig.18.25.

Fig. 18.23 Finger-operated GMM-toy

Fig. 18.24 Double-rotor-type GMM-top

Fig. 18.25 Wooden "planet" tops

18.7 Conclusion

High-speed ultrasonic motors using various types of stator vibrators can be easily realized. Considering the practical applications of these motors, problems related to durability and stability, which are greatly influenced by wearing of friction surfaces, must be solved.

Acknowledgments This research was supported in part by the Ministry of Education, Culture, Sports, Science and Technology through a Grant in-Aid for Scientific Research on Priority Areas (No.438) No.19016001.

References

1. Aoyagi M, Suzuki F, Tomikawa Y and Kano I (2004) Jpn. J. Appl. Phys. 43, 5B: 2873-2878
2. Aoyagi M and Tomikawa Y (1996-06) Electronics and Communications in Japan. 79, 6: 60-67
3. Takano T, Tamura H, Tomikawa Y and Aoyagi M (2006) Proc. 10th Int. Conf. Actuators2006: 453-456
4. Kawai K, Tamura H, Tomikawa Y, Takano T, Hirose S and Aoyagi M (2006) IEICE Tech. 106, 250, US2006-42: 37-42 [in Japanese]
5. Tamura H, Kawai K, Takano T, Tomikawa Y, Hirose S and Aoyagi M (2007) Jpn. J. Appl. Phys. 46, 7B: 4698-4703
6. Tamura H, Iwase M, Hirose S, Aoyagi M, Takano T and Tomikawa Y (2008) Jpn. J. Appl. Phys. 47, 5B: 4034-4040
7. Tamura H, Shibata K, Aoyagi M, Takano T, Tomikawa Y and Hirose S (2008) Jpn. J. Appl. Phys. 47, 5B: 4015-4020
8. Tamura H, Aoyagi M, Takano T, Tomikawa Y and Hirose S (2008) Proc. 11th Int. Conf. Actuators2008: 596-599
9. Kanauchi K and Tomikawa Y (2000) Proc. 7th Int. Conf. Actuators2000: 246-249
10. Tomikawa Y, Kusakabe C, Kanayama K and Takano T (2008) Proc. 11th Int. Conf. Actuators2008: 640-643

Chapter 19
Biochemical Pump with Enzymatic Reaction

- Organic Device with an Active Transportation System -

Yoshihiko WAKABAYASHI[1], Toshiaki OKAMOTO[1], Hirokazu SAITO[1], Hiroyuki KUDO[1], Kohji MITSUBAYASHI[1]

Abstract A biochemo-mechanical pump was constructed with funnel type glass tubes and an enzyme membrane, immobilized catalase onto single-side of dialysis membrane. By applying hydrogen peroxide (H_2O_2) solution into non-enzyme immobilized side tube, the pressure in another tube increased rapidly. Namely, an active transportation of H_2O_2 by the asymmetric enzyme membrane from the funnel area to the tube induced the increase in the tube pressure (max.: 5000 Pa), thus resulting in buffer discharge with non-pulsating and low flow rate. The output pressure was linearly related to the concentration of hydrogen peroxide over a range of 11.8 to 123.6 mmol/l, with good reproducibility.

19.1 Introduction

Many type of Micro-machining devices and MEMS (Micro Electro Mechanical System) with micro-fabrication techniques have been investigated [1]. Particularly, micro-actuator devices such as motor, pump, etc. were required in the medical and analysis fields. The existing devices, however, obtain the mechanical force from electric or heat energy, thus resulting lower energy conversion efficiency and complicate system for energy transfer [2].In living organism, protein molecule such as muscle molecule actin/myosin and ATP synthetase, functioned directly as the devices for energy conversion and transfer with high efficiency [3]. Since some biocatalyst could be catalyze chemical reaction with volume change at room

[1] Yoshihiko WAKABAYASHI, Toshiaki OKAMOTO, Hirokazu SAITO, Hiroyuki KUDO, Kohji MITSUBAYASHI

Institute of Biomaterials and Bioengineering, Tokyo Medical and Dental University temperature and pressure, mechanical force would be expected to be provided directly from chemical energy [4,5].

In this paper, an active transportation system was achieved using asymmetric catalase immobilized membrane with hydrogen peroxide as substrate. A biochemical pump was also fabricated with the enzyme membrane and glass tubes.

19.2 Experimental Section

19.2.1 Construction of Biochemical Pump

Figure 19.1 illustrates the biochemical pump with catalase immobilized enzyme and glass tubes. The chemical pump consisted of a funnel type glass tube (narrow i.d.: 7 mm) and a catalase immobilized membrane.

Catalase (EC 1.11.1.6, activity: 65000U・mg-1, Roche Pharmaceuticals, NJ, USA) was used for constructing the biochemical pump. The enzyme catalyzes specificity for hydrogen peroxide, thus resulting oxygen molecule as one of enzyme products.

For enzyme immobilization, catalase was mixed with PVA-SbQ monomer solution (photocrosslinkable polyvinyl alcohol containing stilbazole quaternized; Type: SPP(styryl pyridinium polymer)-H-13(Bio),Tokyo Gosei Kogyo Co., Tokyo, Japan) and phosphate buffer solution (pH7.0, 50mmol/l) in a weight ratio of 1 : 50: 25, [6-8] to a dialysis membrane (thickness: 15 μm, Part No.157-0144-02., thickness 15μm, Technicon Chemicals Co., S.A., Oceq, Belgium) spread on a glass plate, and then irradiated with a fluorescent lamp for 2 hour in order to photocrosslink the monomer solution and immobilize the enzyme onto single side of the dialysis membrane.

The catalase immobilized membrane was removed from the glass plate and immersed in phosphate buffer. In order to prevent enzyme deactivation when not in use, the membrane was stored in buffer below 10 °C.

The chemical pump was fabricated by sealing the opening mouth of the funnel-type glass tube with phosphate buffer solution, by the asymmetric enzyme membrane, in which the enzyme immobilized side was faced to the inside of the tube (see the enlargement of Figure 19.1). Then, the membrane sealed side of the tube was immersed into a funnel area (filled with buffer solution).

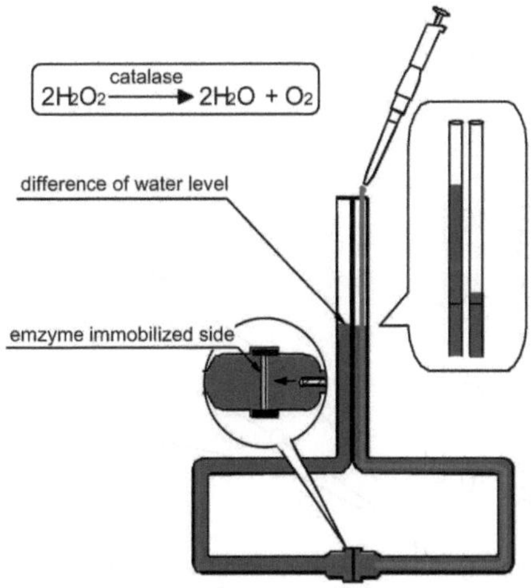

Fig. 19.1 Schematic diagram of a biochemical pump with catalase immobilized membrane

19.2.2 Evaluation of Pump Behavior

Then standard H_2O_2 solution as a substrate of catalase was injected to non-enzyme side of the funnel area by a micro-syringe. A water-head change in the glass tubes was monitored as the height differential of buffer solution between enzyme side and non-enzyme side, thus converting the hydraulic pressure value in the biochemical pump.

19.3 Results and Discussion

Figure 19.2 shows the typical responses of water-head differential in the tubes of the biochemical pump to varying concentrations of hydrogen peroxide (11.8, 45.7, 83.1, and 123.6 mmol/l). As the figure indicates, the differences of water-head increased gradually following addition of H_2O_2.

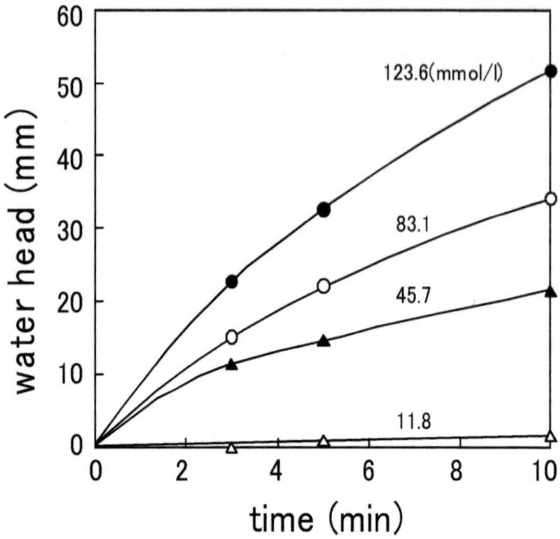

Fig. 19.2 Typical response of the biochemical pump for H_2O_2 (open triangle: 11.5, filled triangle: 45.7, open circle: 83.1, filled circle: 123.6 mmol/l)

The relationship between the final concentration of H_2O_2 in the funnel area and the output pressure (10 minutes after injection) calculated from water-head differential is illustrated in Figure 19.3.

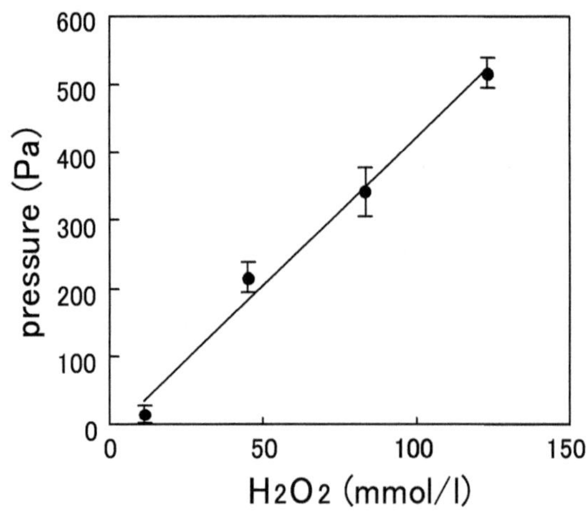

Fig. 19.3 Relationship between final concentration of H_2O_2 and pressure output of the biochemical pump 10 min. after adding H_2O_2 solution

As the figure indicates, the output pressure was linearly related to the concentration of hydrogen peroxide over a range of 11.8 to 123.6 mmol/l, with a correlation coefficient of 0.994 deduced by regression analysis as shown by the following equation:

$$\text{pressure (Pa)} = -15.56 + 4.35\,[H_2O_2\ \text{(mmol/l)}]. \tag{19.1}$$

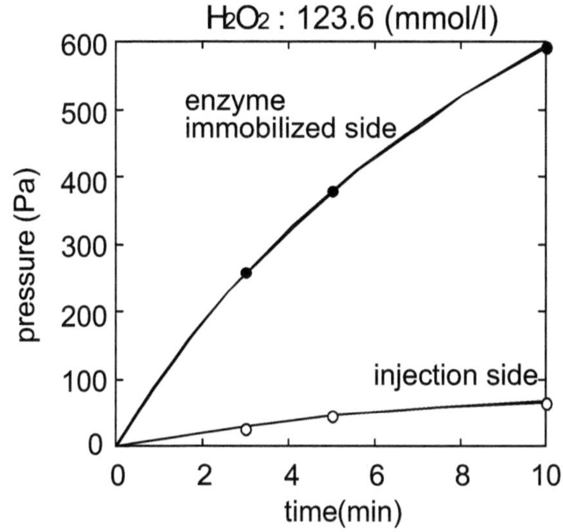

Fig. 19.4 Comparison of pressure change in the tube between enzyme immobilized side and H_2O_2 injection side

Figure 19.4 illustrates the comparison between the pressure changes over time in the enzyme immobilized side and the injection side of the funnel area.

In contrast to the enzyme immobilized side, the pressure output in the injection side increased slightly after H_2O_2 injection

The H_2O_2 molecule penetrated to the dialysis membrane would be catalyzed by catalase to oxygen molecule as enzyme products. And then the oxygen was vaporized in the enzyme immobilized side of the funnel area with volume expansion. As the results, an active transportation of H_2O_2 by the asymmetric enzyme membrane induced the increase in the tube pressure (max.: 5000 Pa) without any electrical and mechanical energy.

The reproducibility of the biochemical pump to H H_2O_2 solution of 123.6 mmol/l is shown in Figure 19.5. Inset figure also illustrates the pressure response of the pump. Sensor performance was reproducible over multiple measurements, showing a coefficient of variation of less than 5 % (n = 5)

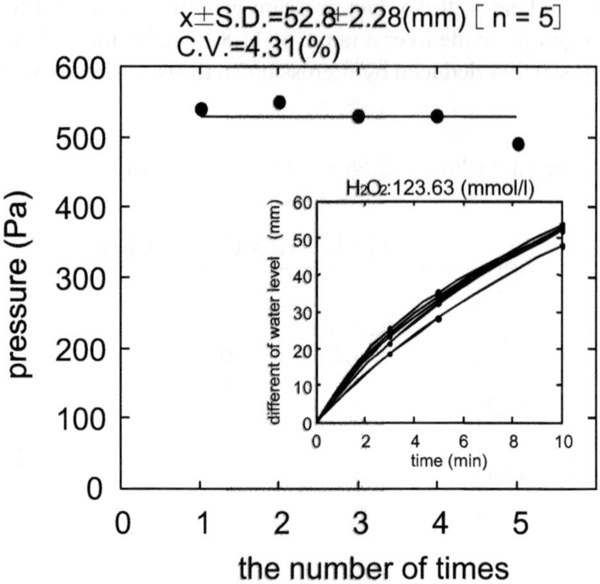

Fig. 19.5 Reproducibility of the biochemical pump to 123.63 mmol/l of H_2O_2. Inset: Response curves after injection (n = 5)

19.4 Conclusions

The biochemical pump with active transportation mechanism was constructed with funnel type glass tubes and asymmetric catalase immobilized membrane.

The output pressure in the enzyme immobilized side of the funnel area increased by injecting hydrogen peroxide (H_2O_2) solution into non-enzyme immobilized side. The active transportation of H_2O_2 by the asymmetric enzyme membrane from the injection side to the enzyme immobilized side induced the increase in the tube pressure (max.: 5000 Pa), thus resulting in buffer discharge with non-pulsating and low flow rate.

The output pressure was linearly related to the concentration of hydrogen peroxide over a range of 11.8 to 123.6 mmol/l, with good reproducibility

Acknowledgments This study was supported in part by MEXT (Ministry of Education, Culture, Sports, Science and Technology) Grant-in-Aid for Scientific Research on Priority Areas (No. 438) "Next-Generation Actuators Leading Breakthroughs", MEXT Special Funds for Education and Research "Advanced Research Program in Sensing Biology", JSPS (Japan Society for the Promotion of Science) Grants-in-Aid for Science Research System.

References

1. Fujimasa I, Nakajima N, Inomachi H, Chinzei T (1991) Notebook for development of micromachining, Shunjun-sha, Tokyo
2. Takamori T (1987) Revolution of actuator, Kogyochosa-kai, Tokyo
3. Jpn Soc. Biophysics (Ed.) (2000) Pump and tranporter, Kyoritsu Publishing, Tokyo
4. Sand TT, Zielinski JE, Arthur C, Bradley D, Wie S (2003) Development of an immunosensor based on pressure transduction, Biosensors and Bioelectronics, 18:797-804
5. Mitsubayashi K, Wakabayashi Y, Okamoto T (2003) Study on a biosensing-actuator. Abstracts of 83rd Japan Chemcal Meeting. 1147
6. Mitsubayashi K, Yokoyama K, Takeuchi T, Karube I (1994) Gas-PhaseBiosensor for Ethanol. Anal Chem 66(20):3297-3302
7. Mitsubayashi K, Hashimoto Y (2000) Development of a gas-phase biosensor for trimethylamine using a flavin-containing monooxygenase 3. Electrochem 68(11): 901-903
8. Ichimura K (1984) A Convenient Photochemical Method to Immobilize Enzymes. J Polym. Sci 22: 2817-2828.

Chapter 20
Domain Wall Engineering in Lead-Free Piezoelectric Materials and Their Enhanced Piezoelectricities

Satoshi WADA[1]

Abstract The engineered domain configurations were induced in barium titanate ($BaTiO_3$) single crystals, and their piezoelectric properties were investigated as a function of the domain size (domain wall density). As a result, it was revealed that the piezoelectric properties were strongly dependent on the domain sizes (domain wall density), i.e., the piezoelectric properties significantly increased with decreasing domain size. The calculated d_{31} of the $[111]_c$ oriented tetragonal $BaTiO_3$ single-domain crystal was –62 pC/N, while the measured value of the $[111]_c$ poled tetragonal $BaTiO_3$ crystal with a domain size of 3 μm was –243.2 pC/N, i.e. an increase of four-fold. When the much finer domain size (below 1 μm) can be induced in the $[001]_c$ poled orthorhombic $BaTiO_3$ crystals, the significantly enhanced piezoelectric properties can be expected.

20.1 Introduction

Domain configurations in the ferroelectric materials can determine ferroelectric and related properties such as piezoelectricity, pyroelectricity and dielectricity. Thus, one of the most interesting investigations for ferroelectric-related applications is the control of the desirable domain configuration. This technique is called domain engineering technique. Domain engineering is the most important technique to obtain the enhanced ferroelectric-related properties for conventional ferroelectric materials. Up to date, the following domain engineering techniques have been proposed and established. The inhibited domain wall motion by the acceptor doping into $Pb(Ti,Zr)O_3$ (PZT) ceramics, i.e., "hard" PZT, is one of the domain engineering techniques, and the hard PZT ceramics are used for the piezoelectric transformer application.1) The enhanced domain wall motion by the do-

[1]Satoshi WADA

Interdisciplinary Graduate School of Medical and Engineering, University of Yamanashi nor doping into PZT ceramics, i.e., “soft” PZT, is also one of the domain engineering techniques, and the soft PZT ceramics are used for the actuator application[1]. The induction of a periodic domain-inverted structure into lithium tantalate ($LiTaO_3$) and lithium niobate ($LiNbO_3$) single crystals is a typical domain engineering technique, and this device is used for the surface acoustic wave application [2,3] and the nonlinear optical application [4,5]. Recently, for the T-bit memory application, the writing and reading techniques of nanodomain on a $LiTaO_3$ single crystal plate were reported by Cho et al., [6,7] and for the 3-D photonic crystal application, the writing and etching techniques of nanodomain of a $LiNbO_3$ single crystal plate were reported by Kitamura et al[8]. These techniques are very important domain engineering techniques, and by using these techniques, enhanced ferroelectric-related properties and new properties can be expected.

Among these domain engineering techniques, the one using the crystallographic anisotropy of the ferroelectric single crystals is known as engineered domain configuration[9-11]. This engineered domain configuration can induce enhanced piezoelectric property in the ferroelectric single crystals. However, there are still many unknown things such as the mechanism of the enhanced property. Especially, it is very important to investigate the most suitable engineered domain configuration for the piezoelectric applications. In this study, the engineered domain configurations were induced in $BaTiO_3$ single crystals, and their piezoelectric properties were investigated as a function of the domain size (domain wall density). Moreover, for the tetragonal $BaTiO_3$ crystals with the engineered domain configurations, it was recently found that the piezoelectric properties significantly improved with decreasing domain sizes (domain wall density) [12, 13]. These results suggested that the domain walls in the engineered domain configuration could contribute significantly to the piezoelectric properties. In this chapter, we try to point out that a significant contribution of domain wall region to piezoelectric, elastic and dielectric properties, and engineered domain configurations can be very helpful to fix the domain wall region in piezoelectric crystals, i.e., to prepare a composite between (a) a distorted domain wall region and (b) a normal tetragonal domain region. Thus, an increase in the volume fraction of the distorted domain wall region can result in crystals with ultrahigh piezoelectric properties.

20.2 Domain Size Dependence of $BaTiO_3$ Crystals with Engineered Domain Configurations

$BaTiO_3$ single crystals have a tetragonal P4mm phase at room temperature. To induce an engineered domain configuration in the tetragonal crystals, the E-field should be applied along the $[111]_c$ direction. The piezoelectric measurements showed that the d_{33} of the $[111]_c$ poled tetragonal $BaTiO_3$ crystal with the engineered domain configuration was almost 203 pC/N, [14] and this value was more than twice as large as the 90 pC/N of the $[001]_c$ poled $BaTiO_3$ single-domain crystal. [15] On the other hand, the d_{33} of the $[001]_c$ poled rhombohedral PZN-PT crystal with the engineered domain configuration was almost 30 times as large as the 83 pC/N of the $[111]_c$ poled PZN-PT single-domain crystal. [16] To explain the origin of this huge difference, we consider the domain size, i.e., the domain wall density, for the engineered domain configuration. This is because the domain structure of the $[001]_c$ poled PZN-PT crystal was composed of the fiber-like domains with a domain length of 130 μm and a domain width of around 1 μm [10,11] while that of the $[111]_c$ poled $BaTiO_3$ crystal was made of very coarse domain with a wide domain width of 300~400 μm [14]. This result suggested that non-180° domain wall region could contribute to piezoelectric properties significantly owing to its structure gradient region. Thus, when the fine domain structure is induced in the $[111]_c$ poled tetragonal $BaTiO_3$ crystals with the engineered domain configuration, it is possible to obtain much enhanced piezoelectric property. Therefore, the piezoelectric properties of $BaTiO_3$ single crystals were investigated as a function of domain size, i.e., domain wall density. Especially, in the $[111]_c$ oriented tetragonal $BaTiO_3$ crystals with an engineered domain configuration, the domain size dependence on E-field and the temperature was investigated in detail.

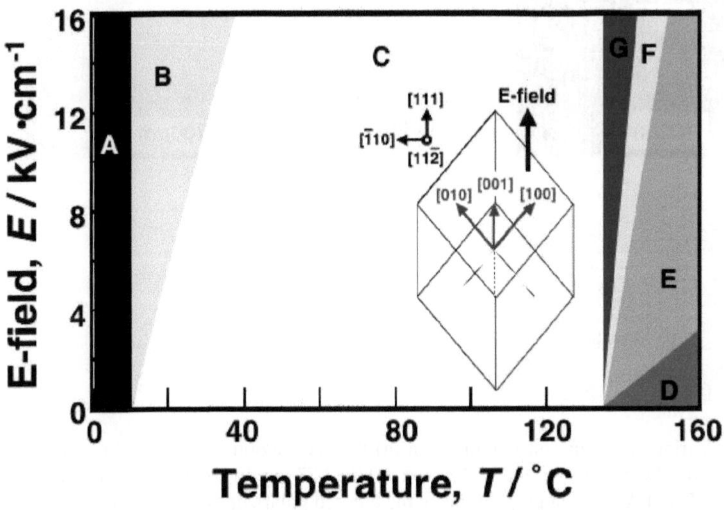

Fig. 20.1 Schematic domain configuration map as a function of temperature and E-field along the $[111]_c$ direction for the $[111]_c$ oriented $BaTiO_3$ single crystals with the engineered domain configuration. Areas A through G stand for a region with different domain configurations related to Fig.20.2, respectively

20.2.1 Domain Size Dependence on E-Filed and Temperature

To understand domain size dependence on E-field and temperature for the $[111]_c$ oriented $BaTiO_3$ crystals, the domain structures were observed at various temperatures from 0 to 200 °C and various E-fields from 0 to 16 kV/cm. Prior to any domain observation, the $BaTiO_3$ crystals were heated up to 160 °C, and then cooled down to the observation temperatures at a cooling rate of 0.4 °C/min without E-field. At the constant temperature, the dc E-fields were applied along the $[111]_c$ direction very slowly.

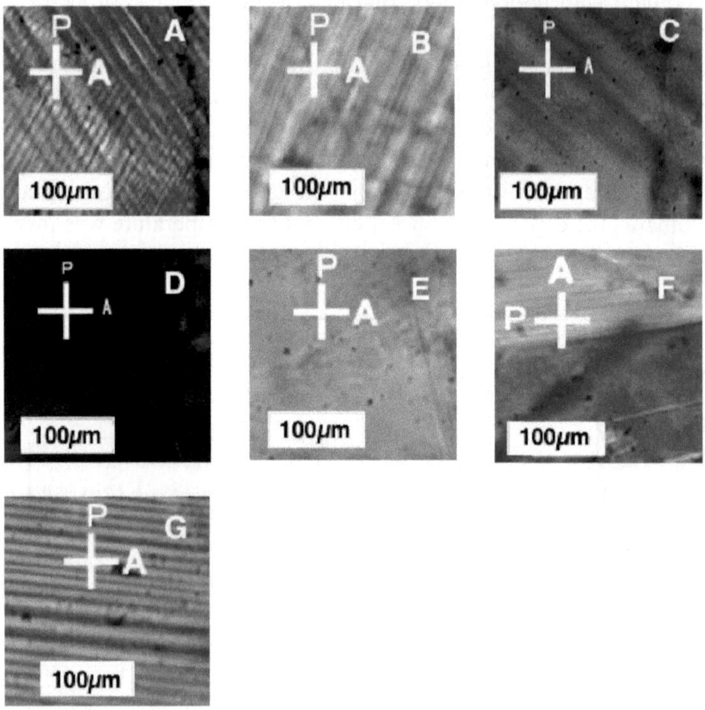

Fig. 20.2 Various domain configurations in (a) the "A" region, (b) the "B" region, (c) the "C" region, (d) the "D" region, (e) the "E" region, (f) the "F" region and (g) the "G" region as shown in Fig.20.1. P and A on each graph indicate the crossed polarizer and analyzer

Figure 20.1 shows the domain size dependence on the E-filed and temperature for the $[111]_c$ oriented $BaTiO_3$ crystals with the engineered domain configuration. Figure 20.2 shows the typical domain structures observed in the corresponding regions A to G presented in Fig.20.1 (P and A indicate the crossed polarizer and analyzer). In Fig.20.1, the "A" and "B" regions were assigned to the orthorhombicmm2 phase. At 25 °C, the dc E-field drive along the $[111]_c$ direction for the tetragonal $BaTiO_3$ crystals resulted in the E-field induced phase transition from

4mm to mm2, and this result was consistent with previous reports [14,17-19]. Figures 20.2-(a) and (b) shows that these domain structures were composed of the fine domains. The fine domains in the "A" region were induced by a normal phase transition from the tetragonal phase to the orthorhombic phase without E-field. On the other hand, the fine domains in the "B" region resulted from the E-field induced phase transition from the tetragonal to orthorhombic phase. Thus, when the E-field was removed, the domain structure in the "B" region easily returned to the normal tetragonal domain configuration as shown in Fig.20.2-(c). Thus, the fine domain structure in the "A" and "B" regions cannot exist at room temperature without E-field. The "C" region was assigned to the tetragonal 4mm symmetry. By the poling treatment in this region, the domain structure was slightly changed, but the observed domain walls were all completely assigned to 90° domain walls of the $\{110\}_c$ planes as shown in Fig.20.2-(c). [20] The "D" region was assigned to the optical isotropic state with the cubic m-3m symmetry as shown in Fig.20.2-(d). However, the "E" region was very unclear and abnormal.

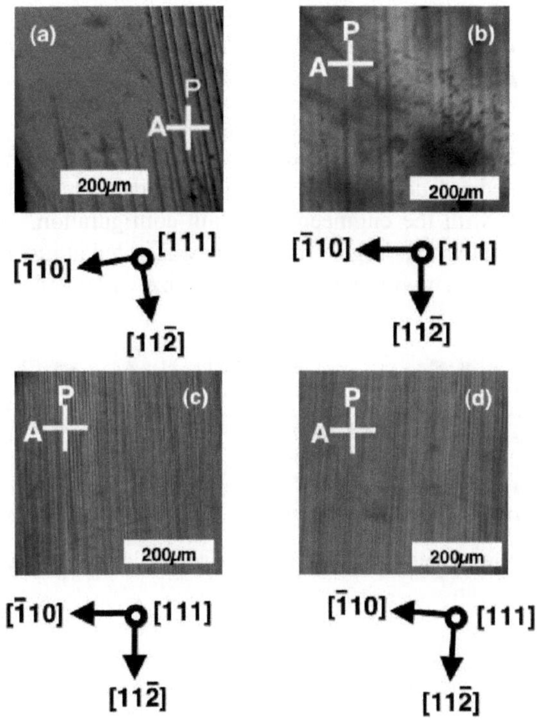

Fig. 20.3 Engineered domain configurations of the [111]c oriented $BaTiO_3$ single crystals with the different average domain sizes of (a) greater than 40 μm, (b) 13.3 μm, (c) 6.5 μm and (d) 5.5 μm

The same brightness in the "E" region under crossed-polarizers suggested that this domain state was not optical isotropic, but an anisotropic state. [21] When the

E-field was applied along the $[111]_c$ direction for the cubic m-3m symmetry, it is expected that the cubic m-3m symmetry can be slightly distorted to become the rhombohedral or monoclinic symmetry. However, at present, the origin of this birefringence is unknown. In the "F" region, the coexistence of the bright state without the domain walls and the fine domain state was observed. In the "G" region, only fine domain structure was clearly observed, all the domain walls of which were assigned to 90° domain walls of the $\{110\}_c$ planes.

In the "A", "B", "C" and "D" regions, these symmetries were assigned on the basis of some reports [14,17-19]. However, there is no information about the "E", "F" and "G" regions. Thus, these symmetries were measured using the in-situ Raman scattering measurement, which was combined with the polarizing microscopic observation. As a result, the change from "D" to "G" by E-field was assigned to an E-field induced phase transition from the cubic to tetragonal phase. [22]

20.2.2 Domain Size Dependence of the Piezoelectric Property Using 31 Resonators

On the basis of the result of the domain size dependence on the E-filed and temperature, various kinds of domain sizes were induced in the $[111]_c$ oriented $BaTiO_3$ single crystals with the engineered domain configuration. The engineered domain $BaTiO_3$ crystal with a large domain size was poled at just below the Curie temperature (Tc) while that with the finer domain size was poled at just above Tc.

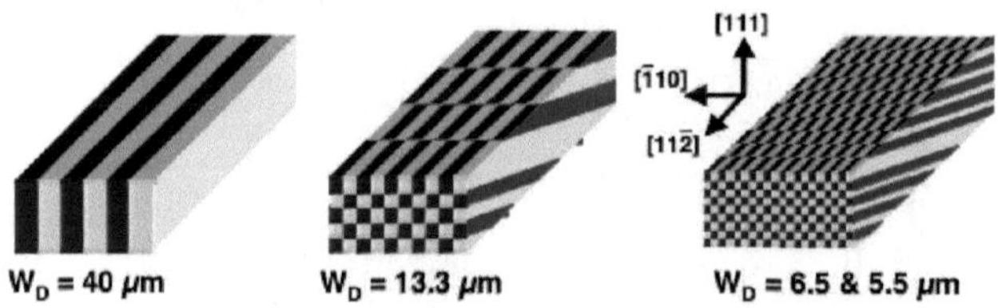

Fig. 20.4 Schematic 31 resonators composed of the engineered domain configuration with different domain sizes

Table 20.1 Domain size dependence of piezoelectric properties for the 31 $BaTiO_3$ crystals resonators

$BaTiO_3$ single crystals	ε_{33}^T	s_{11}^E (pm²/N)	d_{31} (pC/N)	k_{31} (%)
[001] a) (single-domain)	129	7.4	-33.4	---
[111] b) (single-domain)	---	---	-62.0	---
[111], charged (domain size of 80 μm)	1,299	10.9	-85.3	24.1
[111], charged (domain size of 50 μm)	2,117	7.80	-98.2	25.7
[111], charged (domain size of 40 μm)	2,185	7.37	-97.8	25.9
[111], charged + neutral (domain size of 20.0 μm)	2,117	8.30	-102.7	26.0
[111], charged + neutral (domain size of 15.0 μm)	2,186	8.20	-112.5	28.2
[111], charged + neutral (domain size of 13.3 μm)	2,087	7.68	-134.7	35.7
[111], charged + neutral (domain size of 12.0 μm)	1,921	8.20	-137.6	36.8
[111], charged + neutral (domain size of 10.0 μm)	2,239	9.30	-140.5	32.8
[111], charged + neutral (domain size of 8.0μm)	2,238	9.10	-159.2	37.5
[111], charged + neutral (domain size of 7.0 μm)	2,762	9.30	-176.2	36.9
[111], charged + neutral (domain size of 6.5 μm)	2,441	8.80	-180.1	41.4
[111], charged + neutral (domain size of 5.5 μm)	2,762	9.58	-230.0	47.5
"soft" PZT ceramics c) $Pb_{0.988}(Ti_{0.48}Zr_{0.52})_{0.976}Nb_{0.024}O_3$	1,700	16.4	-171.0	34.4

a): measured by Zgonik *et al.* [15]
b): calculated using the values measured by Zgonik *et al.* [15]
c): measured by Jaffe *et. al.* [1]

Figure 20.3 shows the $BaTiO_3$ single crystals with four kinds of domain sizes, i.e., (a) over 40 μm, (b) of 13.3 μm, (c) of 6.5 μm and (d) of 5.5 μm. All of the domains observed in Fig.20.3 were assigned to 90° domain walls of $\{110\}_c$ planes. The domain configurations for all the resonators prepared in this study were composed of the same 90° domain walls[24], and the difference between these domain configurations was just domain size, i.e., domain wall density, as shown in Fig.20.4. Using these 31 resonators with the different domain sizes, the piezoelectric properties were measured at 25 °C by a conventional resonance-antiresonance method [23] using a weak ac E-field of 125 mV/mm. Using the resonance and antiresonance frequencies, the piezoelectric related constants were estimated. Table 20.1 shows the domain size dependence of the properties using these 31 resonators. The d_{31} of the $[001]_c$ oriented $BaTiO_3$ single-domain crystal was reported as –33.4 pC/N while the effective d_{31} of the $[111]_c$ oriented $BaTiO_3$ single-domain crystal calculated using the d_{33} of 90 pC/N, d_{31} of -33.4 pC/N and d_{15} of 564 pC/N reported by Zgonik et al. [15] was –62 pC/N. On the other hand, the d_{31} of the $[111]_c$ poled $BaTiO_3$ single crystal with a domain size over 40 μm was estimated to be -97.8 pC/N while that of the $[111]_c$ poled $BaTiO_3$ single crystal with a domain size of 6.5 μm was found to be –180 pC/N. Especially, the 31 resonator with a domain size of 5.5 μm showed much higher d_{31} of -230 pC/N and k_{31} of 47.5 % than those (d_{31} of -171 pC/N and k_{31} of 34.4 %) reported for soft PZT ceramics [1]. Therefore, the $[111]_c$ oriented engineered domain $BaTiO_3$ crystal exhibit much higher piezoelectric properties than the $[001]_c$ oriented $BaTiO_3$ single-domain crystal.

To date, it was believed that the highest piezoelectric property must be obtained in single-domain crystals, and it is impossible for the material constants to be beyond the values of single-domain crystal. However, this study reveals that the 90° domain walls in the engineered domain configuration significantly enhance the piezoelectric effects, giving rise to much higher piezoelectric constants than those from single-domain crystals.

In general, under a high E-field drive, the domain walls can move very easily, and this domain wall motion made an intrinsic contribution to the piezoelectric properties zero. However, in the engineered domain configuration shown in Fig.20.3, the 90° domain walls cannot move with or without unipolar dc E-field drive. [17,18] This means that in the engineered domain configuration, the 90° domain walls can exist very stably with or without a unipolar E-field drive. Therefore, the contribution of the domain walls to the piezoelectric properties has been clarified for the first time using the engineered domain configuration. In other words, the engineered domain configuration is considered as domain-wall engineering among the domain engineering techniques. [24]

20.2.3 Domain Size Dependence of the Piezoelectric Property Using 33 Resonators

By the poling treatment at various electric-fields and temperatures, [12,13] the 33 resonators of $BaTiO_3$ crystals with different domain sizes were prepared. The average domain sizes in the engineered domain configuration were changed

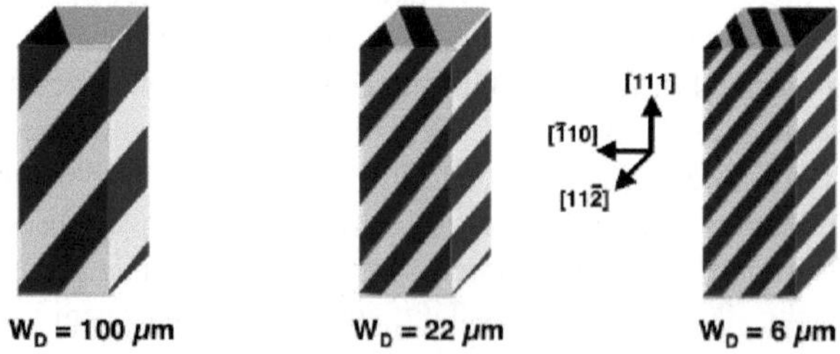

Fig. 20.5 Schematic 33 resonators composed of the engineered domain configuration with different domain sizes

Table 20.2 Domain size dependence of piezoelectric properties for the 33 $BaTiO_3$ crystals resonators

$BaTiO_3$ single crystals	ε_{33}^T	s_{33}^E (pm^2/N)	d_{33} (pC/N)	k_{33} (%)
[001] a) (single-domain)	---	---	90	---
[111] a) (single-domain)	---	---	224	---
[111], neutral (domain size of 100 μm)	1,984	10.6	235	54.4
[111], neutral (domain size of 60 μm)	1,959	10.7	241	55.9
[111], neutral (domain size of 22 μm)	2,008	8.8	256	64.7
[111], neutral (domain size of 15 μm)	2,853	6.8	274	66.1
[111], neutral (domain size of 14 μm)	1,962	10.8	289	66.7
[111], neutral (domain size of 6 μm)	2,679	10.9	331	65.2

a): measured by Zgonik *et al.* [15]
b): calculated using the values measured by Zgonik *et al.* [15]

from 100 μm to 6 μm. The domain configurations for all the 33 resonators prepared in this study were composed of the same 90° domain walls, [20] and the difference between these domain configurations was just the domain size, as shown in Fig.20.5.

The piezoelectric properties were measured at 25 °C using a weak ac E-field of 50 mV/mm. Table 20.2 shows the domain size dependence of piezoelectric-related properties using these 33 resonators. As a reference, the calculated d_{33} piezoelectric constant for the $[111]_c$ oriented $BaTiO_3$ single-domain crystal using the material constants reported by Zgonik et al. [15] was also listed. With decreasing domain size, all the piezoelectric-related properties increased. The most surprising thing is the fact that the piezoelectric properties significantly depend on the domain size regardless the 31 or 33 mode. Thus, we must consider the possible effect of non-180° domain walls on the piezoelectric property. Liu et al. proposed a model for the role of domain walls in the engineered domain configurations.[29] Here, we consider this useful model to explain the enhanced piezoelectric property with increasing non-180° domain wall density.

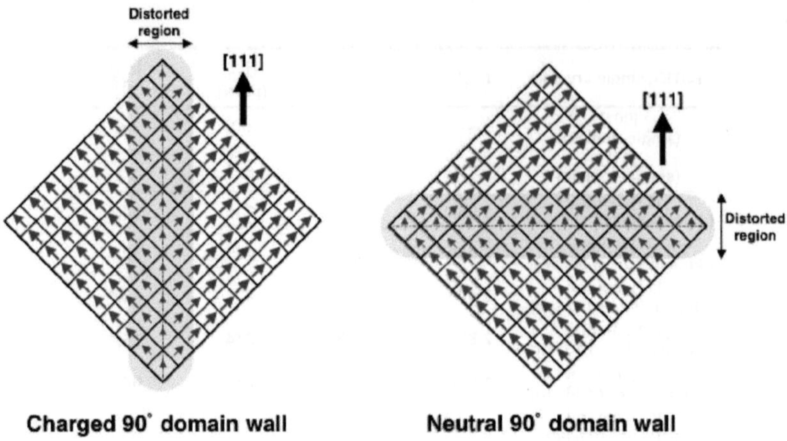

Fig. 20.6 Schematic structure near the 90° domain walls for $BaTiO_3$ crystals

20.3 Role of Non-180° Domain Wall Region on Piezoelectric Properties

It is well known that a region near the 90° domain walls is gradually distorted to relax the strain between domains with different polar directions, as shown in Fig.20.6. [25-29] Moreover, for $BaTiO_3$ crystals, the crystal structure near the 90° domain walls gradually changed from normal tetragonal with c/a ratio of 1.011 to tetragonal with c/a ratio close to 1.0. Thus, it can be expected that the region near the 90° domain walls with pseudo-cubic structure exhibits the material constants of $BaTiO_3$ single crystals near Tc, as reported by Budimir et al. [29] In this study, the $BaTiO_3$ crystals with engineered domain configuration were regarded as a composite of (a) normal tetragonal region and (b) distorted 90° domain wall region. On the basis of this 2-phases model, a volume fraction of the distorted domain wall region over the normal tetragonal region was estimated as follows. For a simplification of calculation, one dimensional model was applied. [30] First, a thickness of the distorted 90° domain wall region (W_{DW}) was assumed using various sizes from 1 to 100 nm. [25-29] Next, using this thickness (W_{DW}) and the domain size (W_D), the volume fraction of the distorted 90° domain wall region (F) was estimated using the following equation,

$$F = \frac{W_{DW}}{W_D} . \qquad (20.1)$$

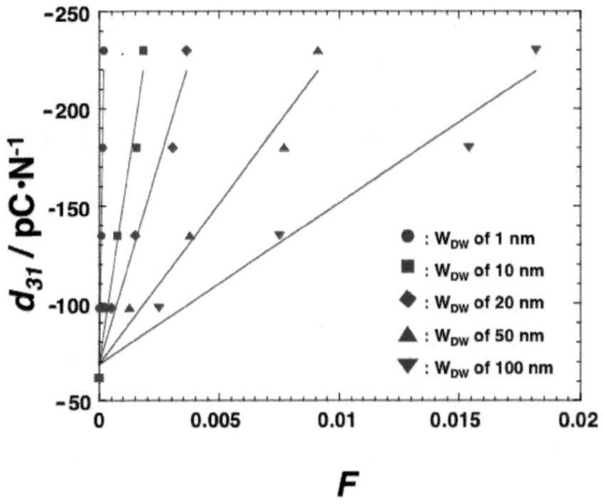

Fig. 20.7 Relationship between d_{31} and F calculated using various W_{DW} values from 1 to 100 nm

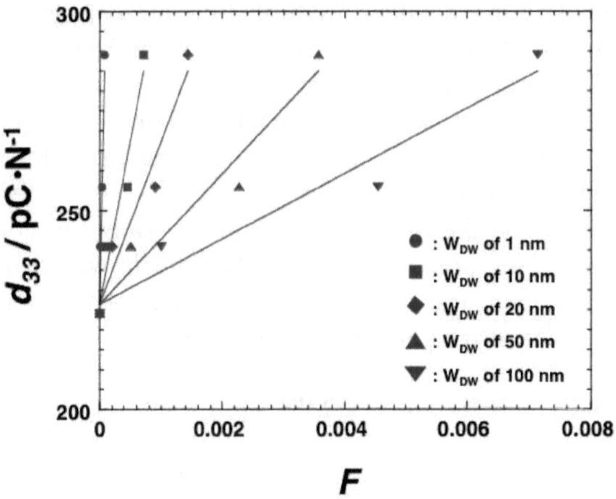

Fig. 20.8 Relationship between d_{33} and F calculated using various W_{DW} values from 1 to 100 nm

W_D can be measured from the experiment in this study while W_{DW} is unknown values. Thus, we must use valid W_{DW} values for the above calculation. To date, many researchers tried to estimate the domain wall thickness using Landau-Ginzburg-Devonshire (LGD) theories and transmission electron microscopic (TEM) observation, and their estimated values from 1 to 10 nm. [25-29] Recently, based on the developed measurement equipments and theories, some new methods were proposed to estimate the domain wall thickness. [30-32] Especially, the domain wall thickness was related to point defect, and the defect was responsible for the broadening of the domain wall thickness. [31] This means that it is very difficult to determine the 90° domain wall thickness for $BaTiO_3$. Thus, in this study, F

values were calculated using the various domain wall thickness from 1 to 100 nm. Using the F values, the relationships between d_{31} and F and d_{33} and F were plotted in Fig.20.7 and Fig.20.8, respectively, in which the slope of a line indicates the piezoelectric constant expected from the corresponding distorted 90° domain wall region. Using the estimated 90° domain wall thickness of 10 nm, d_{31} and d_{33} can be expressed using the following equations,

$$d_{31} = -82{,}676\mathrm{F} - 69, \tag{20.2}$$

$$d_{33} = 81{,}744\mathrm{F} + 227. \tag{20.3}$$

i.e., d_{31} and d_{33} from the distorted 90° domain wall region are calculated to be 82,676 and 81,744 pC/N, respectively. Similarly, if the 90° domain wall thickness is 3 nm, d_{31} and d_{33} from the distorted 90° domain wall region are estimated to be 275,590 and 272,480 pC/N, respectively. If the 90° domain wall thickness is 1 nm, d_{31} of 826,760 and d_{33} of 817,440 can be expected from the distorted 90° domain wall region. The above values reveal that the piezoelectric constants resulting from the distorted 90° domain wall region are significantly high. Based on a recent theoretical calculation, the domain wall width is estimated to be around 3 nm [26], while the maximum domain wall width is found from recent experiments to be 10 nm. [33] Therefore, the d constants arising from domain wall region should higher than 80,000 pC/N.

This study shows that the distorted 90° domain walls can contribute significantly to the piezoelectric properties. When the domain wall density continue to increase, how the piezoelectric constants will increase? The domain size dependences of d_{31} and d_{33} can also be expressed as follows,

$$d_{31} = \frac{-826{,}760}{W_D} - 69 \quad , \tag{20.4}$$

$$d_{33} = \frac{817{,}440}{W_D} + 227 \quad . \tag{20.5}$$

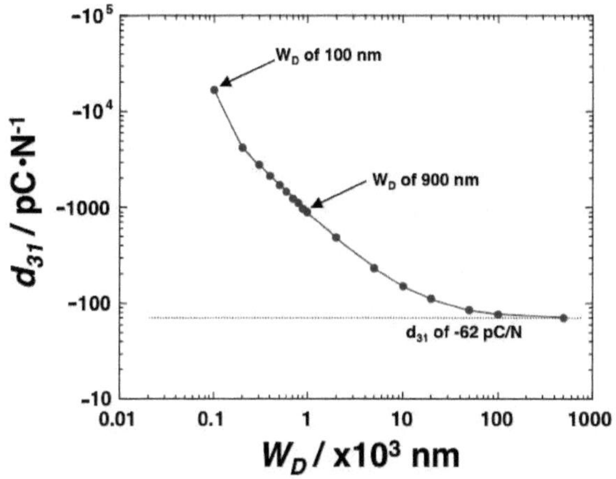

Fig. 20.9 Relationship between d_{31} and W_D calculated using Equation (20.4)

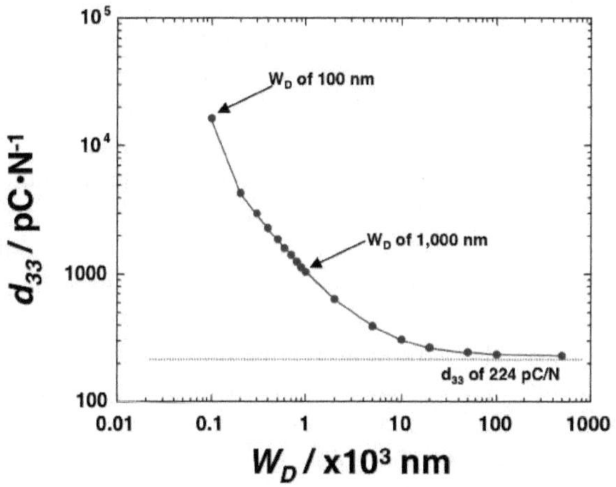

Fig. 20.10 Relationship between d_{33} and W_D calculated using Equation (20.5)

It should be noted that these equations (20.4) and (20.5) are independent of W_{DW}. Thus, using the various W_D values, the relationships between d_{31} and W_D and between d_{33} and W_D were plotted in Fig.20.9 and Fig.20.10, respectively. It can be seen that above the W_D of 20 μm, the piezoelectric coefficients were almost constant at the calculated single-domain values, while below the W_D of 10 μm, piezoelectric constants drastically increased with decreasing domain sizes. Moreover, when the domain size decreased down to 1 μm, both d_{31} and d_{33} reached around 1,000 pC/N.

Park and Shrout reported that $[001]_c$ poled PZN single crystal exhibited the ultrahigh d_{33} of 1,100 pC/N, [34] and the domain size in this PZN crystal was found to be around 1 μm. [10] They also reported that for $[111]_c$ poled PZN single-domain crystal had the d_{33} of just 83 pC/N [34], similar to that of $BaTiO_3$. Therefore, it can be expected that when the domain size of around 1 μm is induced in the $[111]_c$ poled $BaTiO_3$ crystals, high performance lead-free piezoelectrics with ultrahigh piezoelectric constants over 1,000 pC/N can be created.

20.4 What is Domain Wall Engineering?

To enhance piezoelectric property, we must consider two contributions. i.e., intrinsic and extrinsic effects. The intrinsic effect is dependent on unit cell with symmetry and chemical composition. On the other hand, it was revealed that 90° domain wall regions with distorted and structure graduation had ultrahigh piezoelectric constant above 80,000 pC/N. This contribution is an extrinsic effect, and this technique should be called domain wall engineering. In this technique, we must prepare domain wall region fixed in the crystals, and engineered domain configuration is a technique to fix domain wall stably in the crystals. As a result, we can obtain a composite of domain wall regions and normal domain region.

Thus, to obtain the highest piezoelectric crystals, a combination of intrinsic and extrinsic effects is required. To maximize the intrinsic contribution, we should use the $[001]_c$ oriented orthorhombic crystals, and if this crystal has phase transition from orthorhombic to tetragonal at around –10 to 0 °C, we can expect more enhancement of piezoelectric properties. In addition, to maximize the extrinsic contribution, we should induce domain size of around 150 nm in the $[001]_c$ oriented orthorhombic crystals. Arlt et al. reported that in $BaTiO_3$ ceramics, the dielectric constant increased with decreasing grain size and at a grain size of 800 nm, dielectric constant reached a maximum value of 5,000. [35] They also depicted the relationship between the grain size and domain size, and the domain size of $BaTiO_3$ ceramics with a grain size of 800 nm was 150 nm. This result suggested that for domain sizes above 150 nm, the dielectric and piezoelectric properties can increase with decreasing domain sizes, while for the domain size below 150 nm, the both properties can decrease with decreasing domain sizes.

To induce the fine domain size around 150 nm, a combination of patterned electrode and uniaxial stress-field can be used in the new poling system above Tc. We consider that domain wall engineering includes a universal concept of ultrahigh domain wall contribution to piezoelectricity and any techniques to induce fine domain size of about 150 nm in the $[001]_c$ oriented orthorhombic crystals. It is expected that the development of the domain wall engineering will allow us to create new lead-free piezoelectrics with ultrahigh piezoelectric properties.

20.5 Conclusions and Future Trends

The engineered domain configurations were induced in the $BaTiO_3$ single crystals, and their piezoelectric properties were investigated as a function of the domain size (domain wall density). As a result, it was revealed that the piezoelectric properties were strongly dependent on the domain sizes (domain wall density), i.e., the piezoelectric properties significantly increased with decreasing domain size. The calculated d_{31} of the $[111]_c$ oriented tetragonal $BaTiO_3$ single-domain crystal was –62 pC/N, while the measured value of the $[111]_c$ poled tetragonal $BaTiO_3$ crystal with a domain size of 3 μm was –243.2 pC/N, i.e. an increase of four-fold. When the much finer domain size (below 1 μm) can be induced in the $[001]_c$ poled orthorhombic $BaTiO_3$ crystals, the significantly enhanced piezoelectric properties can be expected.

When application of the domain wall engineering is considered in the future, the single crystals have three disadvantages as compared to other piezoelectric materials. One is always very high cost, and this cost can make application limited. Second is very weak mechanical strength, and this weakness also reduces application. Third is an intrinsical difficulty to induce very fine domain sizes below 3 μm. On the other hand, normal piezoelectric ceramics is completely opposite side of the single crystals, and disadvantage of normal ceramics is no crystallographic orientation. Therefore, we must combine of advantage between single crystals and ceramics. I believe that this dream material is grain-oriented ceramics along the engineered domain configuration. In this case, we can obtain four advantages, especially very fine domain sizes. For the piezoelectric ceramics, it is known well that it is very easy to induce fine domain sizes below 100 nm. The application of domain wall engineering for the grain oriented ceramics along engineered domain direction can lead to super piezoelectrics with ultrahigh piezoelectric properties over d_{33} of 1,000 pC/N and k_{33} of 90 %.

Acknowledgments We would like to thank Mr. O. Nakao of Fujikura Ltd. for preparing the TSSG-grown $BaTiO_3$ single crystals with excellent chemical quality. We also would like to thank Dr. S.-E. "Eagle" Park, Dr. T. R. Shrout and Dr. L. E. Cross of MRL, Pennsylvania State University for their helpful suggestion and many discussions about the engineered domain configurations. Moreover, we would like to thank Dr. Y. Ishibashi of Aichi-shukutoku University, Dr. D. Damjanovic of EPFL, Dr. A. J. Bell of University of Leeds and Dr. L. E. Cross of Pennsylvania State University for their helpful discussions about the domain wall contribution to the piezoelectric properties. We would like to thank Dr. J. Erhart and Dr. J. Fousek of ICPR, Technical University of Liberec for their helpful discussions about the analysis of the domain configuration and calculation of the d_{31} surface. This study was partially supported by a Grant-in-Aid for Scientific Research from the Ministry of Education, Culture, Sports, Science, and Technology, Japan.

References

1. Jaffe B, Cook, Jr WR, Jaffe H (1971) Piezoelectric Ceramics. New York, Academic Press
2. Nakamura K, Shimizu H (1983) Poling of Ferroelectric Crystals by using Interdigital Electrodes and Its Application to Bulk-wave Transformer. Proc 1983 IEEE Ultrasonic Symp: 527-530
3. Nakamura K, Ando H, Shimizu H (1986) Partial Domain Inversion in $LiNbO_3$ Plates and Its Applications to Piezoelectric Devices. Proc. 1986 IEEE Ultrasonic Symp: 719-722
4. Lim EJ, Fejer MM, Byer RL, Kozlovsky WJ (1989) Blue Light Generation by Frequency Doubling in Periodically Poled Lithium Niobate Channel Waveguide. Electron Lett 25:731-732
5. Webjorn J, Laurell F, Arvidsson G (1989) Blue Light Generated by Frequency Doubling of Laser Diode Light in a Lithium Niobate Channel Waveguide. IEEE Photonics Technol Lett 1:316-318
6. Hiranaga Y, Fujimoto K, Cho Y, Wagatsuma Y, Onoe A, Terabe K, Kitamura K (2002) Nano-Sized Inverted Domain Formation in Stoichiometric $LiTaO_3$ Single Crystal using Scanning Nonlinear Dielectric Microscopy. Integrated Ferro 49:203-209
7. Cho Y, Hiranaga Y, Fujimoto K, Wagatsuma Y, Ones A, Terabe K, Kitamura K (2003) Tbit/inch^2 Ferroelectric Data Storage Based on Scanning Nonlinear Dielectric Microscopy. Trans Mater Res Soc Jpn 28:109-112
8. Terabe K, Higuchi S, Takekawa S, Nakamura M, Goto Y, Kitamura K (2003) Nanoscale Domain Engineering of a $Sr_{0.61}Ba_{0.39}Nb_2O_6$ Single Crystal using a Scanning Force Microscope. Ferroelectrics 292:83-89
9. Park S-E, Mulvihill KL, Lopath PD, Zipparo M, Shrout TR (1996) Crystal Growth and Ferroelectric Related Properties of $(1\text{-}x)Pb(A_{1/3}Nb_{2/3})O_3 - xPbTiO_3$ ($A=Zn^{2+}$, Mg^{2+}). Proc 10th IEEE Int Symp Applications of Ferroelectrics 1:79-82
10. Wada S, Park S-E, Cross LE, Shrout TR (1998) Domain Configuration and Ferroelectric Related Properties of Relaxor Based Single Crystals. J Korean Phys Soc 32:S1290-S1293
11. Wada S, Park E-E, Cross LE, Shrout TR (1999) Engineered Domain Configuration in Rhombohedral PZN-PT Single Crystals and Their Ferroelectric Related Properties. Ferroelectrics 221:147-155
12. Wada S, Tsurumi T (2004) Enhanced Piezoelectric Property of Barium Titanate Single Crystals with Engineered Domain Configurations. Br Ceram Trans 103:93-96
13. Wada S, Yako K, Kakemoto H, Tsurumi T, Erhart J (2004) Enhanced Piezoelectric Property of Barium Titanate Single Crystals with the Different Domain Sizes. Key Eng Mater 269:19-22
14. Wada S, Suzuki S, Noma T, Suzuki T, Osada M, Kakihana M, Park E-E, Cross LE, Shrout TR (1999) Enhanced Piezoelectric Property of Barium Titanate Single Crystals with Engineered Domain Configurations. Jpn J Appl Phys 38:5505-5511
15. Zgonik M, Bernasconi P, Duelli M, Schlesser R, Gunter P, Garrett MH, Rytz D, Zhu Y, Wu X (1994) Dielectric, Elastic, Piezoelectric, Electro-optic, and Elasto-optic Tensors of $BaTiO_3$ Crystals. Phys Rev B 50:5941-5949
16. Park S-E, Shrout TR (1997) Ultrahigh Strain and Piezoelectric Behavior in Relaxor Based Ferroelectric Single Crystals. J Appl Phys 82:1804-1811
17. Wada S, Kakemoto H, Tsurumi T, Park S-E, Cross LE, Shrout TR (2002) Enhanced Ferroelectric Related Behaviors of Ferroelectric Single Crystals using the Domain Engineering. Trans Mater Res Soc Jpn 27:281-286
18. Park S-E, Wada S, Cross LE, Shrout TR (1999) Crystallographically Engineered $BaTiO_3$ Single Crystals for High-performance Piezoelectrics. J Appl Phys 86:2746-2750
19. Wada S, Tsurumi T (2001) Domain Configurations of Ferroelectric Single Crystals and Their Piezoelectric Properties. Trans Mater Res Soc Jpn 26:11-14
20. Fousek J (1971) Permissible Domain Walls in Ferroelectric Species. Czech J Phys B21:955-968

21. Wahlstrom EE (1979) Optical Crystallography. New York, John Wiley and Sons
22. Wada S, Yako K, Kiguchi T, Kakemoto H, Tsurumi T (2005) Enhanced Piezoelectric Properties of Barium Titanate Single Crystals with the Different Engineered Domain Sizes. J Appl Phys 98:014109
23. (1976) IEEE Standard on Piezoelectricity, American National Standard Institute
24. Fousek J, Litvin DB, Cross LE (2003) Domain Geometry Engineering and Domain Average Engineering of Ferroics. J Phys Condens Matter 13:L33-L38; Fousek J, Cross LE (2003) Open Issues in Application Aspects of Domains in Ferroic Materials. Ferroelectrics 293:43-60
25. Damjanovic D (1998) Rep Prog Phys 61:1267
26. Ishibashi Y, Salje E (2002) A Theory of Ferroelectric 90 Degree Domain Wall. J Phys Soc Jpn 71:2800-2803
27. Setter N (2002) Piezoelectric Materials in Devices. Lausanne, ed. N. Setter. 1
28. Meyer B, Vanderbilt D (2002) Ab Initio Study of Ferroelectric Domain Walls in $PbTiO_3$. Phys Rev B 65:104111
29. Budimir M, Damjanovic D, Setter N (2003) Piezoelectric Anisotropy-Phase Transition Relations in Perovskite Single Crystals. J Appl Phys 94:6753-6761
30. Chaib H, Schlaphof F, Otto T, Eng LM (2003) Electric and Optical Properties of the 90° Ferroelectric Domain Wall in Tetragonal Barium Titanate. J Phys Condens Matter 15:1-14
31. Shilo D, Ravichandran G, Bhattacharya K (2004) Investigation of Twin-wall Structure at the Nanometer Scale using Atomic Force Microscopy. Nature Mater. 3:453-457
32. Tsuji T, Ogiso H, Akedo J, Saito S, Fukuda K, Yamanaka K (2004) Evaluation of Domain Boundary of Piezo/Ferroelectric Material by Ultrasonic Atomic Force Microscopy. Jpn J Appl Phys 43:2907-2913
33. Ishibashi Y (2004) Private Communication
34. Park S-E, Shrout TR (1997) IEEE Trans Ultrason Ferroelectr & Freq Control 44:1140
35. Arlt G, Hennings D, de With G (1985) Dielectric Properties of Fine-grained Barium Titanate Ceramics. J Appl Phys 58:1619-1625

Chapter 21

IPMC Actuator

Next Generation Medical Actuator Using Ion Polymer Metal Compound

Tadashi IHARA[1], Isao YADA[1] and Taro NAKAMURA[1]

Abstract Ion polymer metal compound (IPMC) was extensively studied for its possible medical applications. IPMC works as a soft-actuator which demonstrates unique characters such as light weight, low voltage drive, thin, versatile shaping, and no noise generation. Also, it works as a sensor as well. Force generation of IPMC was studied with isometric tension measurement. Sensor characteristics was carried out with high performance differential amplifier and laser displacement meter. To explore the force generating mechanism, an impedance analysis of IPMC was performed with a precision LCR meter. Also, fabrication procedure of IPMC was examined in order to generate greater force. A prototype medical device that remove phlegm from ventilator airway with IPMC was developed with its driving system.

21.1 Introduction

Ion polymer metal compound features many unique characteristics as a "soft-actuator " material including extremely light weight, ease of control by low voltage, very fast response, large displacement, and versatility in fabrication into various shapes and configurations. It is expected to be one of the most promising materials for artificial muscle and other medical devices. Practical application of IPMC, though, is still limited due mainly to the relative weakness of force generation.

Our preliminary studies [1-5] indicated that the generated force increases with increased thickness of IPMC, repeated plating, and pre-heating process. We examined the change of IPMC characteristic under these conditions in order to clarify what factors contribute to the change in force generation. We have examined isometric force generation because it is one of the standard methods in measuring

[1] Tadashi IHARA, Isao YADA and Taro NAKAMURA

Department of Clinical Engineering, Suzuka University of Medical Science muscle force. A sensing characteristics of IPMC was also examined to explore robotics application as well as sensor-actuator coupled device.

Also, we have fabricated a prototype model of phlegm removing unit with the IPMC which is designed to be used in a ventilator circuit to help patients supported by a ventilator who need frequent suction of phlegm from their airway.

The other advantage of applying Nafion-based hydrophilic membrane onto sensor/actuator for ventilator application is that it circumvents the problem of IPMC that require hydrologic environment. IPMC usually operates in water and deceases its actuating motion when placed in air for about 5 minutes. Since a ventilator generally uses humidifier unit in order to maintain moist air supplied to a patient, IPMC keeps working in mist environment just as in water.

As for the sensor applications, though, IPMC works either in water, vapor, or dry environment and its application domain is greatly enhanced.

21.2 Fabrication of IPMC

IPMC was fabricated using modified standard method [6, 7]. Nafion R-1100 resin was heat-pressed at 185 °C with 20-30 MPa. The thickness of IPMC can be adjusted by changing the amount of resin, pressure, and time to heat-press. Immersion and reduction process was repeated up to 4 times to thicken the thickness of gold plate surface. Disk-shaped membrane was then immersed to hydrolysis solution using a mixture of dimethyl sulfoxide (DMSO), potassium hydroxide (KOH), and water. The pre-processed membrane was permeated in $[Au(phen)Cl_2]^+$ solution. After the immersion, the membrane was reduced with 5% Na_2SO_3 solution to perform gold plating (Fig.21.1). The membrane was cut into a rectangular shape to fit into the respiratory circuit of a ventilator. IPMC membranes of 180 μm (fabricated from Nafion 117), 400 μm, and 800 μm (both fabricated from Nafion R-1100 resin) were cut into 10 mm by 20 mm rectangular slices.

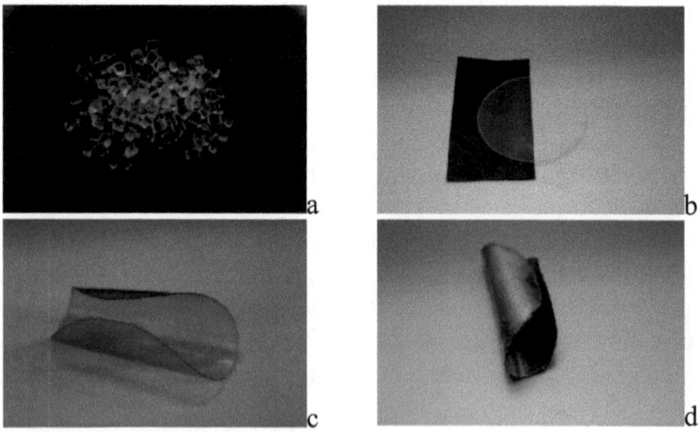

Fig. 21.1 Fabrication of IPMC a) Nafion resin b) after heat-press c) after immersion to $[Au(phen)Cl_2]^+$ solution d) after reduction

21.3 Force Measurement

In order to measure isometric tension, prepared IPMC slice was fixed on an electrode table in water as shown in Fig.21.2. IPMC was driven by a galvanostat unit (Hokuto Denko Co.: HA151) and developed force was measured by an isometric transducer (Nihon Koden Co.: TB-651T) connected to an amplifier (Nihon Koden Co.: EF-601G). Applied voltage ranged 1-3 V.

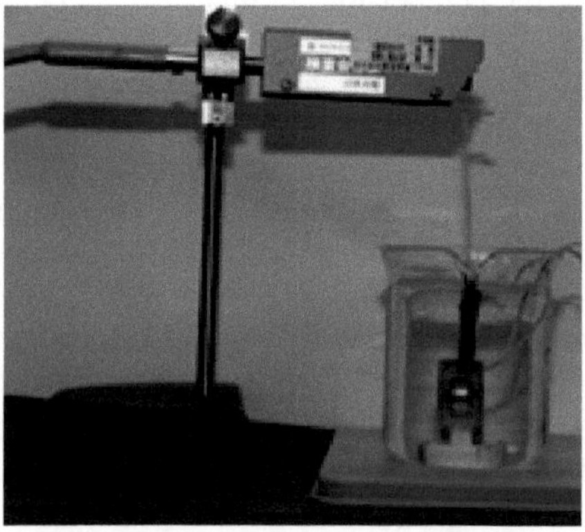

Fig. 21.2 Force measurement of IPMC by the isometric transducer

A typical isometric tension change in increased current with varied thickness of IPMC of width 5 mm and length 20 mm is shown in Figure 21.3.

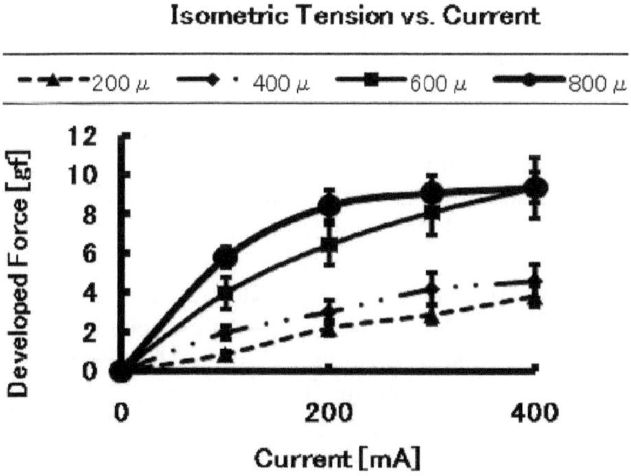

Fig. 21.3 Isotonic tension vs. applied current with varied thickness

Isometric tension increased by 6.7 times at 800 μm as compared to the 200 μm membrane at 100 mA while this ratio decreased as larger current was applied. For example, at 100 mA, the tension was proportional to the power of 1.4 of the thickness h, i.e., $F=0.000545h^{1.4}$ while at 200 mA, it was the power of 0.98; $F=0.0121h^{0.98}$. Tension, as measured by force in gram of force (gf) increased with increased current of up to 300 mA and subsequently dropped the increment rate, and continued to increase up to 500 mA. Similarly, tension increased as the width of the membrane increased but not much change was observed with increased length of the membrane.

Mechanisms of IPMC operation are extensively investigated [8-14]. Asaka et al. proposes a displacement mechanism: the trapped water molecule inside the Nafion electrolyte moves along the counter-ion, in this case, sodium, so that one side of the IPMC membrane swells while the other side shrinks causing bending motion in low frequency driving current. Also, he showed the displacement follows driving current waveform rather than voltage. At higher frequency, another mechanism of surface ion movement is considered to be the case.

Isometric force generation was measured in our study rather than the conventional measurement with load cell. This method measures internal force of IPMC independent of the displacement associated with bending motion. It would measure the internal contractile force in longitudinal direction rather than the developed force tangential to the membrane surface. Also, it is a standard method in measuring isometric contraction in biological muscle and allows the comparison of force generation between the artificial and the natural muscle.

21.4 Impedance Measurement

To examine the factors contributing to the force generation of IPMC, we have measured impedance characteristics of IPMC in conditions of different thickness, number of repeated plating, and duration of pre-heating.

A precision LCR meter (Hewlett Packard 4284A) was used to measure impedance of prepared IPMC. Impedance was measured over a range of 20 Hz to 100 kHz. Using resulted Cole-Cole plot and Bode plot, the values of resistance and capacitance of IPMC were determined.

Developed force measured by isometric transducer, as well as membrane resistance and capacitance derived from impedance measurement are summarized as follows.

In parallel with the force increase, the IPMC capacitance increased with increased IPMC thickness (Fig.21.4). The increase of capacitance, however, was not linear to the thickness. The capacitance of 400 μm IPMC was three times as large as 200 μm IPMC while that of 800 μm IPMC was remarkably large; 9 times as large as 200 μm IPMC. IPMC resistance, as measured by the impedance analysis revealed increased resistance with increased thickness (Fig.21.5). The increase was nearly linear to the thickness.

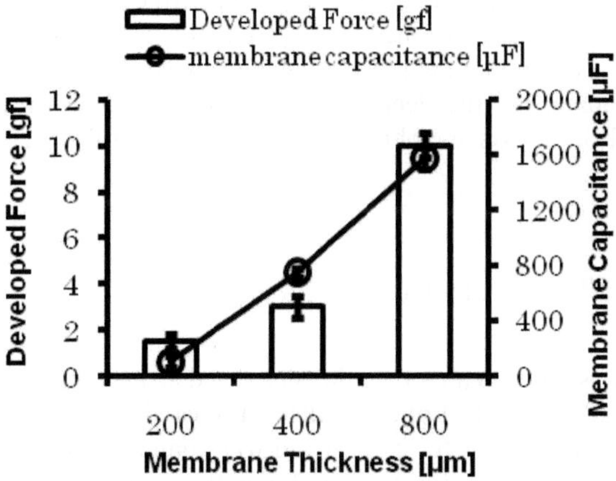

Fig. 21.4 Membrane capacitance by membrane thickness

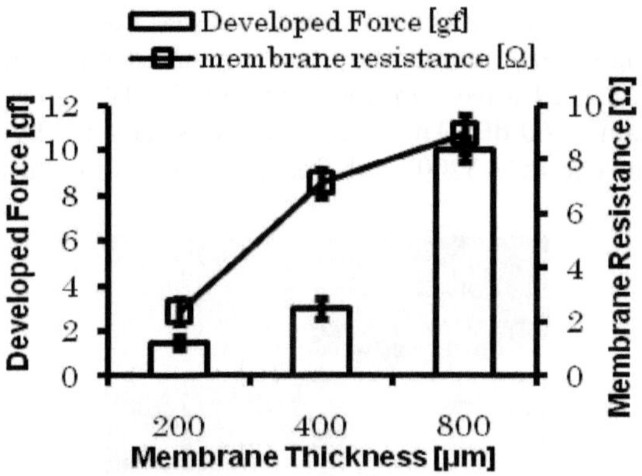

Fig. 21.5 Membrane resistance by membrane thickness

The electrode-Nafion boundaries are described as a complex fractal-like structure. At the boundary surface, the electric double layer is formed and the principal electro-chemical behavior of IPMC is well described with a simple lumped resister-capacitor series circuit which was used in our study.

Takagi et al. [15] developed a distributed constant model of IPMC and examined the effect of counter ion on the impedance of IPMC. They found larger resistance and smaller capacitive element in the IPMC in which TEA (Tetra Ethyl Ammonium) was used as counter-ion than the one with sodium ion as counter-ion.

In our study, the increase of developed force of IPMC with increased membrane thickness is considered to be related to the increased stiffness and increased electric double layer surface. While the increased stiffness could simply be attributed to the increased dimension, the increase in double layer surface may originate from two factors; (1) the more number of counter ions are contained in the thicker the IPMC membrane, (2) the deeper penetration of gold electrode into Nafion layer could result in thicker membrane thereby increasing the total electrode surface area.

21.5 Sensing Characteristics

The membrane was cut into a rectangular shape and arranged to a array of up to 6 IPMCs to fit into the flow duct of a ventilator. Prepared IPMC was fixed as an array on a plastic pipe of diameter 22 mm and was connected to a ventilator circuit and driven by a ventilator (Draeger Evita) with a volume control ventilation (VCV) mode (Fig.21.6). IPMC was connected both to a sensing unit (NF Corp.: SA400F3 differential amplifier) and an actuation unit (Hokuto Denko Co.:HA151 potentiostat / galvanostat). A humidifier unit of the ventilator is located at the center of the ventilator below the control panel. Induced voltage by the IPMC with the air-flow setting of 20 L/min to 100 L/min was measured with volume control ventilation mode. Sensed waveform was amplified by a laboratory made amplifier unit of voltage gain of 60 dB. Amplified signal was recorded by a PC controlled oscilloscope (Kenwood PCS-3200) with frequency range of up to 100 MHz.

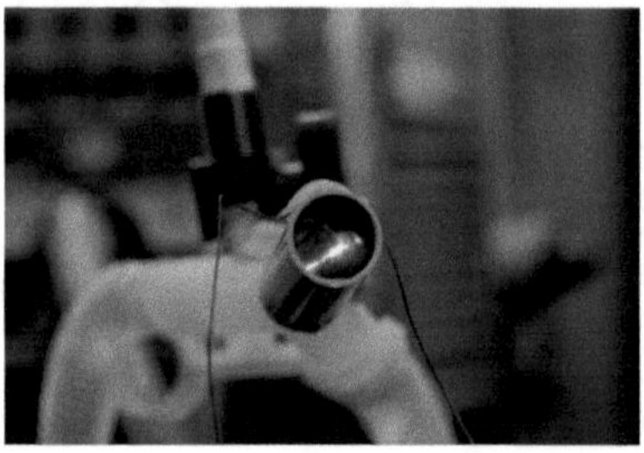

Fig. 21.6 IPMC sensor attached to a ventilator

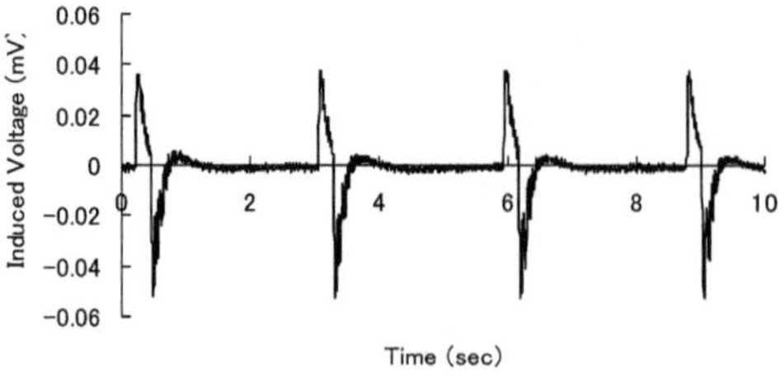

Fig. 21.7 A typical sensor waveform in response to ventilator air-flow

IPMC sensor demonstrated a biphasic response to the ventilator flow (Fig.21.7). In this setting, inspiration flow (ventilator to patient) was detected as positive deflection while expiration flow (patient to ventilator) was negative. The developed voltage was in order of several hundred μV. This sensor unit detects the pressure change associated with respiratory flow. Respiratory flow measurement demonstrated a proportional response to the flow volume (Fig.21.8). This linearity was observed over a range of membrane thickness of 150-250 μm, membrane width of 0.1-1.0 cm, and membrane length of 0.75-1.5 cm.

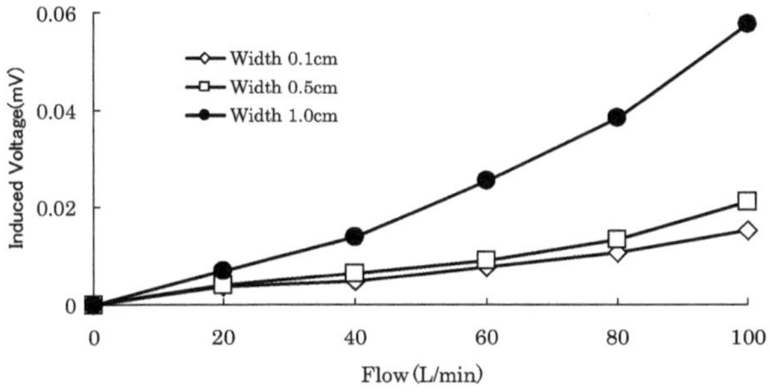

Fig. 21.8 Induced voltage on IPMC sensor as a function of flow

21.6 Driving IPMC Actuator

By exploring the characteristics of IPMC, a new medical device could be developed. We have fabricated a prototype sensor-actuator coupled device for the detection and removal of phlegm from a trachea tube (Fig.21.9). These membranes are driven by signals sequentially initiating the movement in order to carry out the substance placed on top of the membranes.

Further analysis of IPMC actuation mechanism will help bringing IPMC into practical use.

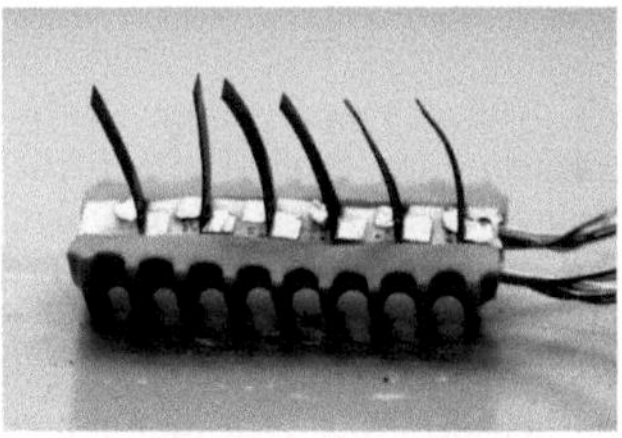

Fig. 21.9 A prototype ventilator phlegm removing unit by IPMC

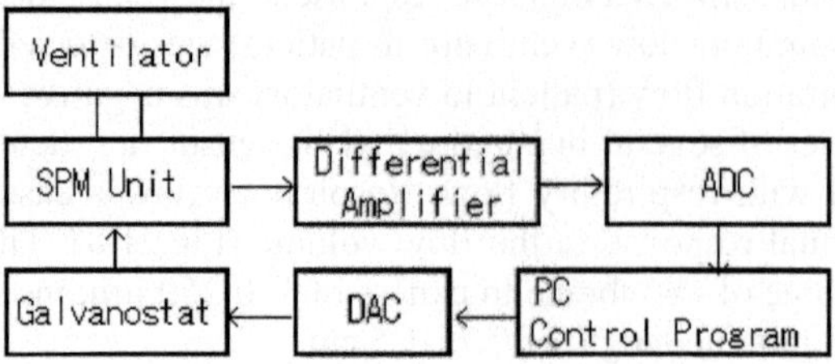

Fig. 21.10 Schematic diagram of sensor-actuator control of IPMC

The sensed signal was filtered and digitized with an A/D-D/A converting unit (Contec, ADA 16/32-2 (PCI)) on a PC board. A real time operating program was used to detect the sensed signal that indicates the narrowing of trachea. The program then activated a driving signal to control the actuation of the membrane. The signal was sent to a D/A converting part of the unit. The output of the D/A unit was sent to galvanostat units which drive the membrane with constant current regardless of the change of the load resistance (Fig.21.10).

In addition to the mechanism of actuation and sensing of IPMC, its actuation-sensing interation is yet to be clarified. Kamamichi et al [16] discusses the interference effect of actuation over sensing signal when the same IPMC was used both as a sensor and an actuator.

In practical application, a major problem would rather be the ventilation flow which hampers actuation motion in the trachea tube. The force required to remove viscous objects such as phlegm is another important issue to be resolved. This application of IPMC as a sensor-actuator coupled device, nevertheless, would be one of the most useful application of IPMC. This prototype study may pave the way for further medical and other applications [17-20].

Acknowledgments This work was support in part by Grants-in-aid for scientific research 16500305, 17040024, 19016021 Ministry of Education, Culture, Sports, and Technology, Japan and by the grant from Asian Office of Aerospace Research and Development, AOARD-03-4037.

References

1. Ihara T, Ikada Y, Nakamura T, Mukai T and Asaka, K (2006) Solid Polymer Electrolyte Membrane Flow Sensor for Tracheal Tube. Proc.SPIE 6167:61670U-1-61670U-8
2. Ihara T, Nakamura T, Muka T and Asaka K (2008) Ion Polymer Metal Compound Actuator for Active Phlegm Remover in Ventilator Circuit. Proc. 11th International Conference on New Actuators:523-526
3. Nakamura T, Ihara T, Muka T, Asaka K and Noritsugu,T (2008) Ion Polymer Metal Compound as a Power Assist Hand Sensor. Proc. 11th International Conference on New Actuators:981-984
4. Ihara T, Nakamura T, Muka T, Asaka K (2007) Heat Press Process of Ion Polymer Metal Compound and its Generating Stress. Proc. 8th SICE System Integration Division Annual Conference:221-222

5. Nakamura T, Ihara T, Muka T, Asaka K (2007) Water Content of Ion Polymer Metal Compound and its Sensor Characteristics. Proc. 8th SICE System Integration Division Annual Conference: 223-224
6. Fujiwara N, Asaka K, Nishimura Y, Oguro K and Torikai E (2000) Preparation of Gold-Solid Polymer Electrolyte Composites as Electric Stimuli-Responsive Materials. Chemistry of Material 12:1750-1754
7. Fujiwara N, Asaka K, Y. Nishimura, Oguro K, Torikai E (2000) Preparation of Gold-Solid Electrolyte Composites as Electric Stimuli-Responsive Materials. Chem. Materials 12:1750-1754
8. de Gennes PG, Okumura K, Shahinpoor M, Kim KJ (2000) Mechanoelectric effects in ionic gels. Europhysics Letters 50:513-518
9. Newbury KM and Leo DJ (2003) Linear Electromechanical Model of Ionic Polymer Transducers – Part I: Model Development, Journal of Intelligent Material Systems and Structures. 14:333-342
10. Newbury KM and Leo DJ (2003) Linear Electromechanical Model of Ionic Polymer Transducers – Part II: Experimental Validation. Journal of Intelligent Material Systems and Structures 14:343-357
11. Asaka K, Oguro K, Nishimura Y, Mizuhata M and Takenaka H (1995) Bending of Polyelectrolyte Membrane-Platinum Composites by Electric Stimuli I. Response Characteristics to Various Waveforms. Polymer Journal 27:436-440
12. Asaka K, Oguro K (2000) Bending of Polyelectrolyte Membrane-Platinum Composites by Electric Stimuli Part II. Response Characteristics to Various Waveforms. Journal of Electroanalytical Chemistry 480:186-198
13. Asaka K, Oguro K (2000) Bending of Polyelectrolyte Membrane-Platinum Composites by Electric Stimuli III: Self-Oscillation. Electrochim. Acta 45:4517-4523
14. Ihara T, Nakamura T, Nakamura T, Ikada Y, Asaka K, Oguro K, Fujiwara N (2004) Application of a Solid Polymer Electrolyte Membrane-Gold to an Active Graft. Proceedings of the first International Congress on Biomimetics and Artificial Muscle 2:60-63
15. K.Takagi, Nakabo Y, Luo ZW, Asaka K (2007) On a Distributed Parameter Model for Electrical Impedance of Ionic Polymer. Proc. SPIE 6524:652416;1-652416;8
16. Kamamichi N, Stoimenov B, Mukai T and Asaka K (2006) A Sensor-Actuator Integrated System with a Patterned IPMC — the interference of actuation to the sensing signal. SICE System Integration Division Annual Conference: 175-176
17. Kim KJ, Shahinpoor M (2002) A novel method of manufacturing three-dimensional ionic polymer-metal composites (IPMCs) biomimetic sensors, and artificial muscles. Polymer 43:797-802
18. Yamakita M, Kamamichi N, Kaneda Y, Asaka K and Luo ZW (2004) Development of an Artificial Muscle Linear Actuator Using Ionic Polymer-Metal Composites. Advanced Robotics 18:383-399
19. Guo S, Okuda Y, Zhang W, Ye X, Asaka K (2006) The development of a hybrid type of underwater micro biped robot. J. Applied Bionics and Biomechanics 3:143-150
20. Hitsumoto S, Ihara T, Morishima K (2007) Study on Cell Stimulating System using Micro fabricated IPMC. Proc. 8th SICE System Integration Division Annual Conference: 231-232

Chapter 22
Pneumatic Rubber Artificial Muscles and Application to Welfare Robotics

Toshiro Noritsugu[1], Masahiro Takaiwa[1] and Daisuke Sasaki[1]

Abstract In the coming advanced age society, an innovative technology to assist the activities of daily living of elderly and disabled people and the heavy work in nursing is desired. A wearable power assist robot is one of effective approaches as such a technology, which is equipped to the human body to assist the muscular force. Since this kind of robot directly acts on the human body, it should be friendly for human, so should be small, lightweight and has to provide a proper softness. A pneumatic rubber artificial muscle is effective for such requirements. We have developed some types of pneumatic rubber artificial muscles and applied them to the wearable power assist robot. In this paper, some types of rubber artificial muscles developed and manufactured in our laboratory are introduced. Further, some kinds of wearable power assist robots, for example, power assist glove, power assist device for standing up motion, power assist wear like clothes are introduced. Experiments clarify the effectiveness of pneumatic rubber artificial muscles for an innovative human assist technology.

22.1 Introduction

The introduction of the robot is expected as one means to send untroubled and comfortable living life in the coming less children and aging society. It leads to a big improvement of QOL (Quality of Life) if a robot supporting elderly and physically handicapped person's daily life and social participation, nursing work, rehabilitation, etc. can be achieved.

The research of such a human assist robot has been advanced also in our laboratory. The robot developed is equipped in the body to support the bodily movement by assisting and enhancing the muscular force. This kind of robot is called a wear able power assist robot, which is being developed in some research laboratories of the United States and Japan [1], but its practical use as the commodity has not been realized.

[1] Toshiro Noritsugu, Masahiro Takaiwa and Daisuke Sasaki

Graduate School of Natural Science and Technology, Okayama University

This paper describes pneumatic rubber artificial muscles and some wearable power assist robots we have developed.

22.2 Background of Research

A physically handicapped elderly and disabled people can expect to live more independent life by using this kind of robots. Although the research of wearable power assist robot has been temporarily executed in the 1960's, it was extremely heavy, large-scale, and its practicality was scarce. The research of this field was hardly executed afterwards. Now, the wearable power assist technology becomes realistic by miniaturizing an electronic component and developing a small-sized and a light-weight actuator. Its application to the welfare robotics is paid to attention.

It is absolutely required for a wearable robot that it is safe and friendly for the human. For this, the equipments composing it should be small-sized, lightweight, soft, and are hoped to have moderate output power. In our laboratory, a pneumatic rubber artificial muscle has been paid to attention as an actuator that satisfies such requirements, some new rubber artificial muscles have been developed, and the wearable power assist robots driven with them have been manufactured.

22.3 Pneumatic Rubber Artificial Muscles

A McKibben type pneumatic rubber artificial muscle is well known. Some companies manufacture commercially available one, for example, Shadow Robot Company in UK, or FESTO in Germany. Also we have developed some new types of pneumatic rubber muscles in addition to the McKibben type.

Figure 22.1 shows a conventional McKibben type rubber muscle, which was manufactured by covering the surrounding of a rubber tube with fiber net in our laboratory. By pressurizing the rubber tube, the muscle generates the axial contraction force. We can manufacture the rubber muscle with an arbitrary size using commercially available rubber tube and polyester fiber net. Fig.22.2 shows the fundamental characteristics of the manufactured rubber muscle comprising a rubber tube with an outer diameter of 11.6mm, an inner one of 8.0mm, and a length of 793mm when not pressurized. The contraction rate is saturated to about 25%. The saturated value depends on the cross angle of fibers. When pressurizing the rubber tube to 600kPa, the contraction force reaches 340N. The performance is equivalent to the commercially available products.

It can generate a large contraction force, but its maximum contraction rate (about 25%) is smaller than that of human muscle (about 50%).

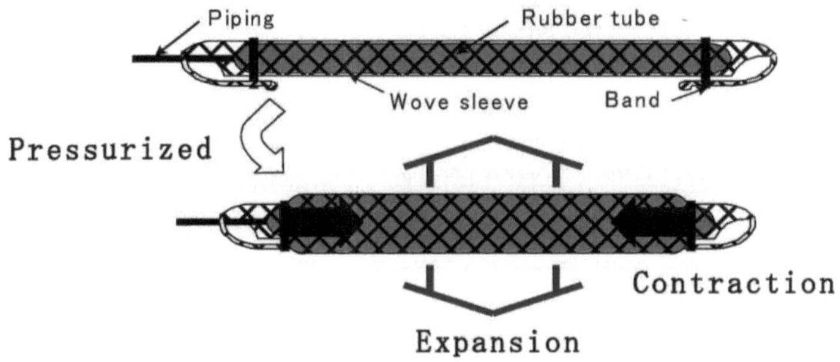

Fig. 22.1 McKibben type rubber artificial muscle

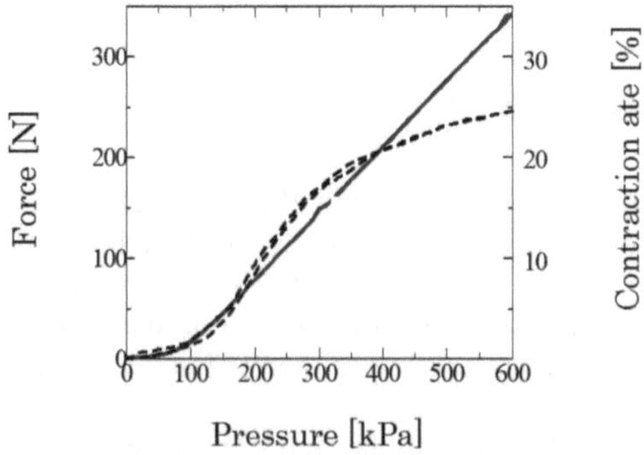

Fig. 22.2 Fundamental characteristics of McKibben type rubber muscle

To assist the rotational motion of human joint with a linear artificial muscle, the mechanism for securing the sufficient motion range and the conversion of the generated force to the joint torque is required. Such a mechanism threatens to damage the small-sized and lightweight advantages of the device. The McKibben type which can generate large force is suitable for the device necessary large assist force such an assist device for standing up motion [2]. However, for the device to assist the motion of finger or upper arm which requires not so large assist force, to secure the small-sized and lightweight properties an assist device unnecessary the torque conversion mechanism is desired.

From that viewpoint, curved type pneumatic rubber artificial muscles doing the curved operation and a twist type rubber muscle doing the rotational operation by supplying compressed air are newly developed.

Figure 22.3 shows the structure of the typical curved type pneumatic rubber artificial muscle. It is composed of a rubber tube covered with a bellows sleeve which extends or contracts only axially. By inhibiting the extension of one side with the fiber reinforcement, the bending motion toward the reinforcement direction occurs by supplying the compressed air to the rubber tube. The curved angle depends on the supplied air pressure.

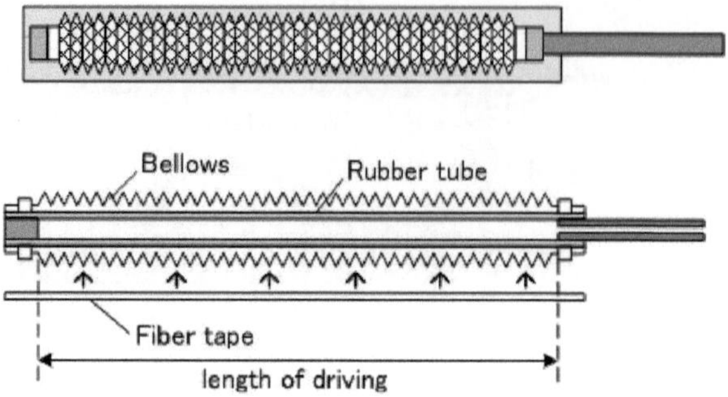

Fig. 22.3 Curved type rubber artificial muscle

22.4 Standing Assist Device

Figure 22.4 shows the structure and the assist effect of standing assist device manufactured in our laboratory, which is composed of leg orthosis and McKibben type pneumatic rubber artificial muscles. Knee and ankle joints are driven with rubber muscles. The knee joint lifts up the human body to stand up. The ankle joint controls the gravity center position of human. To evaluate the assist effect, the human stands up by means of only upper arm force. The upper arm force is measured thorough the rings. In the figure, a top line is for no assist, a middle line for the case only the knee joint is assisted , a lower one is for the case both joints are assisted. When both joints are assisted, the required force for the human can be decreased to about 10% compared with the no assist case [2].

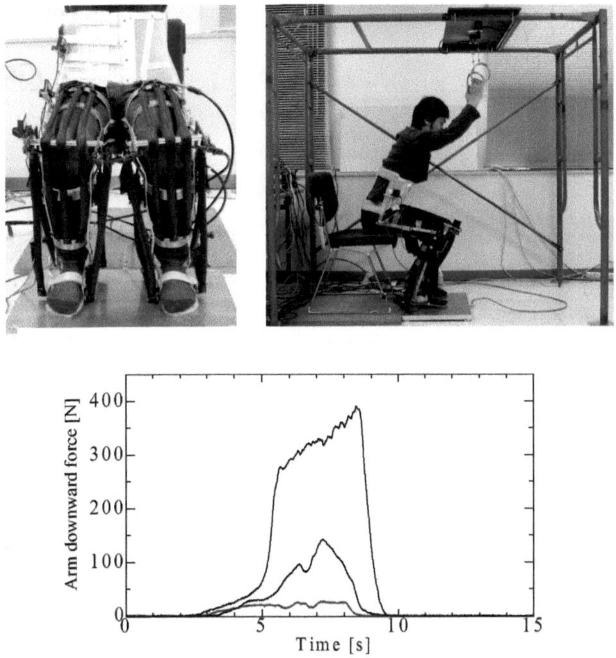

Fig. 22.4 Effect of standing assist device

22.5 Power Assist Glove

It assists the bending motion and increases the grasping and pinching force of hand finger by installing curved type pneumatic rubber artificial muscle in each upper side of five finger parts of a glove. The glove works as a mechanical interface to transmit the generated force of the artificial muscles to the human fingers. It is composed of only soft materials such as rubber, polyester fiber and cloth, and be equipped in the human body like clothes without any hard orthosis. Authors call such clothes like wearable power assist device *power assist wear.*

22.5.1 Two joints type power assist glove

One joint type power assist glove installed one artificial muscle shown in Fig.22.3 in each finger has been developed, and its utility has been confirmed [3]. It can well assist the grasping operation as shown in Fig.22.5. So far, it has a problem of unsuitable to the pinch operation with finger tips.

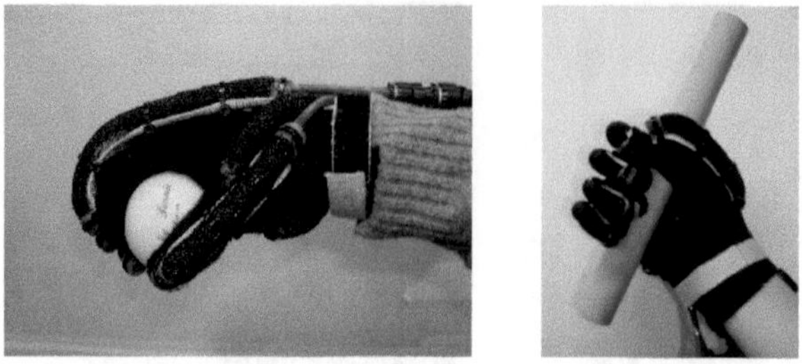

Fig. 22.5 One joint type power assist glove

To deal with this problem, two joints type power assist glove shown in Fig.22.6 is newly developed. Two curved type rubber muscles with a different diameter are connected in series, each of which basic structure is the same as Fig.22.3. The muscle for the finger tip is composed of a rubber tube of 3.0mm in the inside diameter and 4.5mm in the outside diameter and bellows of 7.5mm in the inside diameter and 10.0mm in the outside diameter, and the muscle for the root of finger is composed of a rubber tube of 6.4mm in the inside diameter and 8.4mm in the outside diameter and bellows of 12.0mm in the inside diameter and 16.0mm in the outside diameter. These materials are commercially available usual products.

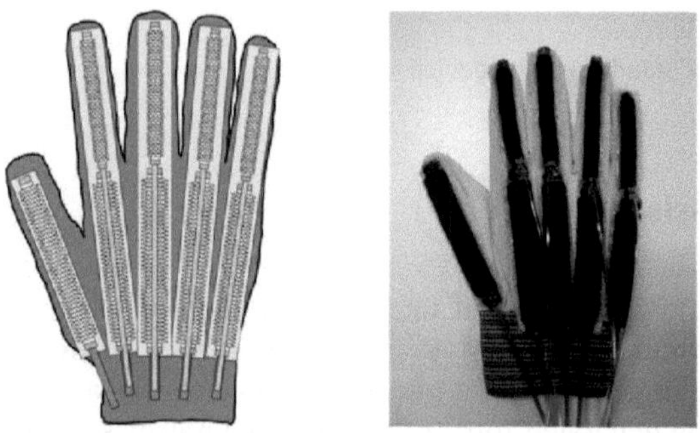

Fig. 22.6 Two joints type power assist glove

A glove made of cloth or leather is used as an interface to transmit the generated torque of the artificial muscle to the human finger. The artificial muscles are equipped along the back of each finger to assist the bending motion. The extended motion of finger is assisted by the restoration torque of the artificial muscle when exhausting the pressure. The total weight of the glove is about 120g.

Figure 22.7 shows the grasp operation and the pinch operation assisted with this glove. The tip and the root of finger can be separately operated. In the grasp operation, both artificial muscles in the tip and root of fingers are pressurized. In the pinch operation, only the muscle in the root of finger is pressurize to hold the finger tip straight as shown in Fig.22.7(b). By controlling the pressure supply to the artificial muscles in the tip and root of fingers individually, various operations can be assisted. Assist force can be controlled by adjusting the supply pressure.

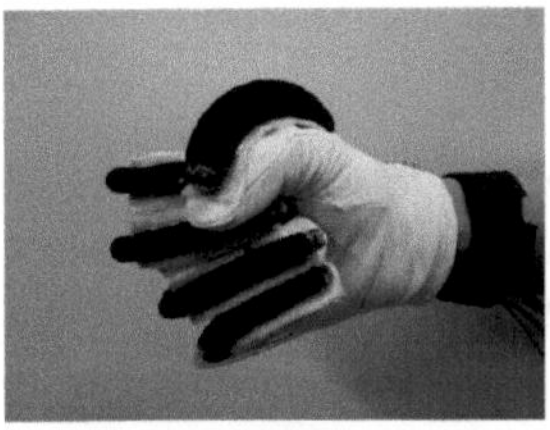

(a) Grasp operation

(b) Pinch operation

Fig. 22.7 Operations of two joints type power assist glove

22.5.2 Generated Force in Pinch Operation

The glove is equipped to the human in the initial state of no pressurizing the rubber artificial muscle. The generated force of forefinger tip in the pinch operation is measured with a thin tactile force sensor equipped at the finger tip as shown in Fig.22.8. PI feedback control of assisting pinch force is executed. The inner pressure of each rubber muscle is adjusted as P_1=P_0-50kPa, P_2=P_0-150kPa, where P_0 is the control input determined by PI controller. In the condition of that the human slightly pinches the object with thickness of 20mm, the step responses are measured for the reference force from 1N to 4N, as shown in Fig.22.8. Although the delay time of about 1s appears, the response is enough fast to assist the human operation. Further, considering the pinch force of the general male is about 20N to 25N, this glove is effective in assisting the pinch operation.

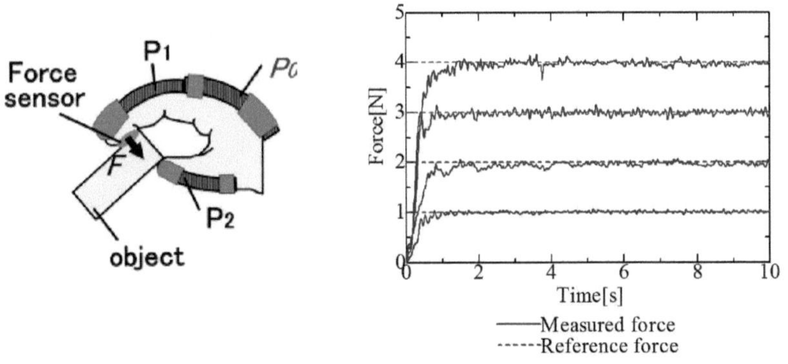

Fig. 22.8 Measurement of assisting pinch force

22.6 Elbow Power Assist Wear

A sheet-like curved type pneumatic rubber artificial muscle is newly developed to realize a thin rubber muscle. By using this muscle, a power assist wear assisting the bending motion of the elbow is developed [4].

22.6.1 Sheet-Like Curved Type Rubber Artificial Muscle

It is composed of the rubber tube sandwiched between upper and lower two sheets sutured surroundings as shown in Fig.22.9. The end of rubber tube is sealed, the other is piped to the pressure control valve. By using the elastic sheet (fabric rubber) extending only to the axial direction of two sheets, when the rubber tube is pressurized, the extending operation in the axial direction is obtained.

Three kinds of operations can be achieved by making the difference in the amount of extension of both sheets due to the difference in the number of sheets or the elasticity of the sheets. Fig.22.9(a) shows the case of using same numbers of sheet in both sides. Since the amounts of extension are same, the linear extension in the axial direction is caused. When there is difference in the number of upper and lower sheets, the difference in the amounts of extension causes the extension and curved motion as shown in Fig.22.9(b). In Fig.22.9(c), the material not to extend in the axial direction (nylon band) is used in the upper sheet, only the curved operation occurs by pressurizing the rubber tube. Due to the combination of these operations, the power assist smoothly along the human body movement can be obtained.

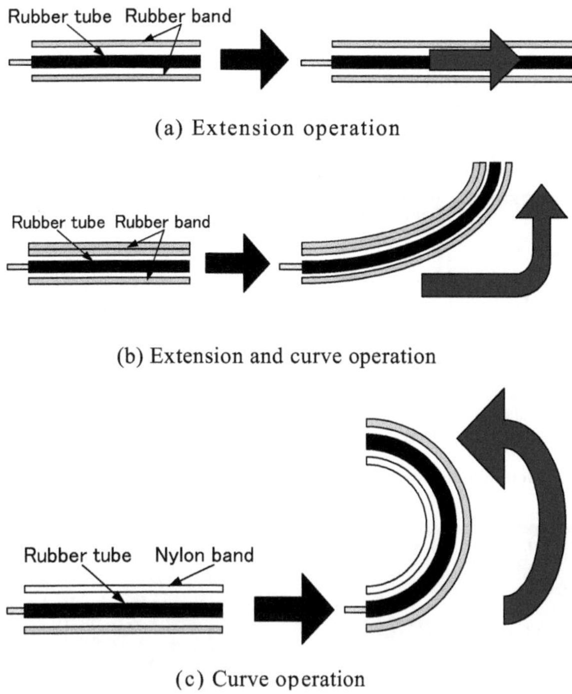

Fig. 22.9 Operation of sheet-like curved type muscle

22.6.2 Operation and Fundamental Characteristics of Power Assist Wear

Figure 22.10 shows the operation of manufactured artificial muscle. The artificial muscle used for the elbow power assist wear is 350mm in the total length, and 100mm in the width. Two rubber tubes are sandwiched between two sheets. The upper sheet is composed of the alternate arrangement of rubber bands and nylon bands along the axial direction. The lower sheet is composed of rubber band except both ends of the nylon band. Such structure is effective to obtain the adjustable movement to the human body. The maximum bending angle of human's elbow joint is about 145°. The manufactured artificial muscle can bend toward this angle under pressurizing the artificial muscle about 120kPa. Moreover, it extends axially about 60mm under the same pressure.

Figure 22.11shows the operation of the elbow power assist wear. A soft equipment composed of cloth and fabric rubber is used to transmit the generated force of the muscle to the human elbow. Moreover, by installing the slide part between the artificial muscle and equipment, the artificial muscle becomes possible to extend axially for the smooth assist operation. The weight of the artificial muscle is

175g, and the total weight including the equipment is 365g. It is enough light-weight for human to wear.

A torque necessary to hold the weight of general adult male's forearm is about 4Nm, when 320kPa is pressurized, the wear can hold the weight and assist the bending of the forearm over the range of about 100° or less in bending angle.

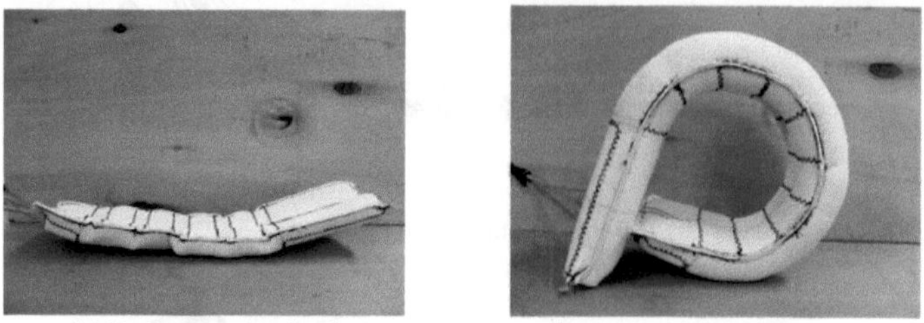

(a) Before pressurized (b) Pressurized 250kPa

Fig. 22.10 Operation of sheet-like artificial muscle

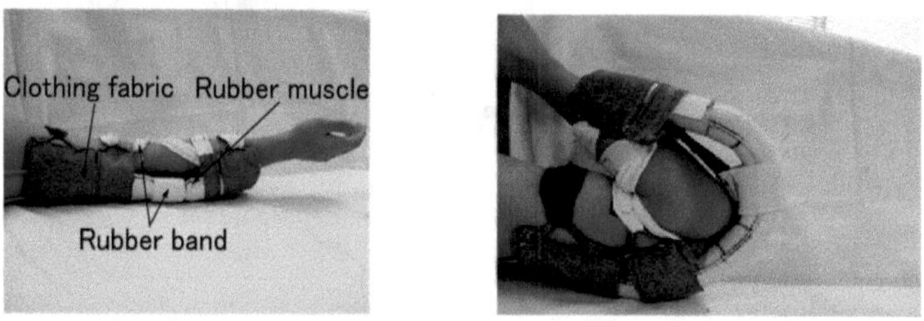

(a) Before pressurized (b) Pressurized 250kPa

Fig. 22.11 Operation of power assist wear for elbow

To evaluate the effect of power assist, the bending angle and the myogenic voltage of the biceps brachii are measured when the wear is non-equipped and equipped. The bending operation of the elbow starts from the posture in which the forearm is extended forward horizontally. The pressure that the artificial muscle begins the operation is set to the initial pressure for the muscle, and the experiment is conducted according to the following processes.

1. From 0s to 5s, the wear is pressurized to the initial pressure of 80kPa.
2. From 5s to 15s, pressurized to 320kPa like a ramp when the human is relaxing.
3. From 15s to 20s, the assist is maintained under relaxing of the human.

Fig.22.12 shows the experimental results, where the solid lines show the myogenic voltage and the broken lines show the bending angle. The maximum elbow bending angle of the human subject is about 120°. A relative bending angle between the forearm and the body is about 210° (The bending angle of elbow joint is about 120°). The assist range of about 100% is possible for the maximum bending angle of the subject. It can be confirmed to be able to reduce the load for the human elbow by using the power assist wear according to the decrease of the myogenic voltage owing to the generated torque of the wear.

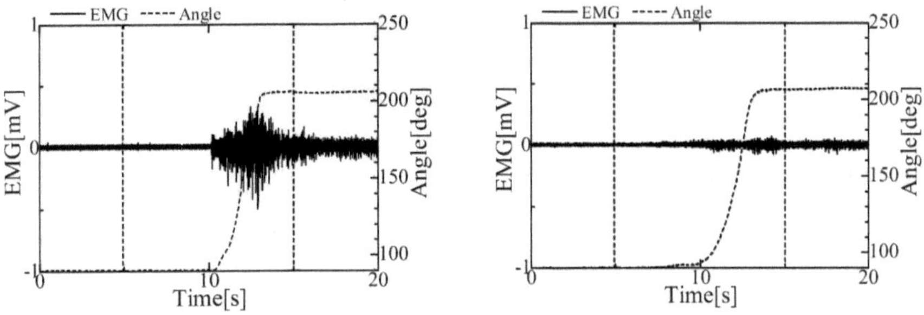

(a) Without assist

(b) With assist

Fig. 22.12 Effect of power assist for elbow bending

22.7 Assist of Rotary Motion

Since there are some rotary motions in the human joints, a rotational rubber muscle is useful. Fig.22.13 shows a newly developed twist type rubber muscle for the assist for the rotation. The inclined fiber reinforcement causes a twist motion when the inner rubber tube is pressurized. In Fig.22.14, four twist type rubber muscles are equipped on the human body to assist the rotary motion of the waist. This type of muscle can be available for the assist of some rotary motions of human body.

22.8 Conclusions

The structure and the characteristics of some types of pneumatic rubber artificial muscles and the power assist robots driven with these rubber artificial muscles have been described. The torque conversion mechanism such as belt and pulley required when the linear type artificial muscle is used becomes unnecessary by using the curved type artificial muscles. As a result, the composition of the small and lightweight power assist wear becomes possible. Moreover, clothes like and soft power assist wear can be achieved by composing the device of soft materials such as rubber and cloth.

To advance a practical use of these power assist wears in the future, a sensing of the human intention is required to control the wear. Additionally, a pneumatic energy source is important issue. Ideally, a small-sized, lightweight, low noisy and wearable air compressor or any other new methods are desired.

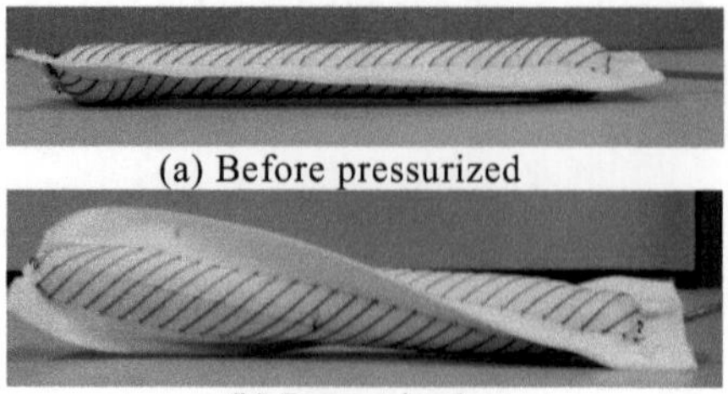

(a) Before pressurized

(b) Pressurized

Fig. 22.13 Operation of twist type rubber muscle

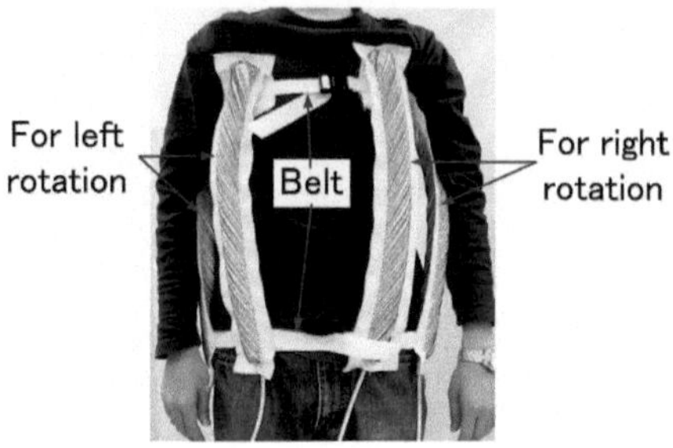

Fig. 22.14 Power assist wear for rotary motion of waist

Acknowledgments This research was executed as 16078210 "Pneumatic Soft Actuator for Human Friendly Mechanism" in the research project "Next-Generation Actuators Leading Breakthroughs" funded by the MEXT Grant-in Aid for Scientific Research on Priority Areas, Japan.

References

1. Kobayahi H (2004) Development of a muscle suit for realizing all motion of the upper limb. Proceedings of 2004 IEEE/RSJ International Conference on Intelligent Robots and Systems, Vol.2:1630-1635
2. Noritsugu T et al (2007) Wearable Power Assist Device for Standing up Motion Using Pneumatic Rubber Artificial Muscles. Journal of Robotics and Mechatronics, Vol.19, No.6: 619-628
3. Sasaki D, Noritsugu T, Yamamoto H, Takaiwa M (2006) Development of Power Assist Glove using Pneumatic Artificial Rubber Muscle. Journal of the Robotics Society of Japan, Vol.24, No.5:640-646
4. ragane M, Noritsugu T et al (2008) Development of Sheet-like Curved Type Pneumatic Rubber Muscle and Application to Elbow Power Assist Wear. Journal of the Robotics Society of Japan,Vol.26,No.6:674-682

Chapter 23
Dynamic Characteristics of Ultrasonic Motors

Nonlinear Dynamic Analysis of Traveling Wave-Type Ultrasonic Motors*

Akira SAITO[1] and Takashi MAENO[2]

Abstract This chapter discusses the dynamic characteristics of ultrasonic motors (USMs) with special attention to a bar-type USM that is one of the traveling wave type USMs. The USMs are known to possess nonlinearities caused by the frictional interface between the stator and the rotor, as well as the amplitude-dependent nonlinearity in the dynamics of the stator. Therefore, the transient dynamics of a bar-type ultrasonic motor, such as starting-up and stopping, can be quite complicated due to such nonlinearities. The transient response of the bar-type ultrasonic motor at starting-up and stopping is discussed in detail by using experimental results measured by a Laser Doppler Velocimeter (LDV), and results of numerical simulations with a second-order nonlinear oscillator model.

23.1 Introduction

An ultrasonic motor (USM) is an actuator with advantageous characteristics such as low-speed/high-torque, rapid response, high holding torque, and low noise emission. In particular, the so-called traveling-wave type USMs are widely used for electronic devices such as autofocus lenses of cameras and driving actuator of photocopying machines (e.g., Ref. [2]). Recently, the increasing demand for understanding the dynamics of the USM has led to many research activities for mod-

* This work was originally presented in Ref. [1].

[1] Akira SAITO
Department of Mechanical Engineering, University of Michigan, Ann Arbor, Michigan 48109-2125, USA

[2] Takashi MAENO
Graduate School of System Design and Management, Keio University, Hiyoshi, Kohokuku, Yokohama 223-8526, Japan

eling the dynamics of the USM for steady and unsteady state operating conditions. The major challenges for such modeling are the nonlinearities due to the contact interface at the rotor-stator interface [3][4], and due to the softening type nonlinearity of the stator vibration [5].

In this work, a bar-type USM (developed by Canon Inc. [6]) is used as a representative of the traveling wave type ultrasonic motors. The structure of the bar-type USM, along with the schematic of the operation principle is shown in Fig.23.1. The bar-type USM is comprised of stator and rotor. A donut-shaped piezoelectric ceramic is embedded and bonded in the stator, and the rotor is in contact with the stator being pressed by a spring along the axial direction. By applying two-phase alternating voltage whose frequencies are close to the natural frequencies of the modes, the first two bending modes of vibration are excited. The planes of vibration for the modes are perpendicular to each other, and the phase difference of these vibration modes is adjusted to be 90 deg. Therefore, the stator shows a *wobbling* motion as shown in Fig.23.1c, and the trajectory of a point on the upper surface of the stator shows an elliptical orbit. Therefore the rotor is forced to rotate by the friction force generated between the rotor and the stator. The contact point changes in time; hence it is a traveling-wave type USM.

In this chapter, the dynamics involved in these processes, including the electromechanical conversion at the piezoelectric ceramics, stator's dynamics, and the rotor's dynamics through the frictional interface are investigated. In particular, the nonlinear transient dynamics at the start-up and stopping of the USM are discussed in detail with measurement results and numerical simulations.

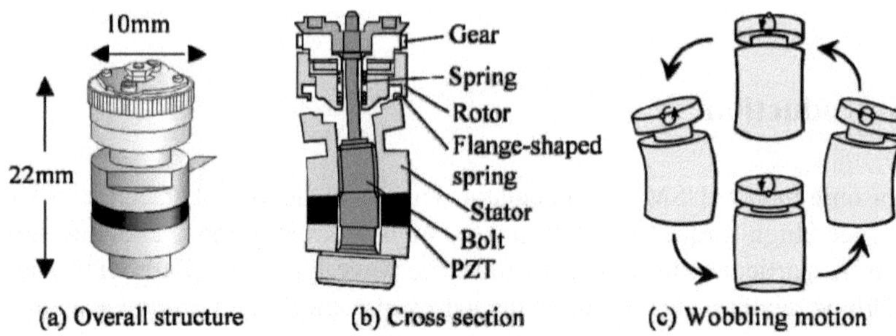

Fig. 23.1 Structure of the USM and working principle of the USM

23.2 Related Work

The pioneering contribution on the investigation of the nonlinear dynamics of the USM was made by Hagood and McFarland [7] where the dynamical model of a ring-type USM was developed considering the electromechanical conversion of the piezoelectric ceramics, and the frictional interface between stator and rotor using Rayleigh Ritz method. The modeling method developed by Glenn *et al.* [8], is an extension of that of Hagood and McFarland, which features stick-slip friction, a priori determination of damping coefficient, and harmonic-balance based mixed-domain solution procedure. By a similar approach, Tsai *et al.* [9] analyzed the unsteady state characteristics of the USM without considering the stick-slip phenomenon. The method proposed by Pirrotta *et al.* [10] employs a similar approach; however the transient effect is neglected. Gudschmidt and Hagedorn [11] developed a dynamical model of a bar-type USM by using a contact interface model that can treat stick-slip phenomenon, with a simplified point-to-point-contact. The modeling method proposed by Kandare and Wallaschek [12] features the simplified equivalent circuit model for expressing the dynamics of the stator as well as the electro-mechanical coupling. The results of numerical simulation for the starting-up and stopping show similar response such as shown in the work by Hagood and McFarland [7]. Pons *et al.* [13] developed a mathematical model of the stator of a USM based on laminated plate theories in conjunction with Ritz method, which can be incorporated with the rotor model, rotor-stator interface model, as well as electromechanical coupling [14]. Furthermore the model was used for the optimization of stator's parameters [15][16]. More recently, several researchers have attempted to construct 3D finite element (FE) models to understand both steady-state and transient dynamics of USMs, such as the work by Duan et al. [17]. As was briefly reviewed so far, there have been many attempts to understand the dynamics of the USMs. However, due to the complications caused by the nonlinearities, the behavior of the USMs in transient states cannot be well predicted by the existing models.

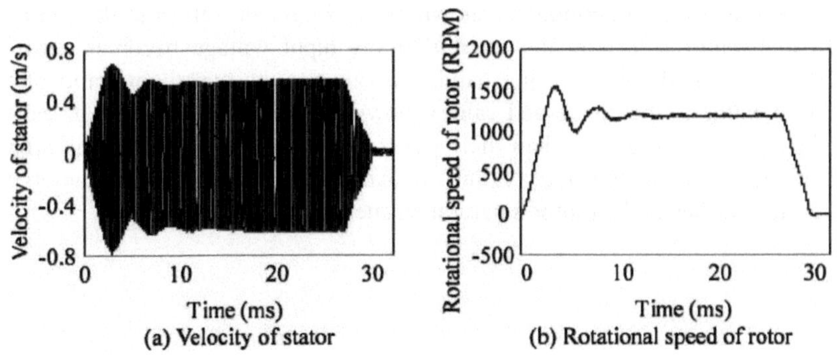

Fig. 23.2 Transient response of the stator and the rotor of the USM

23.3 Measurements of Driving Characteristics

First, in order to analyze the dynamic characteristics of the bar-type USM, the transient response was measured by using Laser Doppler Velocimeter (LDV). In particular, the starting-up and stopping behavior of the motor was measured by applying 1000 periods of two-phase alternating voltage to the piezoelectric ceramics, where the frequency was 37.5kHz, the amplitude was 15 Vp-p, and the phase difference was 90 deg. Figures 23.2a and 23.2b show the time histories of the velocity of the stator along the normal direction and the rotational speed of the stator respectively. First off, we can see that the stator's vibration velocity shows a *beat phenomenon* with decaying amplitude due to damping. Second, the rotational speed shows the trait that is similar to the step response of an under-damped second-order system. It is noteworthy that the rotor appears to follow the *envelope* of the stator's vibration velocity with a slight delay. This also indicates that the USM shows a rapid response with several milliseconds of starting-up and stopping if the motor is used without decelerator.

Next, in order to examine the transient dynamics for different input parameters, similar measurements were performed with different input parameters of applied voltage. The frequency of the voltage was swept from 36.5 to 37.5 kHz, and the amplitude of the voltage was changed from 11 to 17 Vp-p. The measured results are shown in Fig.23.3.

Figure 23.3a shows the steady-state rotational speed of the rotor versus the frequency of the input voltage. As can be seen, as the frequency is swept down from 37.5 kHz with constant amplitude of input voltage, the rotational speed gradually increases and drops to a lower value at a certain frequency. Moreover, the peak frequency decreases as the amplitude of input voltage increases. This *jump phenomenon* is due to the variation of natural frequency of the stator caused by the softening nonlinearity, as discussed in section 1.4. A similar trend can be observed in the delay time (time required to reach the maximum, measured from the moment when the input is applied) as shown in Fig.23.3b, as well as in the stopping time (time required to stop the rotor after the input voltage becomes zero) as shown in Fig.23.3d. On the other hand, the value of the overshoot (ratio of the maximum value to the converged value) shows slightly different trend, as can be seen in Fig.23.3c. Namely, when the input frequency is varied with fixed amplitude of the input voltage, the amount of overshoot decreases as the input frequency approaches to the motor's natural frequency.

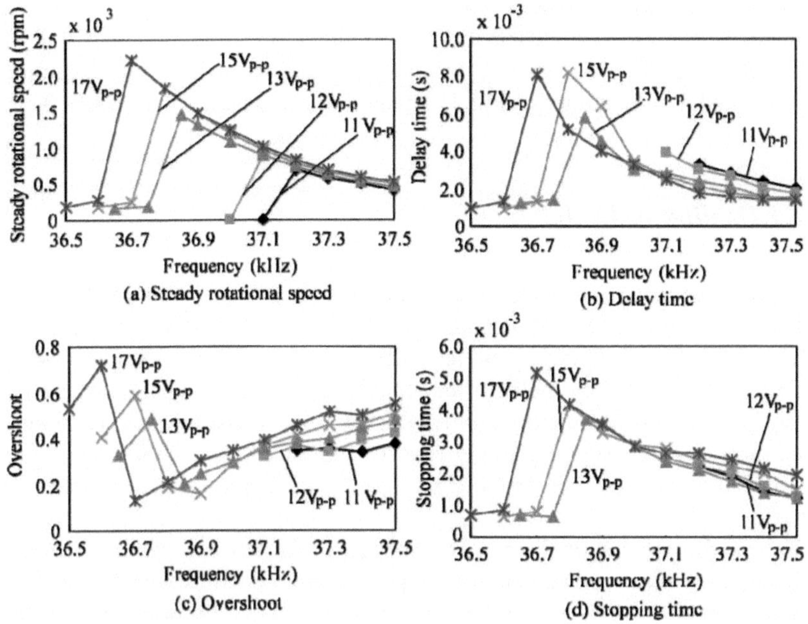

Fig. 23.3 Measurement results of the dynamic characteristics

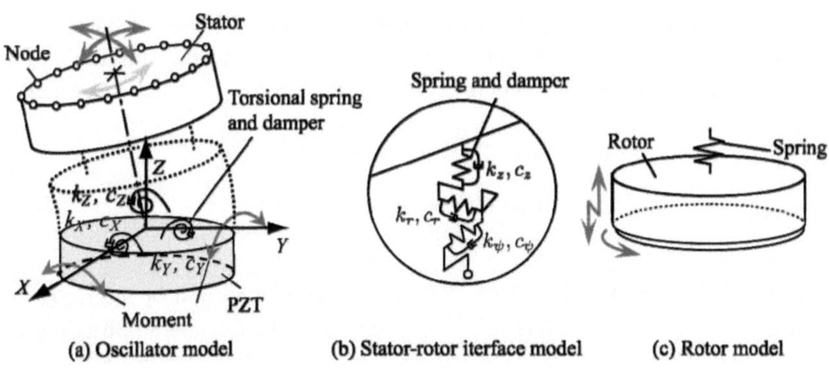

Fig. 23.4 Schematics of the mathematical model

23.4 Mathematical Modeling

The next step is to construct a mathematical model that can predict the measurement results shown in 1.3. It is noted that a detailed derivation of the mathematical modeling, analysis procedure, as well as the parameter identification method can be found in Ref. [1]. The schematics of the model are shown in Fig.23.4. The model consists of three sub-models: oscillator model, stator-rotor interface model, and rotor model as illustrated in Fig.23.4, and they are separately discussed below.

23.4.1 Oscillator Model

In the proposed model, the stator is modeled as a three degrees-of-freedom (DOF) oscillator consisting of a rigid disk, torsional springs and dampers, which is an extension to the model proposed in Ref. [11]. Namely as shown in Fig.23.4a, the bending modes excited by the deformation of the piezoelectric ceramics are modeled as the vibrations due to the moments about X and Y axes, which are generated by the piezoelectric material. This yields the vibration distribution along the perimeter of the upper surface of the rigid disk, which agrees well with that of the real stator.

Now, the formulation on the modeling of the oscillator is divided into two parts: (i) electromechanical conversion at the piezoelectric ceramics, and (ii) the vibration of the stator excited by the moment. It is known that the force generated by the piezoelectric ceramics is governed by a piezoelectric equation, and is proportional to the amplitude of the applied voltage. For the USM however, there are frictional losses at the bolted-joint between the stator and the piezoelectric ceramics. Therefore in this study, considering that the frictional losses at the interface between the stator and the piezoelectric ceramics increase as the amplitude of vibration increases, the moment generated from piezoelectric ceramics M_{PZT} is modeled by an empirical formula:

$$M_{PZT} = \frac{V}{\alpha_{PZT} + \beta_{PZT} A^2} \tag{23.1}$$

where V is the applied voltage, A is the amplitude of vibration of the oscillator, α_{PZT} and β_{PZ} are constants. The Eq. (23.1) represents that as the amplitude of vibration of the stator increases, the moment transmitted from the piezoelectric ceramics decreases due to the energy loss at the interface between the piezoelectric ceramics and the metal part of the stator. This formulation is based on the measurements of the dynamic characteristics of the motor and was confirmed by the authors that Eq. (23.1) can be used as a good empirical formula to accurately express the moment generated by the piezoceramics.

The relationship between the torsional angle θ of one of the torsional springs attached to the rigid disk, and the restoring moment M_E is modeled as a cubic equation, i.e.,

$$M_E = k\theta - d\theta^3 \tag{23.2}$$

where k and d are positive constants. The cubic term is employed in order to express the softening-spring type nonlinearity. Now, considering that the oscillator is subject not only to the moment from the piezoelectric material and the restoring moment from the stator, but also to the tangential and normal contact forces from the rotor through contact interface, the oscillator equation can be formulated as:

$$\mathbf{I}\dot{\omega}_s + \omega_s \times (\mathbf{I}\omega_s) = \mathbf{M}_P - \mathbf{M}_E - \mathbf{M}_C - \mathbf{M}_D \tag{23.3}$$

where $\mathbf{I}$ is the inertia tensor of the stator, ω_s is the angular velocity of the stator, $\mathbf{M}_P$, $\mathbf{M}_E$, $\mathbf{M}_C$, and $\mathbf{M}_D$ are the moments due to the force generated by the piezoelectric ceramics, elastic restoring force of the stator, contact force at the contact interface, and the damping force generated by the dampers.

23.4.2 Contact Interface Model

In the proposed model, the frictional contact force developed at the stator-rotor interface is calculated using a finite number of three-dimensional springs and dampers that are attached to the equally-spaced *nodes* in the circumferential direction along the rim of the stator [18], and the Coulomb friction model. The schematic of the springs is shown in Fig.23.4b. The algorithm for calculating the frictional force is described as follows. If the frictional force at each node does not exceed the maximum static frictional force, the stator contacts with the rotor at the node without slipping. The maximum static frictional force f^i_{max} at the ith node is calculated as

$$f^i_{\max} = \mu_s f^i_n \tag{23.4}$$

where f^i_n is the force normal to the contact surface, and μ_s is the static friction coefficient. When the frictional force reaches the maximum static frictional force, slipping occurs between the node and the spring; hence a dynamic friction force f^i_d is generated. The dynamic friction force is expressed as

$$f^i_d = \text{sgn}(V^i_{rel})\mu_s f^i_n \tag{23.5}$$

where V^i_{rel} is the relative velocity between rotor and stator at the ith node, and sgn represents the signum function. Moreover, the viscous force generated at the node is expressed as

$$f^i_{\text{viscosity}} = c\Delta z^i V^i_{rel} \tag{23.6}$$

where Δz^i is the non-negative displacement of the spring along the normal direction. This enables the model to include the amplitude dependence of the viscous force. It is noted that the viscous force expressed by the Eq. (23.6) represents the viscoelastic effect of the contact interface of the rotor.

23.4.3 Rotor Model

In the proposed model, the rotor is modeled as a rigid disk with two DOF; rotation about and translation along Z-axis. The Newton's equations of motion for the rotation and translation of the rotor are written as

$$I_r \dot{\omega}_r + c_z \omega_r = M_{cz} - \mathrm{sgn}(\omega_r) M_f \tag{23.7}$$

$$m_r \ddot{z} + c_r \ddot{z} = F_c + F_s \tag{23.8}$$

where I_r is the moment of inertia about z-axis, c_r is the damping coefficient of the rotor, M_{cz} denotes the z-component of $\mathbf{M}_C$, ω_r is the angular velocity of the rotor, M_f is a constant moment due to friction transmitted from the flange, m_r is the mass of the rotor, F_c is the frictional force transmitted from the contact interface, and F_s is the restoring force due to the spring for preloading.

23.5 Numerical Simulations

The state variables used in the analysis are the torsional angles of the oscillator, rotational angle of the rotor, the displacement along Z-direction of the rotor. For the simulation results shown in this section, the fourth-order Runge-Kutta method was used for the numerical time integration of the equations of motion. In conjunction with the numerical integration, an iterative method was employed to calculate the hysteretic property of the contact states, i.e., stick or slip (See [1] for details.)

With the constructed mathematical model, numerical simulations for the starting-up and stopping were performed. In order to make the comparisons with the measured results, the alternating voltage was applied for 1000 cycles and the rotor's response was calculated. The number of nodes, n, at the contact interface was chosen to be 256, which was sufficiently large such that further division yielded no noticeable improvements in the solution. Furthermore, the time step Δt was set to 0.5 μs, which was small enough to obtain the converged solution.

An example of the results of numerical simulations is shown in Fig.23.5(a), where the results were obtained with the amplitude of input voltage of 15 Vp-p and phase difference of 90 deg. The results of measurements are also shown in Fig.23.5(b) for comparison. As can be depicted from Fig.23.5, at both starting up and stopping, the proposed model produces very similar characteristics in rotational speed of the rotor to those of the measured data. In particular, the numerical simulations can predict the steady-state rotational speed, as well as the changes in rotational speeds at stopping very accurately. However, although the qualitative characteristics of rotational speeds at starting-up can be captured by the simulations, there are slight differences between the results of numerical simulations and experiments. However, since the nonlinearity of the oscillator shown in Fig.23.5 was not considered in the conventional modeling approaches, it was difficult to predict the characteristics of the rotational speed of the rotor when the amplitude of vibration of the oscillator changes. In particular, Fig.23.5 shows the qualitative agreement in beating phenomenon accompanied by frequency variation, which is impossible to be predicted with linear oscillator models. Namely, the proposed model can describe the transient dynamics of the motor more appropriately than conventional models.

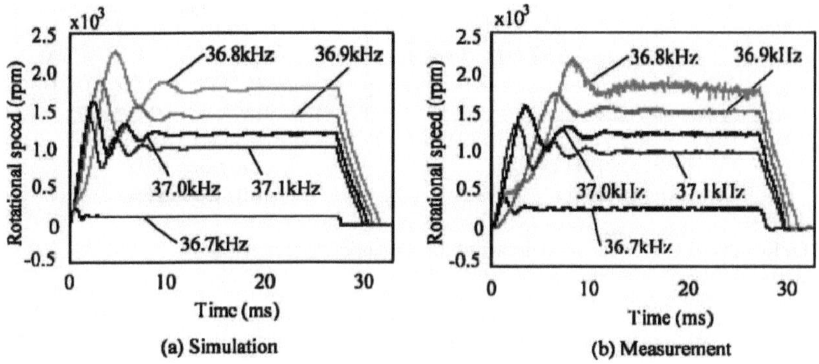

Fig. 23.5 Transient response by numerical simulation and measurements

Next, similar calculations were carried out with various values of applied voltage, and the relationship between the driving parameters and the driving characteristics of the motor in transient states were examined. The results of calculations are shown in Fig.23.6. The results shown in Figs. 23.6a-d correspond to the experimental results shown in Fig.23.3a-d respectively. First, as seen in Fig.23.6a, a good agreement is achieved in the steady-state rotational speed of the motor between measured and calculated results. This indicates that the interaction between the stator and the rotor in steady-state operation is properly modeled. Second, as shown in Fig.23.6d, the stopping time also shows a good agreement between calculated and measured results. This implies that the proposed contact model between the stator and the rotor can predict the contact interaction even under the transient motion. However for the delay time and the overshoot shown in (b) and (c), the agreement is limited to the qualitative one. The main causes for these slight discrepancies could be the nonlinearity of the parameters such as the damping coefficients in the contact interface model and the rotor model, temperature variations, and spring pressure variations, the identifications of which are beyond the scope of this work.

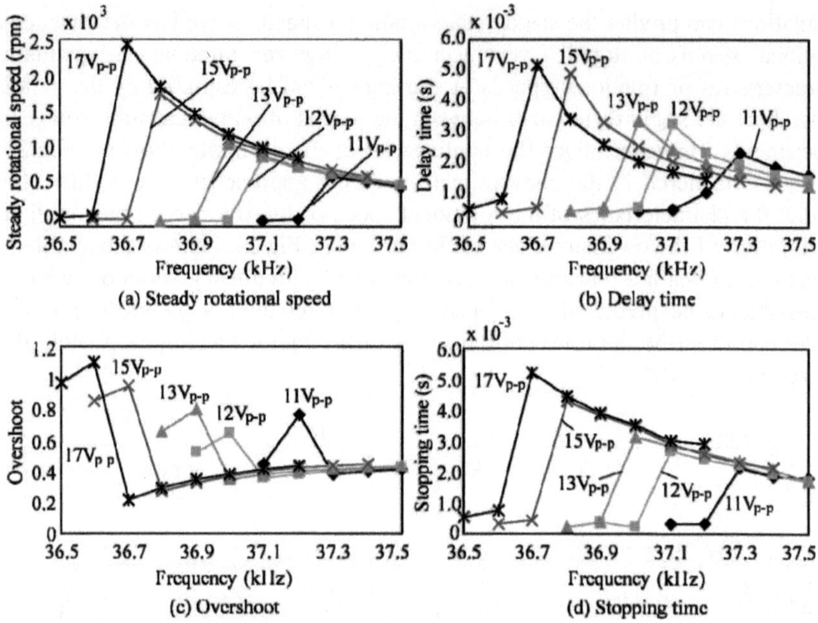

Fig. 23.6 Results of numerical simulations of the dynamic characteristics

23.6 Conclusion

In this chapter, the nonlinear dynamic response of the traveling-type USM was discussed using the results of measurements and the numerical simulations. First, the transient response at starting-up and stopping was measured by using the LDV, and the effects of input parameters to the dynamic characteristics were investigated. Second, a dynamical model of the USM was constructed as a second-order nonlinear oscillator, and it was verified that the proposed model could be used for predicting the transient dynamics of the USM. Furthermore, for steady-state velocity and the stopping time, a quantitative agreement was achieved between the measurements and numerical simulations. Finally, it is noted that the authors believe that the proposed model can be generalized and applied to the prediction of the nonlinear transient dynamics of the other traveling-type USMs.

Acknowledgments This work was supported in part by MEXT Grant-in-Aid for Scientific Research on Priority Areas, No. 438 Next-Generation Actuators Leading Breakthroughs.

References

1. Nakagawa Y, Saito A, Maeno T (2008) Nonlinear Dynamic Analysis of Traveling-Wave-Type Ultrasonic Motors. IEEE Trans Ultrason Ferroelectr Freq Control 55(3):717–725
2. Maeno T (2005) Recent Progress of Ultrasonic Motors. In Proc First Int Workshop Ultrason Motors Actuators:15-18
3. Maeno T, Tsukimoto T, Miyake A (1992) Finite-Element Analysis of the Rotor/Stator Contact in a Ring-Type Ultrasonic Motor. IEEE Trans Ultrason Ferroelectr Freq Control 39(6):668-674
4. Wallaschek J (1998) Cont act mechanics of piezoelectric ultrasonic motors. Smart Mater Struct 7(3): 369–381
5. Parashar SK, von Wagner U, Hagedorn P (2005) Nonlinear shear-induced flexural vibrations of piezoceramic actuators: experiments and modeling. J Sound Vibr 285(4-5):989–1014
6. Okumura I (1992) A Designing Method of a Bar-Type Ultrasonic Motor for Autofocus Lenses. In Proc IFToMM-jo Int Symp Theory of Machines and Mechanisms:836-841
7. Hagood NW, McFarland AJ (1995) Modeling of a Piezoelectric Rotary Ultrasonic Motor. IEEE Trans Ultrason Ferroelectr Freq Control 42(2):210-224
8. Glenn TS, Ghandi K, Atalla MJ, Hagood NW (2001) Mixed-domain traveling-wave motor model with lossy (complex) material properties. In Proc SPIE Smart Struct Mater 4326:525-537
9. Tsai M, Lee C, Hwang S (2003) Dynamic Modeling and Analysis of a Bimodal Ultrasonic Motor. IEEE Trans Ultrason Ferroelectr Freq Control 50(3):245-256
10. Pirrotta S, Sinatra R, Meschini A (2007) A novel simulation model for ring type ultrasonic motor. Meccanica 42(2):127–139
11. Gutschmidt S, Hagedorn P (2004) Modeling the steady and unsteady operation of a piezoelectric bar type motor. In Proc Int Congr Acoust:I405-I408.
12. Kandare G, Wallaschek JO (2002) Derivation and validation of a mathematical model for traveling wave ultrasonic motors. Smart Mater Struct 11(4):565–574
13. Pons JL, Rodriguez H, Ceres R, Calderon L (2003) Novel modeling technique for the stator of traveling wave ultrasonic motors. IEEE Trans Ultrason Ferroelectr Freq Control 50(11):1429–1435
14. Pons JL, Rodriguez H, Seco F, Ceres R, Calderon L (2004) Modelling of piezoelectric transducers applied to piezoelectric motors: a comparative study and new perspective. Sens Actuator A-Phys 110 (1-3):336–343
15. Pons JL, Rodriguez H, Fernandez JF, Villegas M, Seco F (2003) Parametrical optimisation of ultrasonic motors. Sens Actuator A-Phys 107(2):169–182
16. Rodriguez H, Pons JL, Fernandez JF, Ceres R, and Villegas M (2002) On the simulation of mechanical behaviour of travelling wave ultrasonic motors. Ferroelectrics 273(6):65–70
17. Duan WH, Quek ST, Lim SP (2007) Finite element solution for intermittent-contact problem with piezoelectric actuation in ring type USM. Finite Elem Anal Des 43(3):193–205
18. Maeno T (1998) Contact Analysis of Traveling Wave type Ultrasonic Motor considering Stick/Slip Condition. J Acoust Soc Jpn 54(4): 305-311 (in Japanese)

Chapter 24
Actuator with Multi Degrees of Freedom

Tomoaki YANO[1]

Abstract The latest research development of the project "Actuator with multi degrees of freedom (DOF)" is reported. At first, advantages of the multi DOF actuator are presented. Next, the design concept of the multi pole spherical synchronous motor is proposed. The specifications and the drive system are presented. The experimental results of the positioning control, the high speed control, the trajectory control, and the output torque control are shown. Then, the design concept of the hexahedron-octahedron based spherical stepping motor is proposed. The specifications and the drive system are presented. The experimental results of changing the rotational direction, the speed control, the output torque control are shown. The developed spherical actuators will be a high torque actuator by using iron cored, back yoked armature coils.

24.1 Introduction

In recent years, the study of actuator with multi degrees of freedom (DOF) has increased worldwide [1].

Once actuator with multi DOF becomes a reality, the following changes become possible:

1. A decrease in the number of actuators necessary for the realization of the same degrees of freedom is expected. Thus, in a system with actuators connected in series, such as the serial link manipulator, the total weight of the system can be reduced and the system can become compact in size.
2. Among the actuators with multi DOF, especially, spherical actuators are structurally designed so that the centers of rotation of two or three degrees of freedom coincide. Therefore, the kinematics equation is simple and can mostly be solved geometrically, leading to the simplification and speeding-up of control.

[1] Tomoaki YANO

Advanced Manufacturing Research Institute

National Institute of Advanced Industrial Science and Technology

3. The actuator of the robot eye and the active joints of the truss structure can be configured with no addition of extra mechanisms.
4. A mechanism with multiple degrees of freedom can be easily realized in the field of micro-machinery into which it is difficult to build sophisticated mechanisms.
5. A marked energy-saving effect as well as system size reduction can be expected, because the number of actuators used in the system can be reduced. Actuator is one of the basic devices in various products of industry. The energy consumed within the actuator accounts more than 50 percent in Japan. Therefore, energy conservation within the actuator will greatly contribute to carbon dioxide reduction, global environment protection, cultural level improvement, and economic development.
6. In systems which require high-precision positioning, it is difficult to eliminate assembly errors in combining one-degree-of-freedom actuators, and constant maintenance for eliminating secular error is necessary. Thus, in applications requiring high precision, actuators with multi DOF seem to be advantageous.

In view of advantages discussed above, I have manufactured by way of trial and conducted research into a wide range of electromagnetic spherical actuators, including induction motors, synchronous motors, stepping motors and AC servo motors; I have examined their characteristics and performances [2-5], and found new applications thereof [6]. Meanwhile, the output torque of these spherical actuators is not enough for driving the robot joints. So, I set the goal of the project "Actuator with multi degrees of freedom" to propose high-torque actuators with multi DOF. One is a multi pole spherical synchronous motor and the other is a hexahedron-octahedron based spherical stepping motor.

24.2 Multi Pole Spherical Synchronous Motor

24.2.1 Design

Figure 24.1 shows the structure of the developed multi-pole spherical synchronous motor. The motor has a rotor supported by a gymbal mechanism and a stator with four armature winding units. 260 small Ne-Fe-B permanent magnets are attached to the rotor surface so that the North and the South poles are located alternately.

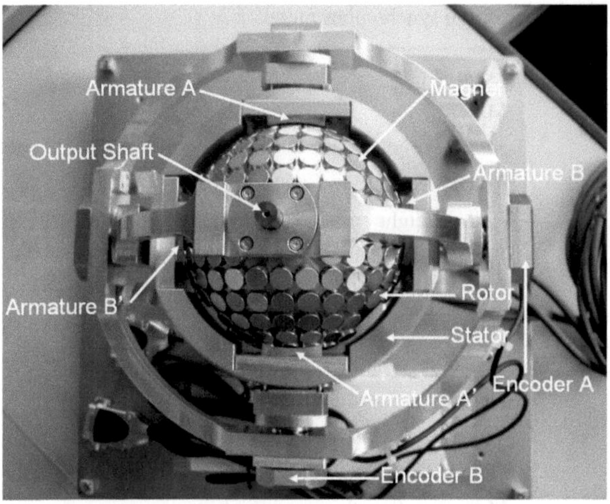

Fig. 24.1 Multi pole spherical synchronous motor

Four armature winding units are positioned on the stator. Each armature winding unit has three windings. The pitch of the windings is 2/3 of the pitch of permanent magnets. Therefore, locally, the relation between the armature windings and the permanent magnets is almost the same as that of the conventional one DOF linear synchronous motor.

Armatures A and A' drive the rotor in the same direction and Armatures B and B' drive the rotor in the direction perpendicular to the first one.

Figure 24.2 shows the positions of permanent magnets on the rotor. The output shaft is at the vertical position in Fig.24.2 (a) and at the position rotating around Y axis in Fig.24.2 (b). Permanent magnets are attached on the rotor so that the North and the South poles are located alternately on the concentric circles. The centers of the concentric circles are the X axis and the Y axis.

Even when the rotor rotates around Y axis as in Fig.24.2 (b), the relation between the permanent magnets and the armature windings are the same on the vertical direction as in Fig.24.2 (a). In a similar way, when the rotor rotates around X axis, the relationship between the permanent magnets and the armature windings are the same on the vertical direction as in Fig.24.2 (a). Therefore, the rotor can rotate around each axis simultaneously.

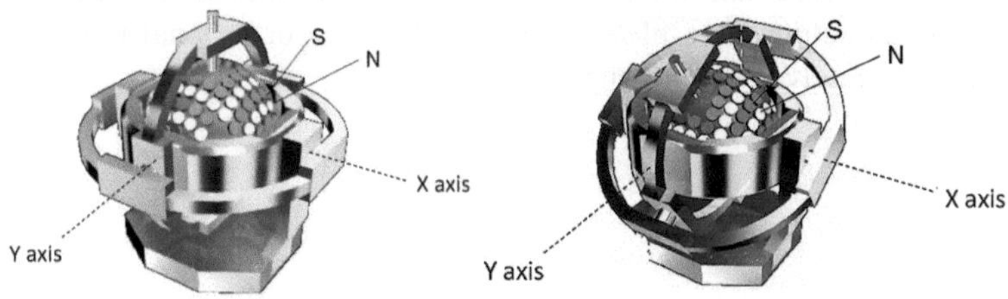

(a) vertical position (b) tilt position

Fig. 24.2 Permanent magnets and movement

Table 24.1 Specifications of the synchronous motor

Term	Value
Motor Size	265 x 265 x 266 mm
Motor Weight	8.6kg
Rotor Size	ϕ130 mm
Rotor Weight	4.43 kg
Permanent Magnet	ϕ11.5 mm x 5mm
Magnet Material	Ne-Fe-B (50MGOe)
Number of Magnet	260
Armature Windings	3 phase (153 turn)
Degrees of Freedom	2
Working Area	± 45 deg
Output Torque	0.49 Nm
Maximum Speed	90 deg/s
Encoder Resolution	163840 pulse/rev (0.0022 deg/pulse)

Air-cored coils are used to avoid torque pulsation with the cogging torque. The rotor position is measured by the encoders A and B. The specifications are shown in Table 24.1. The motor took on a structure in which all electromagnetic forces directly drive the rotor. Therefore, the motor of this type is expected to be a high torque.

24.2.2 Drive System

Fig.24.3 shows the block diagram of the motor driving system. Basic drive concept is the same as that of the conventional one DOF linear synchronous motor. Position signals of the encoders A (rotation angle around X axis) and B (rotation angle around Y axis) are compared with the target position signals from the control software respectively; the DSP board calculates the appropriate amplitudes of the three phase currents of each pairs of armature winding units from each position signal difference; and the PWM Generator commands the Inverters A and B to apply the armature three phase currents. The two orthogonal movements are controlled separately yet simultaneously.

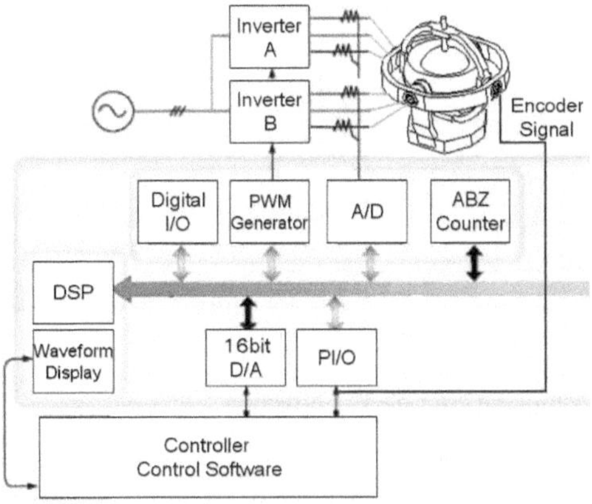

Fig. 24.3 Drive system

24.2.3 Experimental Results

The repeated positioning error was measured by using two laser interferometers. The end of the output shaft, which is 112 mm apart from the center of the rotor, was measured by two laser interferometers after moving the output shaft in an X-shape, a square shape, and a circular shape within ±20 degrees. The evaluation was done by repeating the movement 30 times. Every time, the end of the output shaft was measured 50 times.

Fig.24.4 shows the repeated positioning error of A'A-direction. The maximum, minimum, and average position points were plotted. The maximum positioning error was 6.8 μm, which corresponds to the rotation angle of 0.00348 deg. It is near the encoder resolution. Therefore, the repeated positioning resolution was small enough, within the expected range.

The moving speed was tested by rotating the output shaft from a 32 degrees tilted position to the vertical point from four directions by a 95 deg/sec command. Fig.24.5 shows the experimental results from rotating the output shaft from (32, 32) to (0, 0). It shows that the maximum rotating speed is more than 90 deg/sec. After reaching the maximum speed, the rotor decreased its speed and reached to the vertical point with overshooting. The target maximum speed was achieved at the other eight trials too.

The output torque is measured at a vertical position. Table 24.2 shows the measured output torques at the armature current 1.62 A RMS.

The developed motor was tested by moving the output shaft in an X shape, a square shape, and a circular shape. An X shape movement takes 10 seconds, a square movement takes 10 seconds, and a circular movement takes 8.3 seconds.

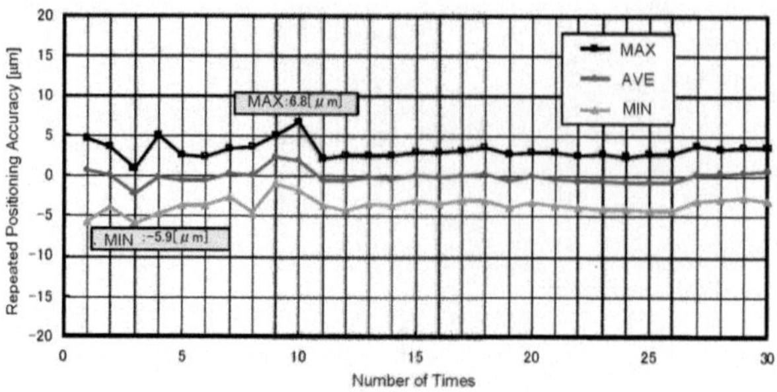

Fig. 24.4 Repeated positioning error of A'A-direction

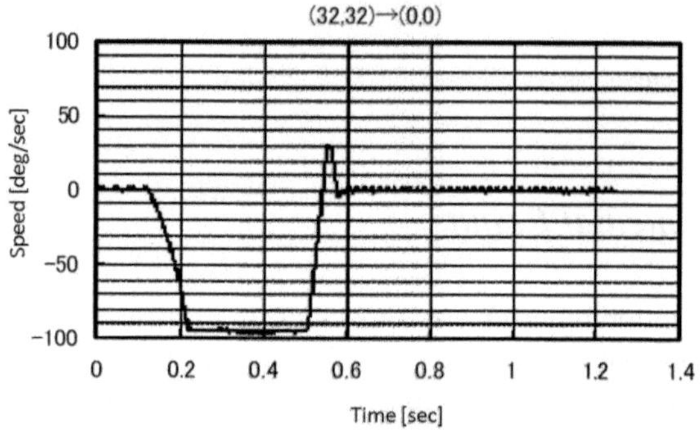

Fig. 24.5 Rotating speed from (32, 32) to (0, 0)

Table 24.2 Output torque

Direction	Torque [Nm]
B' to B	0.500
B to B'	0.502
A to A'	0.498
A' to A	0.491

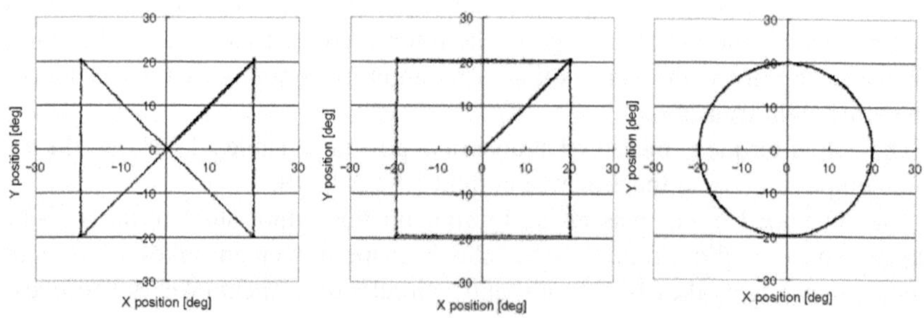

(a) X shape (b) Square shape (c) Circle shape

Fig. 24.6 Trajectory control

Encoder signals are measured and the X-Y positions of the output shaft are plotted on the X-Y plane. Plotted figures are shown in Fig.24.6.

According to the plotted figures of the trajectory control, the developed motor moves with a relatively small positioning error.

The developed motor showed some good performances, but the output torque is still too small to drive a robot's joint. I will increase the output torque by using iron cored back yoked armature coils.

24.3 Hexahedron-Octahedron Based Spherical Stepping Motor

24.3.1 Design

Figure 24.7 (a) shows a picture of the developed hexahedron-octahedron based spherical stepping motor (6-8 spherical stepping motor) and Fig.24.7 (b) shows the structure of it. Both the rotor and the stator are sphere shaped. The rotor is supported by six spherical bearings. Compressed air is supplied for the three bearings positioned on the bottom to reduce the friction of the balls. Figure 24.8 (a) shows the structure of the rotor and Fig.24.8 (b) shows the structure of the stator. Eight NdFeB permanent magnets are attached on the spherical shell at the vertexes of the virtual hexahedron inscribed in the rotor so that the North and the South poles are located alternately. Six iron cores are also attached on the spherical shell at the center of the faces of the virtual hexahedron inscribed in the rotor. The spherical shell is made of iron, the inner diameter is 52mm and the thickness is 5mm. The rotor is covered with an acrylic spherical shell to make the rotor surface sphere. The outer diameter of the acrylic shell is 78mm and the thickness is 4mm. The shape of the permanent magnets and iron cores are cylinder which diameter is 20mm and the thickness is 5mm.

The base of the stator is an acrylic spherical shell. Twenty-five armature coil units are attached through the acrylic spherical shell. Six of them are attached at the vertexes of the virtual octahedron inscribed in the stator. Twelve of them are attached at the center of the edges of the virtual octahedron inscribed in the stator. Seven of them are attached at the center of the faces of the virtual octahedron inscribed in the stator without the upper face. The upper face is open for the output shaft. The number of the turns of each armature coil is 153. The inner diameter and the outer diameter of the acrylic shell are 90mm and 110mm respectively.

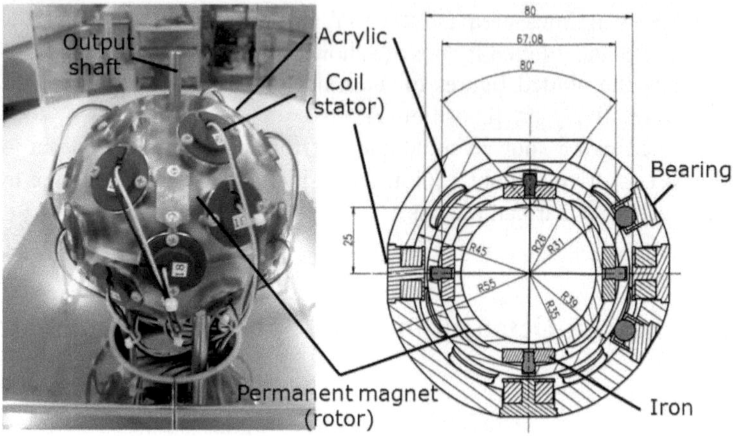

(a) Picture (b) Structure

Fig. 24.7 Design of the 6-8 spherical stepping motor

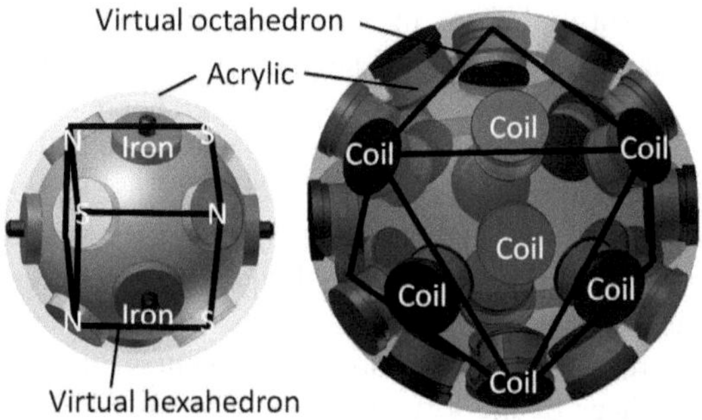

(a) Rotor (b) Stator

Fig. 24.8 Rotor and Stator

The developed motor is air-cored type without back yoke to avoid the cogging torque. The output torque of the developed motor is small but motor control is easy.

24.3.2 Drive Principle

Figure 24.9 is the upper view of the designed 6-8 spherical stepping motor when the output shaft is at the vertical position. The permanent magnets are positioned at every 90 degrees around the output shaft and North Poles and South Poles are positioned alternately. The pairs of coils (1,1'), (2,2') and (3,3') are positioned at every 120 degrees around the output shaft. The relationship between the permanent magnets and the armature coils is similar to that of the conventional three phase planer stepping motor with two pairs of permanent magnets. Therefore, when I supply three phase sinusoidal currents to the armature coils (1,1'), (2,2') and (3,3'), the rotor will rotates around the output shaft.

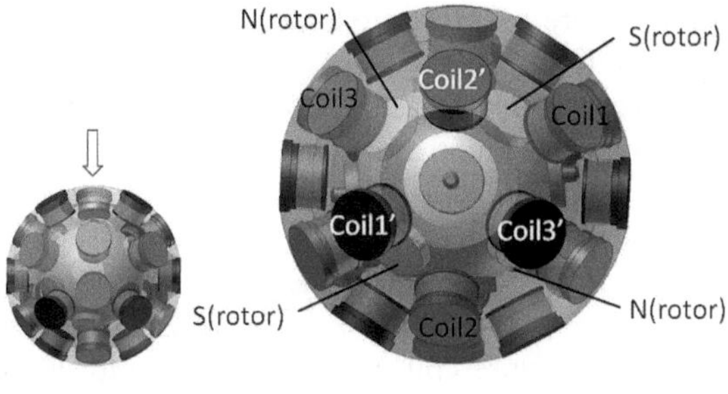

Fig. 24.9 Top view

Figure 24.10 is the bottom view of the designed 6-8 spherical stepping motor when the output shaft is at the vertical position. The positions of the permanent magnets and the coils (4,4'), (5,5') and (6,6') are same as Fig.24.9. Therefore, when I supply three phase sinusoidal currents to the armature coils (4,4'), (6,6') and (5,5'), the rotor will rotates around the output shaft in the same direction as Fig.24.9. When I supply three phase sinusoidal currents to the armature coils (1,1'), (2,2'), (3,3') and (4,4'), (6,6'), (5,5') simultaneously, the output torque will be doubled.

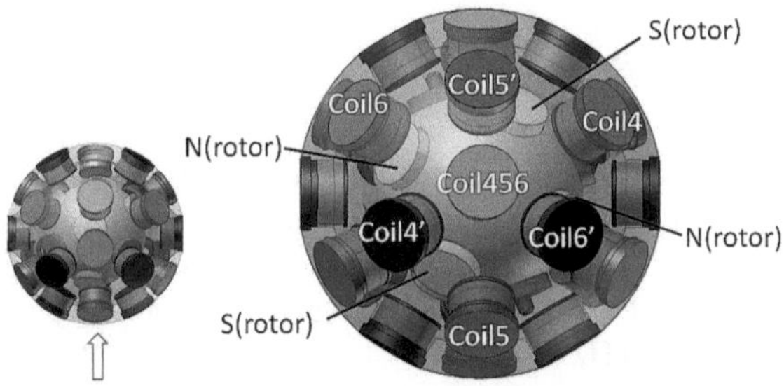

Fig. 24.10 Bottom view

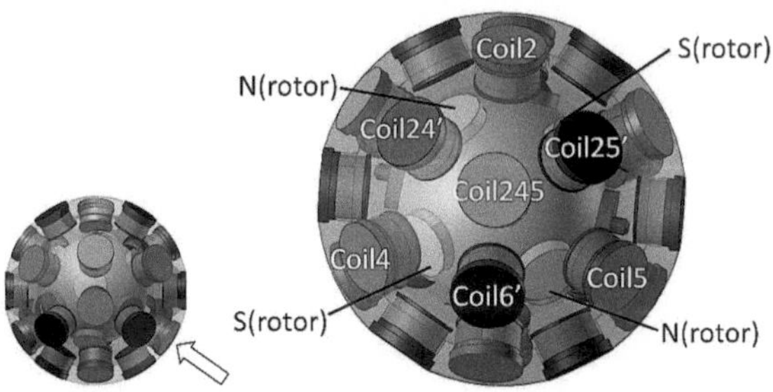

Fig. 24.11 Tilt view

Figure 24.11 is the view of the designed 6-8 spherical stepping motor from the direction of the arrow. One of the iron cores is just under the armature coil 245. The position of the armature coil 245 is the center of the triangle formed from coil2, coil4 and coil5. The positions of the permanent magnets and the coils (2,6'), (5,24') and (4,25') are same as Fig.24.9. Therefore, when I supply three phase sinusoidal currents to the armature coils (2,6'), (5,24') and (4,25'), the rotor will rotates around coil245. At the back view of Fig.24.11, the relationship between the permanent magnets and the armature coils is same as Fig.24.11. Therefore, when I supply three phase sinusoidal currents to the appropriate armature coils, the rotor will rotates around coil245.

The rotor has six iron cores and the stator has seven armature coils at the center of the face of the octahedron inscribed in the stator. When one of the iron cores is positioned just under the armature coil, I can rotate the rotor around the noticed armature coil by supplying three phase sinusoidal currents to the armature coils around the noticed armature coil.

From the discussion above, I can rotate the rotor in almost any direction by the following algorithm;

1) Attract appropriate iron core by the armature coil which will be the rotational center.

2) Supply three phase sinusoidal currents to the six pairs of armature coils around the rotational center.

3) When I change the rotation axis, go to step 1).

24.3.3 Experimental Results

The rotor rotates around vertical axes when supply three phase currents to armature coils (1,1'), (2,2'), (3,3') and (4,4'), (6,6'), (5,5') simultaneously. The rotor also rotates around coil 245 when supply three phase currents to armature coils (2,6'), (5,24'), (4,25') and (6,2'), (1,36'), (3,16') simultaneously. Each armature current is 1A RMS and 3.0 Hz.

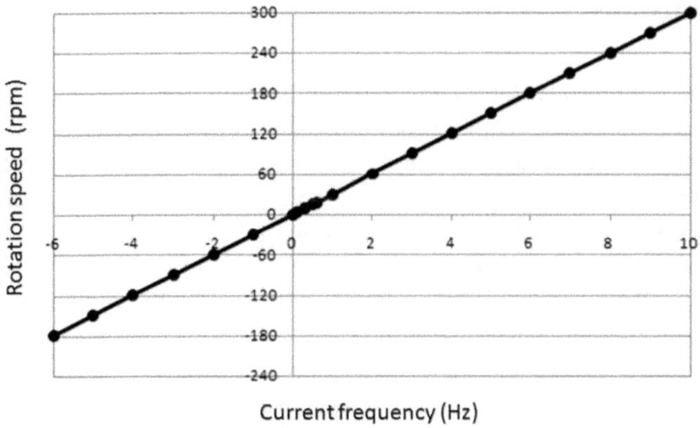

Fig. 24.12 Rotation speed to current frequency

Figure 24.12 shows the rotation speed around vertical axes when the current frequency changes from -6 Hz to 10 Hz. Minus means reverse rotation. The rotor rotates from -180 rpm to 300rpm. When the current frequency is 0.00167 Hz, the rotor rotates at 0.05 rpm. When the current frequency is under -6 Hz or over 10 Hz, the rotor cannot be controlled well.

Figure 24.13 shows the output torque for the angle between the permanent magnet on the rotor and a coil on the stator. Supply DC 1 A to the coils (1,1') and (4,4'), and DC-0.5A to the coils (2,2'),(3,3'),(6,6'), and (5,5'). The maximum output torque is 0.013Nm.

The developed motor rotates around different axis and the rotational speed is controlled from -180rpm to 300rpm. The motor is also controlled at 0.05rpm. The maximum torque is 0.013Nm. As the developed motor is air-cored type without back yoke, the output torque is small but control is easy. Iron-cored, back yoked 6-8 spherical stepping motor will be an omnidirectionaly high torque spherical motor.

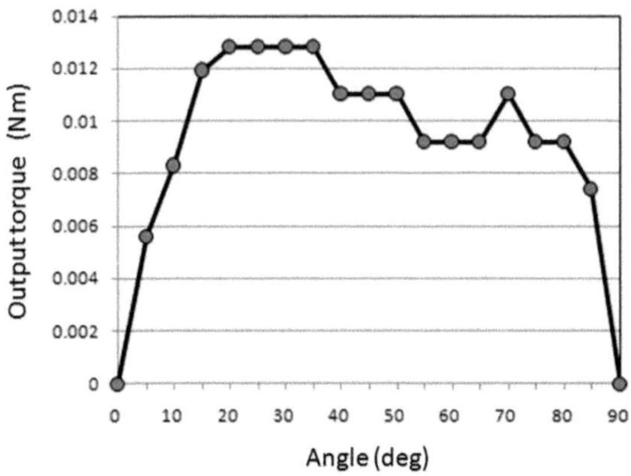

Fig. 24.13 Output torque to the angle between magnet and coil

24.4 Conclusions

Aimed for the high torque multi DOF actuator, a multi pole spherical synchronous motor and a 6-8 spherical stepping motor are developed and tested. A multi pole spherical synchronous motor is 2 DOF and a 6-8 spherical stepping motor is 3 DOF. Each motor moves by the designed principle and the experimental results shows the possibility of making the high torque multi DOF actuator. As the armature coils of both motors are air-cored type without back yoke, the output torques are still too small to drive a robot's joint. I will increase the output torque by using iron cored back yoked armature coils.

Acknowledgments This study was conducted partly with the support of the Grant-in-Aid for Scientific Research in Priority Area No. 438, for which I would like to express my deepest gratitude.

References

1. Yano T (2006) Multi dimensional drive system and their applications, Proc. Speedam06: S16-1-S16-6
2. Yano T, Kaneko M (1995) Basic Consideration of Actuators with Multi DOF Having an Identical Center of Rotation, J. Robotics and Mechatronics, 7, 6 :458-466
3. Yano T, Kaneko M, Sonoda M (1995) Development of a Synchronous Motor with Three degrees of Freedom, Theory and Practice of Robots and Manipulators - Proc. ROMANSY10, Springer-Verlag :275-280
4. Yano T, Suzuki T, Sonoda M, Kaneko M (1999) Basic Characteristics of the Developed Spherical Stepping Motor, Proc. IROS99, 3 :1393-1398
5. Yano T, Suzuki T (2002) Basic Characteristics of the Small Spherical Stepping Motor, Proc. IROS02 :1980-1985
6. Yano T, Takatuji T, Osawa S, Motomura Y, Itabe T, Suzuki T (2005) Development of a Spherical Motor Type Laser Tracker for the Portable 3D Position Measurement System, Proc.LDIA2005 :254-257

Chapter 25
Segment-Structured Diamond-Like Carbon Films Application to Friction Drive of Surface Acoustic Wave Linear Motor

Masaya TAKASAKI[1]

Abstract A surface acoustic wave (SAW) linear motor, which is a kind of ultrasonic motors, has many merits such as thin structure, large force, high speed, precise positioning and so on. Wear, however, is one of critical problems in the motor due to friction drive principle. On the other hand, diamond-like carbon (DLC) films have been recognized as a wear resistant material. In this chapter, application of Diamond-like-carbon (DLC) film as a wear-resistant film for its friction drive is introduced. In the application, projections on the conventional silicon slider can be replaced with segment-structured DLC (S-DLC) films. The installation of the films to slider/stator surface is reported. Improvement of preparation method of the films is proposed. Measurement results of driving characteristics of the motors with the films are also described.

25.1 Introduction

In recent years, development of high-performance actuator is demanded with progress in techniques for production of semiconductor materials, industrial robots and so on. Ultrasonic motor has many merits such as large force/torque, gearless, large holding force/torque, less magnetic field, precise positioning and so on. A surface acoustic wave (SAW) linear motor [1][2] is a kind of the ultrasonic motors, which has additional merits such as thin structure, easy installation, and so on. The motor has employed a silicon slider [3], and acquired stable driving to achieve large output force of 10 N [4] and high no-load speed of 1 m/s [5].

The SAW linear motor is a expecting motor as mentioned above. On the other hand, wear is one of problems in the motor due to friction drive principle. Diamond-like carbon (DLC) film has been recognized as a wear-resistant material.

[1] Masaya TAKASAKI

Graduate School of Science and Technology, Saitama University

The film was tried to be installed as the wear-resistant material to the SAW linear motor (The silicon slider was coated.) and confirmed its feasibility [5]. In our study, we focus protection of the stator transducer from the wear.

The stator transducer consists of a $LiNbO_3$ substrate and surface electrodes on its surface. The material is piezoelectric single crystal and very brittle. Therefore, the transducer was easy to wear out due to friction drive. Moreover, the silicon projections enhanced wearing. Lifetime of the SAW motor was reduced by the wear of the stator transducer. To extend the lifetime, a mechanism to avoid the wear, such as coating of a wear-resistant film, will be expected on the transducer surface. In this paper, we discussed application of DLC film as the wear-resistant film for the transducer and employed segment-structured DLC (S-DLC) films.

In this chapter, installation of the S-DLC films to the friction surface of the SAW linear motor. At first, a silicon piece with the S-DLC films is described to confirm feasibility of the films as friction material for the SAW motor. Second, the stator transducer with S-DLC films on its surface is investigated. Finally, a new method to install the films on the $LiNbO_3$ substrate is introduced.

25.2 SAW Linear Motor

Figure 25.1 shows a schematic view of the SAW linear motor. The motor consists of a stator transducer and a slider. The transducer is a $LiNbO_3$ 128° Y-cut X-prop substrate. When alternating current is applied to an interdigital transducer (IDT) on the substrate, Rayleigh wave, which is a kind of SAW, is generated and propagated to the direction indicated by the black arrow in the figure in the case of IDT (a) to apply the current. In progressive Rayleigh wave, micro-scopically, the surface of the substrate moves along an elliptical locus. The slider arranged on the elastic substrate surface is driven by friction force to the direction described by the grey arrow in the figure. The slider is required to be preloaded for the enough friction force to drive. The driving direction is the reverse of Rayleigh wave propagation direction. Selecting the IDT to apply the driving current, the driving direction can be switched. At the operating frequency of 9.6 MHz, vibration amplitude of the stator transducer surface is only few ten nm. To realize stable contact, a silicon slider, which has the distribution of projections, has been employed [3].

The material of the transducer is piezoelectric single crystal and very brittle. In the contact with the silicon slider under the friction drive, the material was easy to wear out. Moreover, the silicon projections enhance the wear. As a result, lifetime of the motor was reduced. To solve this problem, the S-DLC films were expected to avoid the wear in the friction surface.

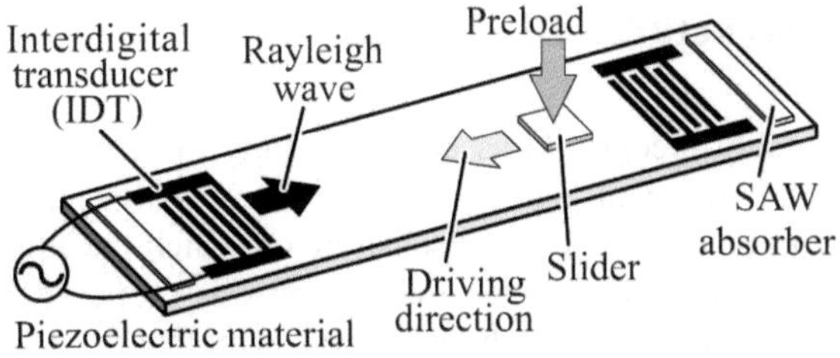

Fig. 25.1 Schematic view of a SAW linear motor

For the following experiments, new experimental apparatus was fabricated. Figure 25.2 is a photograph of the apparatus. In this apparatus, the $LiNbO_3$ transducer was guided by a linear guide. The slider was fixed on a preload mechanism. The moving part of the motor is called "moving table". For standardization the name in our research, the slider is called "slider", though it was fixed. The preload mechanism consisted of the half of a steel ball, a ring and a plate spring, as illustrated in the figure. The couple of the ball and the ring kept the slider parallel to the transducer surface. The upper part of the ring was pressed by the plate spring with a load cell. The pressing force, namely preload, was measured by the cell.

To measure output forces of the motor, the mass of the moving table was multiplied with the maximum acceleration of the table. For the measurement, the mass of the moving table was dynamically determined by the following experiment. A voice coil motor (VCM) vibrated the table under PD control. The shift of resonance frequency of the system with the change of additional mass to the table was observed. The moving part mass including the VCM was determined from the shift. In the following experiments, the moving table contained the $LiNbO_3$ substrate with its weight of 7 g. Their total mass of 86 g was used for calculation of the force.

Fig. 25.2 Experimental apparatus

25.3 Segment-structured DLC Films

DLC has been widely recognized as a tribo-coating material acting like a high wear-resistant solid lubricant with a low friction coefficient. The DLC film, however, is easy to be damaged, if the substrate on which the film is deposited is deformed. To avoid such damage, S-DLC coating was proposed [6]. As a result, the wear-resistance was significantly improved in comparison with that of the conventional continuous DLC films. In this research, we had focused on the segment structure of the S-DLC films. The structure was similar to projections of the conventional silicon slider.

Figure 25.3 shows the preparing method of the S-DLC films. The target substrate was placed on a mesh wire. S-DLC films were prepared on the downside of the substrate, because the downside is masked by a grid pattern of the mesh electrode during deposition. After the sputter etching with argon plasma, S-DLC films were deposited by pulse plasma CVD method with deposition parameters described in Table 25.1. Under these conditions, the deposition rate was estimated to be approximately 8 nm/min and the produced DLC films were approximately 1.0 μm thick after deposition. The prepared S-DLC films on a silicon wafer is shown in Fig.25.4 as an optical microscopic view. The S-DLC films with the segment size of 20 x 20 μm^2 and interval of 20 μm between segments were obtained. The structure is very similar to the silicon slider.

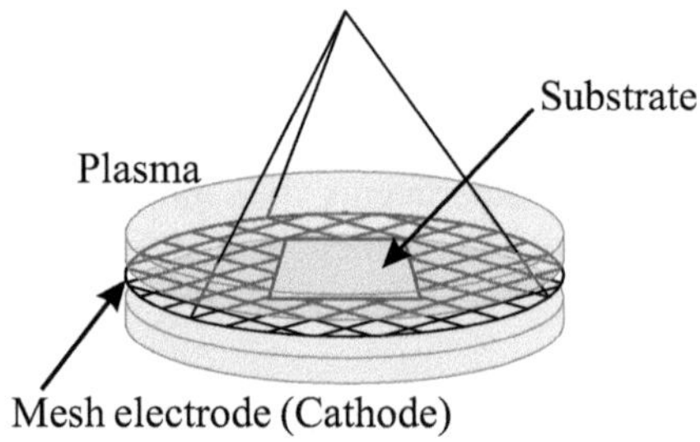

Fig. 25.3 Preparing method of S-DLC films

Table 25.1 Deposition condition

Preparation	Pulse plasma
Method	CVD
Source gas	C_2H_2
C_2H_2 flow rate	14.1 x 10^{-3} L/min
Pressure	3 Pa
Biasing voltage	-10 kV
Pulse frequency	2 kHz
Deposition time	2 h

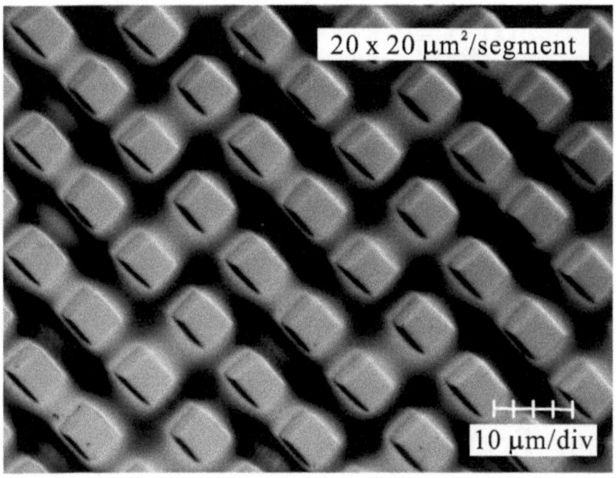

Fig. 25.4 Segment-structured diamond-like-carbon films on silicon wafer

25.4 S-DLC Slider

To indicate feasibility of the S-DLC films as a friction material for the SAW linear motor, a silicon wafer with the S-DLC films was prepared. The wafer was cut into pieces in the same size as the conventional silicon slider. Here, the piece is denoted as "S-DLC slider." The SAW linear motor employing the S-DLC slider worked successfully. Under the preload of 20 N, step responses of velocity with the change of the applied current were plotted in Fig.25.5. Traverse velocity around 700 mm/s was observed. Faster speed was expected if the slider stroke would not be limited. Maximum thrust force of 4 N was obtained at the preload of 33 N with applied current of 1.2 A. These results show the S-DLC films were effective for SAW linear motor friction drive as well as the conventional silicon slider projections.

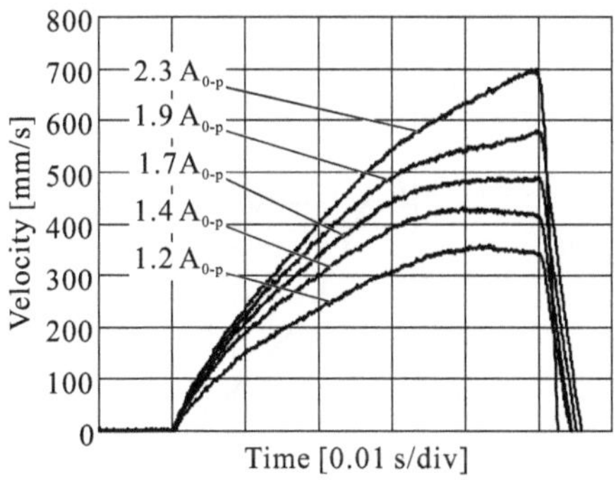

Fig. 25.5 Step responses of SAW linear motor using S-DLC slider

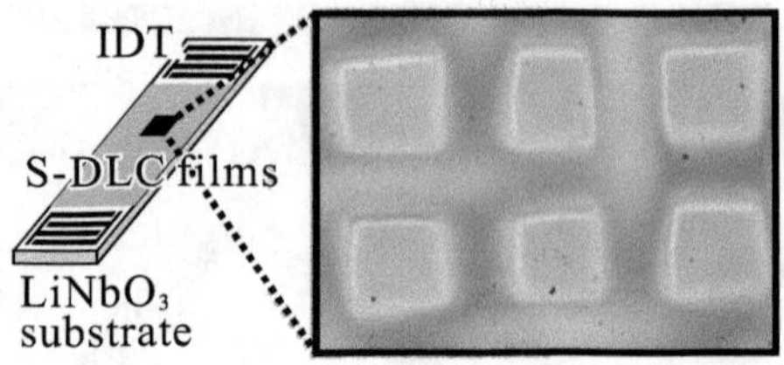

Fig. 25.6 S-DLC films on $LiNbO_3$ stator transducer

25.5 S-DLC Stator Transducer

As mentioned above, our purpose is to protect the $LiNbO_3$ stator transducer. For the purpose, the S-DLC films were deposited on the $LiNbO_3$ transducer. Figure 25.6 shows the films deposited on the $LiNbO_3$ substrate. Deposition condition was same as the S-DLC slider, except for deposition time. It is seen that the segment structure can play the role of the projection of the silicon slider. Therefore, a flat piece like a silicon wafer is available as the slider. The transducer with the S-DLC films is called "S-DLC transducer" in the following part. In the case of the S-DLC transducer, the slider is easy to wear out. Replacing the worn slider with new one, lifetime of the motor can be extended.

Influences of the DLC deposition on $LiNbO_3$ substrate to piezoelectric/acoustic characteristics were observed. Frequency response of conductance of the S-DLC transducer is plotted on Fig.25.7 with that of conventional transducer. It can be seen that S-DLC deposition process influence to piezoelectricity of $LiNbO_3$ transducer can be ignored. Amplitude distributions were measured with the applied current of 0.56 $A_{0\text{-}p}$. The measurement result is plotted with the comparison of conventional one in Fig.25.8. In the plot, origin means the edge of the IDT. It can be seen that attenuation due to the distribution of the S-DLC can be ignored. These experimental results mean that same acoustic characteristics as the conventional transducer can be expected on the processed stator transducer.

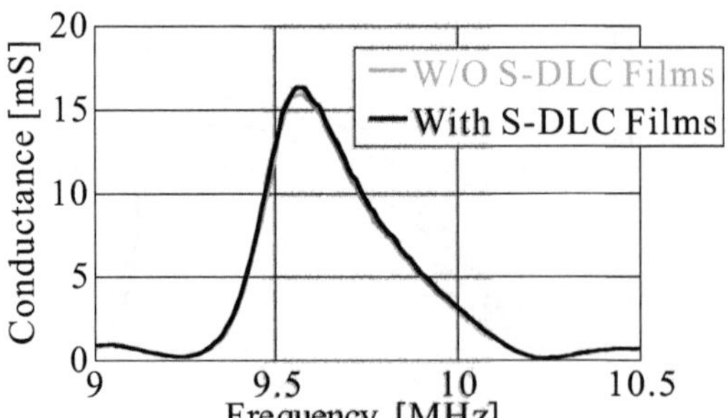

Fig. 25.7 Comparison of admittance characteristics

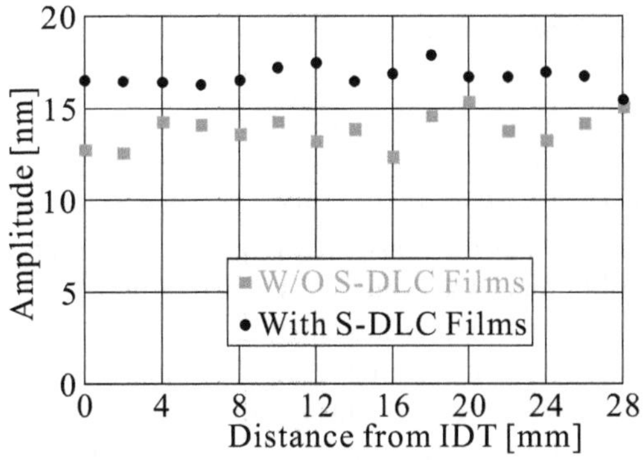

Fig. 25.8 Results of amplitude measurement

The S-DLC stator transducer with the silicon piece worked successfully. Performances of the motor were observed in the same manner as the previous driving experiment. Maximum output force was 0.8 N with applied current of 1.5 $A_{0\text{-}p}$. Much lower force was observed. To discuss the reason, profile of the S-DLC films was measured. Figure 25.9 shows the measurement result. It can be seen that the top of each segment was shaped in a curved surface. Additionally, height (thickness) of segments seems not to be unique. Under the contact between the stator and the wafer, total contact area was reduced rather than the case of conventional silicon slider and contact pressure of each contact point was not constant. As a result, total friction force was reduced.

Repeated driving experiment was also carried out. After 200 trips, some blank segment areas due to omission of local DLC films were observed. It seems that adhesion strength between the $LiNbO_3$ substrate and DLC film was not enough. Finally, the stator transducer would not be protected from the wear problem.

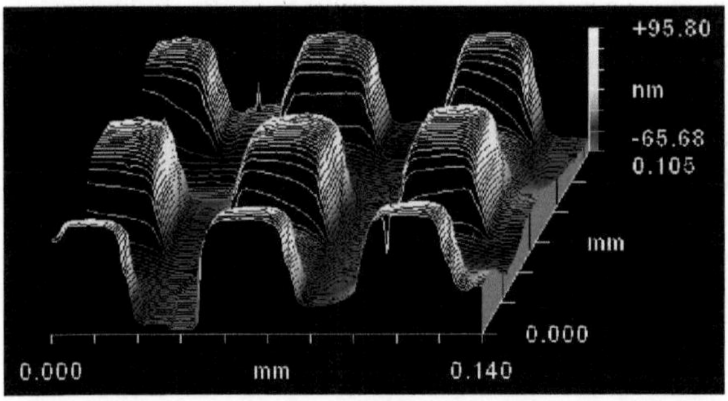

Fig. 25.9 Profile of the segments of S-DLC stator transducer

25.6 New Segment Structure

To solve the problem pointed above, we discussed a new method to prepare segment structure of DLC film on $LiNbO_3$ substrate. It seemed that the mask during DLC deposition influenced the shape of the segments. Therefore, the deposition technique without the mask was required for the new method. On the other hand, the distribution of the projections was required for the friction drive of the motor and easy process can have advantage.

It is known that adhesion strength between chromium and DLC film as well as that between $LiNbO_3$ substrate and chromium. According to the strength, we proposed segment-structured chromium films under DLC (S-Cr/DLC) film as shown in Fig.25.10. The stator transducer with the film is denoted as "S-Cr/DLC film stator." In this configuration, enough adhesion strength can be acquired. The deposition mask is not required during the DLC deposition. Additionally, the deposition of segment-structured Cr is easy process, as mentioned below.

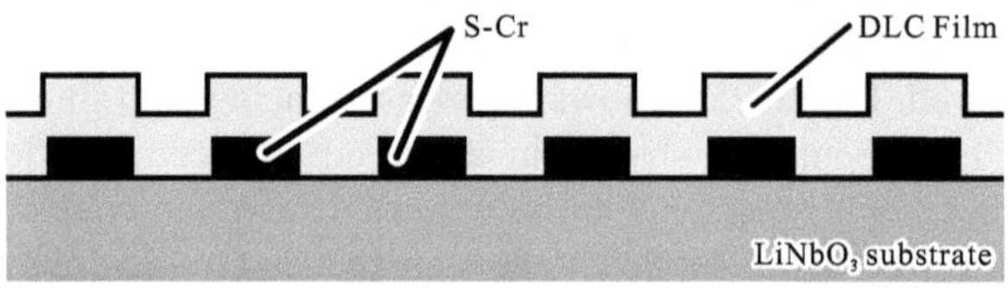

Fig. 25.10 Structure of Segment-structured Cr/DLC film

Figure 25.11 shows fabrication process of S-Cr/DLC film stator. At first, chromium film and aluminium film were deposited on the $LiNbO_3$ substrate by vacuum evaporation process, and a resist layer was coated on the aluminium film (Fig.25.11 (a)). Secondly, the resist layer was formed into same pattern of the IDT and the S-Cr by photolithography, and unnecessary metal films were removed by wet etching (Fig.25.11 (b)). Then, the aluminium films over S-Cr part was also removed (Fig.25.11 (c)). Finally, a continuous DLC film was deposited only on S-Cr part with the deposition condition described in Table 25.2 (Fig.25.11 (d)). Figure 25.12 shows a photograph of the fabricated transducer compared with a conventional stator transducer.

Table 25.2 Deposition condition for S-Cr/DLC film

Preparation	RF Pulse plasma
Method	CVD
Source gas	C_2H_2
C_2H_2 flow rate	100 x 10^{-3} L/min
Pressure	3 Pa
Biasing voltage	-540 Vdc
Pulse frequency	13.56 MHz
Deposition time	45 min

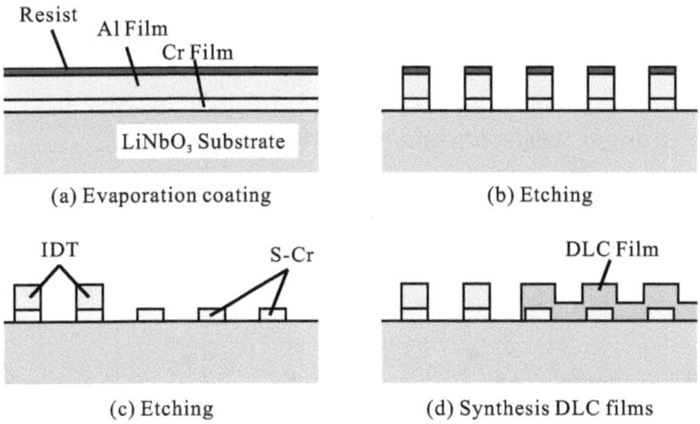

Fig. 25.11 Fabrication process of S-Cr/DLC films stator

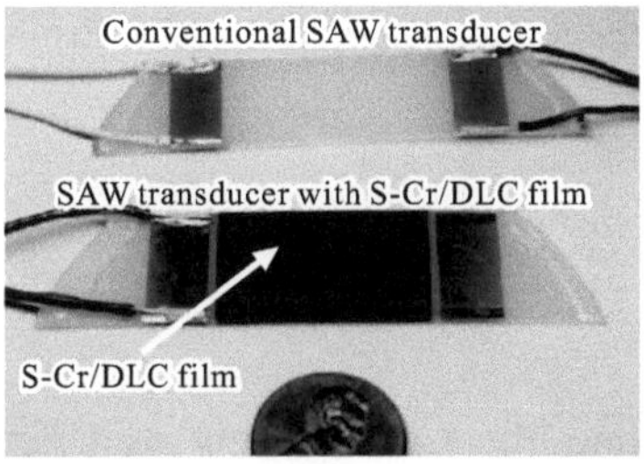

Fig. 25.12 Photograph of the S-Cr/DLC film stator

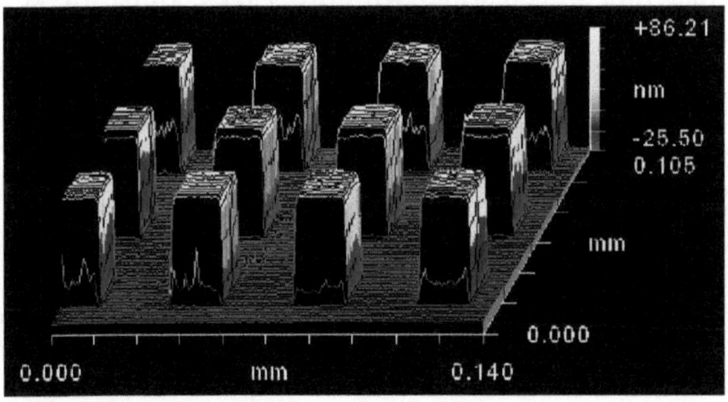

Fig. 25.13 Profile of the segments of S-Cr/DLC stator transducer

Profile of the DLC film of the fabricated transducer was measured. The result is illustrated in Fig.25.13. It can be seen that the top of each segment had flat surface comparing with the S-DLC stator. Surface roughness of the top area was Ra 0.4 nm. Larger contact area during the motor driving can be expected.

The S-Cr/DLC film stator was applied to the SAW motor driving. The motor worked successfully. Maximum speed of 200 mm/s was observed with applied current of 0.84 $A_{0\text{-}p}$. Output force characteristic was observed with the change of the preload. Figure 25.14 shows the characteristic. It can be seen that the new stator could perform large output force rather than the S-DLC stator. Repeat driving was also conducted. After 200 trips, no blank area was observed. The new stator can have stronger wear-resistant property.

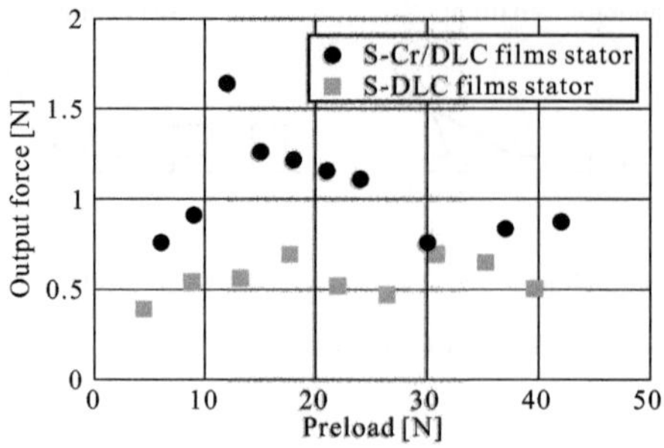

Fig. 25.14 Output forces with the change of preload

25.7 Conclusion

Segment-structured diamond-like carbon films were applied as wear-resistant material for friction drive of a surface acoustic wave linear motor. For such application, their segment structure was focused. To confirm feasibility of DLC film as friction material, silicon wafer with the S-DLC films was tested. Then the films were installed on $LiNbO_3$ stator transducer. To acquire enough performance as a motor and extend its lifetime, we proposed segment-structured Cr films under DLC film. A stator transducer equipped with the film was fabricated and applied to the friction drive. As a result, larger output force than that of S-DLC stator and wear-resistant property were observed. We expect that the DLC film can extend lifetime of the SAW linear motor with keeping its driving performance.

Acknowledgments This research was partially supported by the Ministry of Education, Science, Sports and Culture, Grant-in-Aid, for Scientific Research on Priority Areas (Actuator), 19016005, 2007-2008.

References

1. Kurosawa M, Takahashi M, Higuchi T (1996) Friction drive surface acoustic wave motor. Ultrasonics 34:234-246
2. Kurosawa M, Takahashi M and Higuchi T (1996) Ultrasonic linear motor using surface acoustic waves, IEEE Trans Ultrasonics 43:901-906
3. Osakabe N, Kurosawa M, Higchi T et al (1998) Surface acoustic wave linear motor using silicon slider. Proc. IEEE MEMS 390-395

4. Kurosawa M, Itoh H, Asai K (2003) Elastic friction drive of surface acoustic wave motor. Ultrasonics 41:271-275
5. Nakamura Y, Kurosawa M, Shigematsu T (2003) Effects of Ceramic Thin Film Coating on Friction Surface for Surface Acoustic Wave Linear Motor. Proc IEEE Ultrasonic Symp 1766-1769
6. Aoki Y, Ohtake N (2004) Tribological properties of segment-structured diamond-like carbon films. Tribology International 37:941-947
7. Fujii Y, Kotani H, Takasaki M et al (2007) Surface Acoustic Wave Linear Motor Using Segment-Structured Diamond-Like Carbon Films on Contact Surface. Proc 2007 IEEE Ultrasonics Symp 2543-2546
8. Kotani H, Takasaki M, Mizuno T (2007) Glass Substrate Surface Acoustic Wave Linear Motor. Proc IEEE Ultrasonics Symp 2547-2550
9. Takasaki M, Fujii Y, Kotani H et al (2008) Segment-Structured Diamond Like Carbon Films application to Friction Drive of Surface Acoustic Wave Linear Motor – Driving Characteristics –. Proc International Conference and Exhibition on New Actuators and Drive Systems 627-630
10. Kotani H, Takasaki M, Ishino Y et al (2008) Ultra Low-Velocity Control of a Surface Acoustic Wave Linear Motor. J System Design and Dynamics 2:497-506
11. Kotani H, Takasaki M, Mizuno T (2008) Surface Acoustic Wave Linear Motor Using Glass Substrate. Proc 9th International Conference on Motion and Vibration Control CD1-1310
12. Kotani H, Fujii Y, Takasaki M et al (2008) Surface Acoustic Wave Linear Motor Using Segment-Structured Diamond-Like Carbon Films on Contact Surface (1st report) – Deposition on Driving Surface and Driving Experiment -. J Japan Society for Precision Engineering 74724-729

Chapter 26
Mechanism of Electroactive Polymer Actuator

Multi-Scale Analysis Using Computational Techniques

Kenji Kiyohara[1], Takushi Sugino[1], and Kinji Asaka[1]

Abstract Mechanism of bending motion of the bucky-gel actuator was investigated at three length scales; the millimeter scale, the micrometer scale, and the nanometer scale. When voltage is applied, the cathode and the anode layer of the actuator were found to expand and contract, respectively, by measurements performed at the millimeter scale and a symmetrical analysis. The stress and the strain associated with the bending motion were calculated at the micrometer scale by the elasticity theory. The mechanism of the stress generation was attributed to the ionic structure formed at the nanometer scale in the electrode layers of the actuator by Monte Carlo simulation. Computational techniques are shown to be powerful tools in designing high performance electroactive polymer actuators.

26.1 Introduction

Electroactive polymer (EAP) actuators are expected to be the actuators of the next generation after the metal or ceramic based actuators, which have been utilized in our daily life. EAPs are flexible, lightweight, and driven by low voltage, in contrast to the conventional actuators 0. These properties are considered to be complementary to those of the conventional actuators and, for some applications, even advantageous over the conventional ones. Among the prototypes of the actuators based on EAPs are the nafion-metal composite 0-0, the carbon nanotube films 0, and the bucky-gel composites 0. All of these have the three-layer (electrode layer-separator layer- electrode layer) structure. The electrodes are made of metal, carbon nanotube, or the bucky-gel while the separators are made of insulating polymers containing aqueous solutions or ionic liquids. Those prototypes show bend-

[1] Kenji Kiyohara, Takushi Sugino, and Kinji Asaka

Research Institute for Cell Engineering, National Institute of Advanced Industrial Science and Technology (AIST) ing motion when voltage is applied between the electrode layers. In order to improve the performance of the EAP actuators and use them for practical applications, it is essential to understand the mechanism of the EAP actuators. For this purpose, we discuss the EAP actuators at three length scales; the millimeter scale, the micrometer scale, and the nanometer scale.

The EAP actuators function at the millimeter scale. The bending motion of an EAP actuator can be measured by the displacement of the tip of the actuator at this length scale using a laser displacement meter. The size of the actuator is often in this length scale, too. The design of the geometrical structure of the actuator in this length scale should be determined in such a way that the performance of the actuator is maximized for the given set of material properties composing the actuator. Therefore, in order to design the structure of the actuator, we need to know the material properties composing the actuator.

The bulk material properties such as the elastic constants are defined at the micrometer scale. The expansion or contraction is described by the elastic deformation of the polymer materials composing the actuators. Each volume element of the micrometer scale in the actuator deforms in response to applied voltage according to its bulk material properties such as the elastic constants and the stress exerted by the applied voltage. The deformation at each volume element of the actuator in turn results in the bending motion of the actuator as a whole with the constraints which are determined by the geometrical structure of the actuator. The connection between the micrometer scale deformation and the millimeter scale deformation can be described by the elasticity theory [8]. The conventional analyses of the elasticity theory [5], however, have been using various assumptions such as those on the symmetry of the deformation. They have been introduced rather for convenience in deriving analytical solutions and their validity has not always been proven to be reasonable. We show in the following that those assumptions can be removed by employing a computational method for solving the problems of elasticity. The computational method enables us to estimate the stress and the strain at each volume element of the actuator at the micrometer scale with the data obtained by measurements of the bending motion at the millimeter scale. This is a significant advance in the analysis of the actuator because there is no direct way to measure the stress and the strain inside the actuator experimentally. Making the connection between the micrometer scale properties and the millimeter scale properties, however, cannot explain the mechanism of the deformation in each volume element in response to the applied voltage. It must be explained at the nanometer scale because in the end the origin of the stress exerted by the applied voltage must be sought in the change of molecular interaction and molecular structure.

The molecular interaction is described at the nanometer scale. Since change of molecular interaction at applying voltage is the origin of stress inside the actuator and, in turn, of the expansion and contraction of the layers of the actuator, clarification of the mechanism at the molecular level would be the most fundamental in designing the actuators of high performance. Molecular simulation [9] is a suitable technique for analyzing the mechanism of stress generation in the actuator at molecular level. It has been known that, at applying voltage, the ions accumulate near the surface of the molecules that work as electrodes and there form the electrical double layers. The theory of the electrical double layers has been developed as early as the beginning of the 20th century [10, 11] but there has been no theory that can completely explain the ionic structure of the electrical double layers, although some advances have recently been made [12-14]. Therefore, molecular simulation is the most reliable technique in studying ionic structure and thermodynamic properties near the electrodes [15-19]. There are, however, only few studies that have been carried out on electrodes with porous structure as the ones in the EAP actuators. We show the development of computational technique specific for porous electrodes and application of it to studying the mechanism of the EAP actuators.

26.2 Structure of the EAP Actuator

The EAP actuator that we consider here is the bucky-gel composite [7], which has the three layer (electrode layer-separator layer-electrode layer) structure. The typical dimension of the actuator is 2 cm in length, 0.5 cm in width, and a few hundred μm in thickness. For both the electrode layers and the separator layer, the base polymer is poly(vinylidene fluoride-co-hexafluoropropylene) (PVdF(HFP)) and an ionic liquid, 1-ethyl-3-methylimidazolium tetrafluoroborate ($EMIBF_4$), is distributed in the base polymer. For the electrode layers, single walled carbon nanotube is also distributed in the base polymer so that they have electronic conductivity. The concentration of carbon nanotube is chosen high enough so that the carbon nanotubes are partially contacted with one another within the electrode layer. In this way, it is ensured that the electrical potential at the surface of the electrode layer and that in the middle are the same at equilibrium.

One separator layer was sandwiched by two electrode layers and they were heat pressed to form an actuator as a whole. Metal electrodes were attached to the electrode layers at the both sides (see Fig.26.1). When voltage is applied between the metal electrodes, the ions in all the three layers redistribute in response to the electric field induced by the voltage. At the molecular level, the ions accumulate near the surface of the carbon nanotube molecules in the electrode layers. Electrical double layers form at the surface of all the carbon nanotube molecules that are electrically connected with the metal electrode. Note that the surface area that is available for the electrical double layers to form is much larger than the surface area of the electrode layer that is exposed to the air, because of the entangled network of carbon nanotubes inside the electrode layer. Also note that there is no large potential barrier for an ion in passing through the interface between an electrode layer and the separator layer because the base polymers are the same for the both layers. There is ionic current but no electronic current in the separator layer.

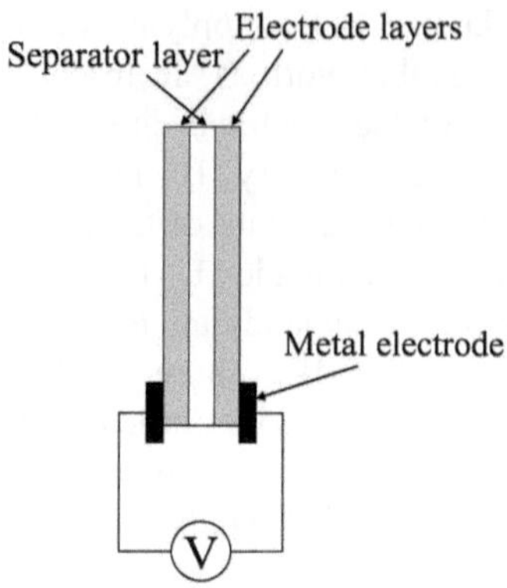

Fig. 26.1 Schematic picture of the structure of an EAP actuator with the three layer structure

When voltage is applied between the electrode layers, the actuator shows bending motion. This is explained by the anisotropic expansion or contraction of the two electrode layers when voltage is applied. The detail of the mechanism is, however, not well understood.

26.3 Symmetrical Analysis: a Millimeter Scale Analysis

The bending motion of the three layer actuator can be characterized by measuring the horizontal displacement of the tip of the actuator when the actuator of the other end is fixed so that it is vertically placed. If the distance between the fixed point in one end and the point at which the displacement is measured is denoted A and the horizontal displacement is denoted D (see Fig.26.2), then the curvature of the actuator, $1/R$, is calculated by [20]

$$\frac{1}{R} = \frac{2D}{A^2 + D^2} \tag{26.1}$$

On the other hand, if the length of the cathode is elongated by δ_C and that of the anode by δ_A, then the length and the thickness of the actuator, L and T, respectively, are related by

$$\frac{\delta_C - \delta_A}{L} = \frac{T}{R} \tag{26.2}$$

Combining these, we find

$$\frac{\delta_C - \delta_A}{L} = \frac{2DT}{A^2 + D^2}. \tag{26.3}$$

This equation shows that, by measuring the horizontal displacement of the tip of the vertically fixed actuator, we are able to determine the difference of the elongation of the cathode layer and the anode layer. This is the standard method for analyzing the deformation of the three layer actuators. However, the absolute values of δ_C and δ_A or even the sign of those cannot be determined separately by this method. For the actuator that we studied, we know that $\delta_C - \delta_A > 0$, which means that the expansion rate of the cathode is larger than that of the anode.

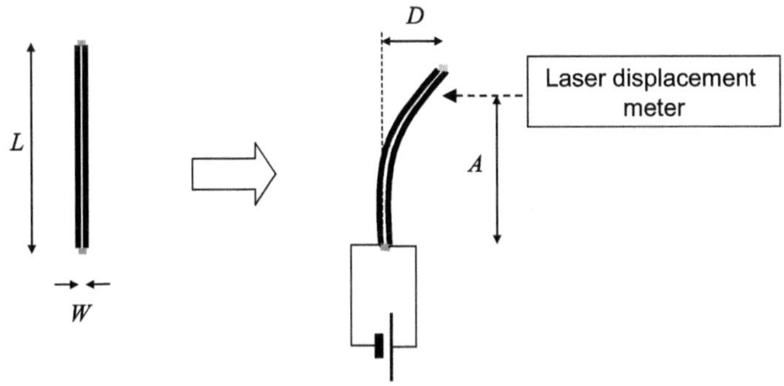

Fig. 26.2 Bending motion of the EAP actuator. Left: before applying voltage. Right: after applying voltage

In order to determine if the cathode and the anode are expanding or contracting, we used a five layer actuator and the experimental measurement was combined with a symmetrical analysis (see Fig.26.3). The five layer actuator is fabricated in the same method as the three layer actuator except that it has three electrode layers and two separator layers in the order electrode-separator-electrode-separator-electrode. By using the five layer actuator, we can induce the stress on the electrode layers in an asymmetrical fashion in contrast to the three layer actuator. Let us consider the following two cases.

In case I, the left and right electrode layers are chosen to be the electrode layers. The middle layer is electrically left open (see Fig.26.3, Left). This structure is similar to that for the three layer structure: it is symmetric with respect to the plane parallel to the layers and placed at the center of the actuator. By flipping the sign of the electrical potential of the two electrode layers, it should behave in a similar fashion to the three layer structure, both electrically and mechanically, except that the elastic property of the middle layer is different. Therefore, when the left layer is chosen to be the anode and the right layer the cathode, the five layer actuator will bend towards left, in the same way as the three layer actuator. There are three possibilities for this to happen; 1. $\delta_C > \delta_A > 0$ (both the cathode and the anode expand), 2. $\delta_C > 0 > \delta_A$ (the cathode expands and the anode contracts), and 3. $0 > \delta_C > \delta_A$ (both the cathode and the anode contract). We cannot distinguish which

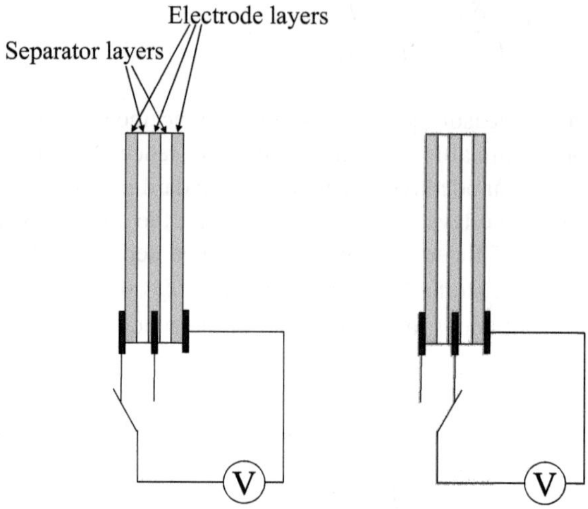

Fig. 26.3 EAP actuator with the five layer structure. Left: the left and the right layers are chosen to be the electrodes (case I). Right: the middle and the right layers are chosen to be the electrodes (case II)

of the three actually occurs only by measuring the motion of this actuator.

In case II, the middle and right electrode layers are chosen to be the electrode layers (see Fig.26.3, Right). The left layer is electrically left open. Unlike in case I, this structure has a different symmetry from that of the three layer structure: it is not symmetric with respect to the plane parallel to the layers and placed at the center of the actuator. By measuring the motion of this actuator when the sign of the electrical potential is flipped, we are able to distinguish the three possibilities above. For possibility 1, $\delta_C > \delta_A > 0$, the actuator should bend toward left, regardless of the choice of the cathode and the anode out of the middle and right electrode layers. It should bend more when the right layer is chosen to be the cathode. For possibility 2, $\delta_C > 0 > \delta_A$, the actuator should bend toward left when the right electrode layer is chosen to be the cathode and toward right when the right electrode is chosen to be the anode. For possibility 3, $0 > \delta_C > \delta_A$, the actuator should bend toward right, regardless of the choice of the cathode and the anode out of the middle and right electrode layers. Therefore, by measuring the motion of the actuator in case II, we are able to distinguish whether the cathode and the anode are expanding or contracting, separately.

We performed experiments of the two cases above and found that possibility 2 of the above discussion turned out to be the reality [21]. Therefore, the cathode expands with a positive expansion rate and the anode contracts with a negative expansion rate for the bucky-gel actuator: that is, $\delta_C > 0 > \delta_A$.

26.4 Elasticity Theory: a Micrometer Scale Analysis

By performing a quantitative analysis based on the elasticity theory for the bending motion of the two cases, we are also able to determine the strain and the stress at the micrometer scale. The free energy functional of the actuator for the bending deformation due to the applied voltage, ΔF, can be written as

$$\Delta F = \Delta F_{el} + \Delta F_{ex} \tag{26.4}$$

where ΔF_{el} and ΔF_{ex} denote the free energy change associated with the elastic deformation and that associated with the isotropic expansion due to the applied voltage, respectively. Those are written as

$$\Delta F_{el} = \int dr \left[\mu \left(u_{ij} - \frac{1}{3}\delta_{ij} u_{ij} \right)^2 + \frac{1}{2} K u_{ll}{}^2 \right] \tag{26.5}$$

$$\Delta F_{ex} = \int dr [-p u_{ll}] \tag{26.6}$$

where μ and K denote the elastic constants, u_{ij} denote the strain tensor, and p denotes the stress exerted by the applied voltage [8]. Note that μ, K, u_{ij}, and p are functions of position of the volume element defined at the micrometer scale, r. The integrations are performed over the whole actuator.

The strain tensor u_{ij} and the exerted stress p should be determined in such a way that ΔF takes the minimum value for the given conditions such as temperature, voltage, etc. The elastic constants μ and K are measured by other experiments. However, the stress p, which is exerted inside the actuator, cannot be measured directly by experiment. Therefore, the distribution of p and that of u_{ij} are determined at the same time by an iteration process in the following way. First, some distribution of p is assumed. Then ΔF is minimized by changing the values of u_{ij} at each position. Once ΔF is minimized, the distribution of u_{ij} is determined and the resulting bending curvature is calculated by specifying the geometry of the actuator. The calculated bending curvature is then compared with the experimentally measured curvature. If they do not agree, then the distribution of p is modified and the same process is repeated until the estimated value and the measured value of the curvature of the actuator agree with each other.

Using this procedure, the strain or the expansion rate of the cathode and that of the anode have been determined separately [21]. The strain of the cathode was estimated to be the order of +1% and that of the anode to be -1%. The pressure exerted inside the electrode layers was estimated to be $+10^5$ MPa for the cathode and -10^5 MPa for the anode [21]. These results are important in designing the structure of the EAP actuators.

As we described above, the micrometer scale analysis clarifies the response of the materials constituting the EAP actuator. This alone, however, does not tell us why the electrode layers expand or contract. Since it is almost certain that the expansion or contraction of the electrode layers are due to redistribution of the ions in response to the applied voltage, we should look at the change of molecular interaction and molecular structure in the EAP actuator.

26.5 Monte Carlo Simulation: a Nanometer Scale Analysis

In order to understand the expansion and contraction of the electrode layers of the actuator at the level of molecular interaction, Monte Carlo simulation [9] is a useful technique. Monte Carlo simulation is a technique of molecular simulation by which thermodynamic properties are numerically calculated based on the statistical mechanics, using the potential model chosen for the molecular interaction. By using Monte Carlo simulation, the ionic structures near the electrically charged molecules and the stress exerted by applying voltage are calculated.

Various potential models can be used in molecular simulation. Some are simple ones that represent the qualitative physical characters of real molecules and others are more refined ones that are designed to represent the chemical characters according to the quantum mechanics. Since we are interested in the basic mechanism of the molecular interaction, modeling the complicated molecular interaction in the composite material of the actuator by a refined model would be too costly for our purpose. Therefore, we used simple potential models: the ions interact with the hard sphere potential and the Coulomb potential.

In order to represent porous nature of the polymer, chargeable hard plates were placed parallel to each other with a separation of a molecular scale, typically a few to several times the diameter of the ions, in both the cathode and the anode. In the electrode layers of the actuator, the carbon nanotube molecules are crudely modeled to form such a porous structure inside the electrode layers. If one pair of chargeable hard plates parallel to each other with a molecular scale separation is picked up from the cathode and another pair from the anode, they represent the porous nature of the whole cathode and that of the anode, respectively. Therefore, it suffices to use one pair of chargeable plates for the cathode, another pair for the anode in the Monte Carlo simulation, and an electrode (separator) layer that separates the cathode and the anode (Fig.26.4). The Monte Carlo simulation was performed in the constant-voltage grand-canonical ensemble [22]. In this ensemble, the voltage or the electrical potential difference between the anode and the cathode is specified as an external field.

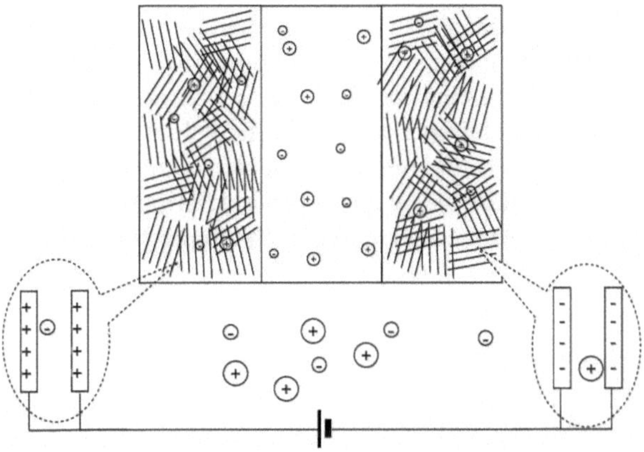

Fig. 26.4 Schematic picture of the model system. The anode and the cathode are composed of porous electrodes. These are represented by a pair of planar electrodes. The middle layer is the separator layer

During the simulations, we calculated the density profile, the stress profile, and the electrical potential profile as functions of the distance from the chargeable plates [23]. The main findings are the following. Before applying the voltage, the anions and the cations are distributed uniformly throughout the system. When voltage is applied between the two pairs of the chargeable plates, one pair of the charged plates are charged positively and the other pair negatively. The anions accumulate near the positively charged plates and the cations near the negatively charged plates. Thus the electrical double layers form. The thickness of the electrical double layers formed by the counter ions is typically one to two times the diameter of the ions. In the region away from the charged hard plates by more than a few ionic diameters, the electrical fields due to the charged plates are neutralized and the density of anions and that of cations are the same.

In the porous electrodes, the ionic structure near the electrode plates is significantly affected by the size of the pore for the following reason. The layers of the counter ions near the charged hard plates become thicker as the magnitude of the voltage is raised. When the voltage becomes so large that the thicknesses of the electrical double layers developing from the both sides of the pore become close to half of the pore size or the distance between the charged plates, a qualitative change occurs. In this case, there is no neutral region between the two charged hard plates and there starts an interaction between the layers of the counter ions from the both sides: the two layers of counter ions are no longer isolated. From this point on, the interaction between these electrical double layers becomes stronger as the voltage is raised.

The critical voltage above which the two layers of counter ions in a pore start to interact varies depending on the ion size, the pore size, and the strength of the Coulomb interaction determined by the temperature and the dielectric constant of the environment. We calculated the transition of density profile and the stress within the pore by varying those parameters and found that they show quite complicated behaviors. For example, the stress exerted by voltage within the pore can be positive or negative depending on the condition. Such complicated behaviors of interacting ions are known as the correlation effect [24-25]: a subtle change of the balance between the Coulomb interaction and the volume exclusion effect of ions can result in a drastic, and sometimes even counter-intuitive, change of thermodynamic behaviors. The details of the quantitative description of the thermodynamics of the ions in porous electrodes are left for the future studies.

26.6 Conclusions

EAP actuators are expected to function most likely at the millimeter scale. On the other hand, the material properties of the components of the actuators are determined at the micrometer scale and the most fundamental mechanism must be found in the change of molecular interaction by applying voltage at the nanometer scale.

The computational techniques useful for bridging our understandings at these length scales were developed. By a symmetrical analysis of the five layer bucky-gel actuator, it was unambiguously found that the cathode expands and the anode contracts at applying voltage. The strain and the stress were determined by utilizing the elasticity theory. The mechanism of stress generation at molecular scale was studied by Monte Carlo simulation, which indicated that the stress generation is closely related to the structural change of electrical double layers that formed in the porous structures of the actuator. Computational analyses such as those shown here are expected to be powerful tools for solving unanswered basic problems of EAP actuators and for developing high performance EAP actuators.

Acknowledgments The authors would like to thank Mr. I. Takeuchi and Mr. K. Mukai for helpful discussions.

References

1. Bar-Cohen Y Ed (2001) Electroactive polymer actuators as artificial muscles. SPIE, Washington
2. Oguro K, Kawami Y, Takenaka H (1992) Bending of an ion-conducting polymer film-electrode composite by an electric stimulus at low voltage (in Japanese). J Micromachine Soc 5:25-30
3. Oguro K, Asaka K, Takenaka H (1993) Polymer film actuator driven by a low voltage. Proceedings of 4th International Symposium on Micro Machine and Human Science at Nagoya, 39
4. Asaka K, Oguro K, Nishimura Y, Mizuhata M, Takenaka H (1995) Bending of polyelectrolyte membrane-platinum composites by electric stimuli I. Polymer J 27:436-440
5. Asaka K, Oguro K (2000) Bending of polyelectrolyte membrane platinum composites by electric stimuli Part II Response kinetics. J Electroanal Chem 480:186-198

6. Baughman RH, Cui C, Zakhidov AA et al (1999) Carbon nanotube actuators. Science 284:1340-1344
7. Fukushima T, Asaka K, Kosaka A, Aida T (2005) Fully plastic actuator through layer-by-layer casting with ionic-liquid-based bucky gel. Angew Chem Int Ed 44: 2410-2413
8. Landau LD, Lifshitz EM (1986) Theory of elasticity, 3rd edn. Butterworth-Heinemann, Oxford
9. Allen MP, Tildesley TD (1987) Computer simulation of liquids. Clarendon, Oxford
10. Gouy M (1907) Sur la constitution de la charge électrique à la surface d'un électrolyte. J. Phys. Théor. Appl. Sér. 4 (Paris) 9:457–468
11. Chapman DL (1913) A contribution to the theory of electrocapillarity. Phil. Mag. 25:475–481
12. Morreira AG, Netz RR (2000) Strong-coupling theory for counter-ion distributions. Europhys Lett 52:705-711
13. Morreira AG, Netz RR (2002) Simulations of counterions at charged plates. Eur Phys J E 8:33-58
14. Chen YG, Weeks JD (2006) Local molecular field theory for effective attractions between like charged objects in systems with strong Coulomb interactions. Proc Nat Acad Sci 103: 7560-7565
15. Torrie GM, Valleau JP (1980) Electrical double layers. I. Monte Carlo study of a uniformly charged surface. J Chem Phys 73:5807-5816
16. Torrie GM, Valleau JP, Patey GN (1982) Electrical double layers. II. Monte Carlo and HNC studies of image effects. J Chem Phys 76:4615-4622
17. Wennerström H, Jönsson B, Linse P (1982) The cell model for polyelectrolyte systems. Exact statistical mechanical relations, Monte Carlo simulations, and the Poisson-Boltzmann approximation. J Chem Phys 76:4665-4670
18. Quesada-Pérez M, Martín-Molina A, Hidalgo-Álvarez R (2004) Simulation of electric double layers with multivalent counterions: ion size effect. J Chem Phys 121:8618-8626
19. Boda D, Henderson D, Rowley R, Sokolowski S (1999) Simulation and density functional study of a simple membrane separating two restricted primitive model electrolytes. J Chem Phys 111:9382-9388
20. Pei Q, Inganas O (1992) Electrochemical applications of the bending beam method 1. Mass transport and volume changes in polypyrrole during redox. J Phys Chem 96:10507-10514
21. Kiyohara K, Sugino T, Takeuchi I, et al (2009) Expansion and contraction of polymer electrodes under applied voltage. J Appl Phys 105:063506-063513; Erratum (2009) J Appl Phys 105:119902
22. Kiyohara K, Asaka K (2007) Monte Carlo simulation of electrolytes in the constant voltage ensemble. J Chem Phys 126: 214704-214717
23. Kiyohara K, Asaka K (2007) Monte Carlo simulation of porous electrodes in the constant voltage ensemble. J Phys Chem C 111:15903-15909
24. Levin Y (2002) Electrostatic correlations: from plasma to biology. Rep Prog Phys 65: 1577-1632
25. Grosberg AY, Nguyen TT, Shklovskii BI (2002) Cooloquium: The physics of charge inversion in chemical and biological systems. Rev Mod Phys 74:329-345

Chapter 27
Development of a Polymer Actuator Utilizing Ion-Gel as Electrolyte

Hisashi KOKUBO[1] and Masayoshi WATANABE[1]

Abstract Polymer ion-gel-actuators comprising carbon materials as electrodes and ion-gels as electrolytes have been prepared and demonstrated to be operable at low voltages under atmospheric conditions. In order to develop a new ion-gel actuator as an alternative to a conventional polymer actuator, we modified the electrode gel by changing the types of carbon materials and polymer network consisted of ion-gel synthesized by living-like radical polymerization route.

The actuator utilizing activated carbon as an electrode material exhibited the maximum displacement of ca. 0.35 mm at ± 1.5 V. The difference in the surface area and electric conductivity of the carbon materials greatly affected the actuator performance. The ion-gel composites were prepared by mixing the obtained triblock copolymer (SMS) and ionic liquid ([C_2mim][NTf_2]). The ion-gel actuator using this composite exhibited improved performance in terms of displacement in comparison with the conventional ion-gel actuator.

27.1 Introduction

27.1.1 EAP Actuators

The electroactive polymers (EAPs) [1][2] used in transducers that convert electrical energy into mechanical energy are classified as ionic EAPs [3][4] and electronic EAPs [5]. Conducting polymers such as polypyrrole represent ionic EAPs driven by migration and diffusion of ions in the polymer. Electronic EAPs driven by electrostatic interaction are popularly known as dielectric elastomer actuators or liquid crystal elastomer actuators. As shown in Table 27.1, both ionic and electronic EAPs have drawbacks: ionic EAPs cannot be operated under atmospheric conditions, thereby limiting the conditions under which they can be used, and the electronic EAPs suffer from safety problems because they need to be driven under a high-applied voltage.

[1] Hisashi KOKUBO and Masayoshi WATANABE

Department of Chemistry & Biotechnology, Yokohama National University, 79-5 Tokiwadai, Hodogaya-ku, Yokohama 240-8501, Japan

Table 27.1 Characteristics of the electroactive polymer (EAP) actuators

	Ionic EAP	Electronic EAP
Operational mechanism	Migration and diffusion of ions	Electrostatic Coulomb force
Advantage	Driven at low voltage	Driven under atmosphere
Disadvantage	Driven in solutions	Driven under high voltage
Example	·Conducting polymers ·Ionomeric polymer-metal composites	·Dielectric elastomers ·Liquid crystal elastomers

27.1.2 Ionic Liquid and Ion-gel

In recent years, ionic liquids, which consist entirely of cations and anions, have received considerable attention. Ionic liquids have characteristic features such as low melting points compared with those of conventional ionic crystals, negligible volatility, non-flammability, thermal and chemical stability, and high ionic conductivity. In previous studies, we proposed ion-gels (polymer-ionic liquid hybrids) as a new class of polymer gels [6][7]. An ion-gel, which contains an ionic liquid in a network polymer, is a soft, lightweight, and transparent polymer electrolyte film. We have developed EAP actuators that comprise ion-gel films sandwiched between two electrodes.

27.1.3 Ion-gel Actuators

We developed polymer ion-gel actuator using a mixture of activated carbon (AC) and acetylene black (AB) as electrodes, poly(methyl methacrylate) (PMMA) as the network polymer, and 1-ethyl-3-methylimidazolium bis(trifluoromethanesulfonyl)imide ([C_2mim][NTf_2]) as the ionic liquid. The structural formulae of these materials are depicted in Fig.27.1. The polymer ion-gel actuators used in this study, as well as those used in previous ones, comprised electrode layers and an electrolyte layer, as depicted in Fig.27.2. A strip of the ion gel-carbon electrode composite immediately changes its shape when the polarity of the applied voltage is changed. We observed the tip of a 4 mm-long strip underwent a displacement of ca. 100 μm toward the anodic side when the other side was clamped and a voltage of ±1.5 V was applied. Thus, we successfully fabricated EAP actuators that can be driven at low voltages under atmospheric conditions and that overcome drawbacks of conventional EAP actuators [8]. In addition, since we utilize of commercially available carbon materials such as AC, we can fabricated this type of actuator at low cost. The actuation mechanism of the polymer ion-gel-actuator is currently unclear. However, the difference between the anodic and cathodic interfaces in terms of the in the electric-double-layer capacity may be one of the driving forces that govern the actuation mechanism.

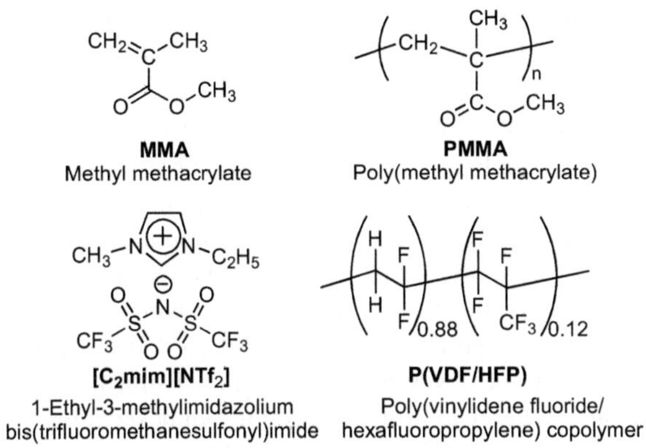

Fig. 27.1 Chemical structures of the monomer, polymers, and ionic liquid used in this study and in our previous study

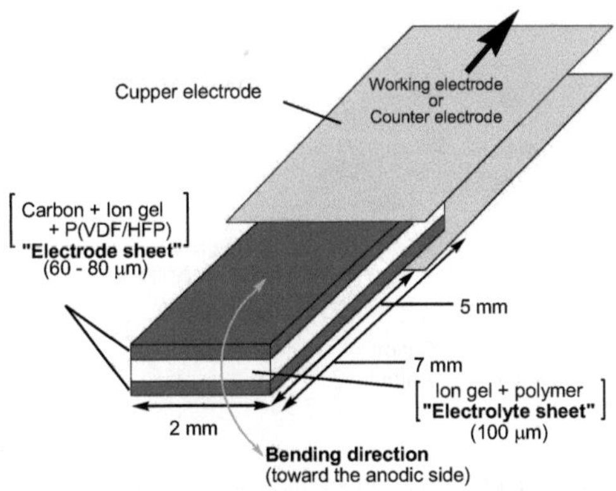

Fig. 27.2 Schematic structure of polymer-ion-gel actuator in this study. Actually, two glasses to support at fixed part are sandwiched on the cupper electrodes (not shown)

27.1.4 Selection of the Polymers Constituted of Ion-gel

Crystalline poly(vinylidene fluoride/hexafluoropropylene) copolymer (P(VDF/HFP)) was used as the network polymer of the ion-gel instead of PMMA, which we used in our previous study [5]; this is because ion-gels that comprise P(VDF/HFP) have better mechanical strength than those that comprise PMMA [7].

It is worth nothing that many studies have described the gelation behavior of ABA-type triblock copolymers in organic solvents [8][9]. A-units and B-units are insoluble and soluble in a solvent, respectively. Fig.27.3 shows the schematic model of the formation of an ion-gel, which comprises an ABA-type copolymer and an ionic liquid. While B-units such as PMMA dissolve in an ionic liquid, the A-units aggregate and serve as cross-linking points in the ion-gel. The mechanical properties of these physically cross-linked gels can be controlled by changing (1) the weight fraction of the polymer in an ionic liquid, (2) the molecular weight of the A- and B-units, and (3) the ratio of the molecular weights of the A- and B-units. We synthesized an ABA-type triblock copolymer (SMS) comprising polystyrene (solvatophobic; A-unit) and PMMA (solvatophilic; B-unit), and we investigated the thermal and electrochemical properties of the synthesized ion-gels with a view to developing application of the fabricated ion-gel actuator.

We investigated the effect on the performance of the polymer ion-gel actuator of the following two factors: (i) the type of carbon material used and (ii) the type of network polymer used, i.e., P(VDF/HFP) and an ABA-type triblock copolymer.

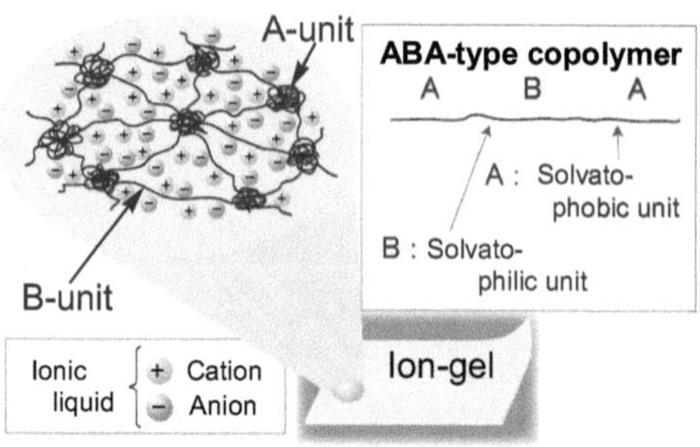

Fig. 27.3 Schematic model of formation of the ion-gels. A-unit (polystyrene) and B-unit (PMMA) are insoluble and soluble in a ionic liquid, respectively. Although B-unit dissolves in an ionic liquid, A-unit aggregates each other, and has a role as cross-link point of the ion-gel

27.2 Experimental Part

27.2.1 Materials and Measurements

[C_2mim][NTf_2] and 50 nm inverse-opal-carbon (OC) were prepared according to literature [10]. AC, multi-walled carbon nanotubes (MWCNTs), vapor growth carbon fiber (VGCF), and poly(vinylidene fluoride/hexafluoropropylene) copolymer (P(VDF/HFP)) were purchased and used without further purification. Scanning electron microscope (SEM) images, X-ray diffraction (XRD) patterns, and actuator displacements were obtained using a Hitachi Science Systems S-2600 SEM, a MAC Science MX-Labo powder X-ray diffractometer, and a KEYENCE LC-2400 laser displacement meter, respectively. Thermal gravimetric analysis (TGA), differential scanning calorimetry (DSC), and complex impedance measurements were carried out an SII TG/DTA6200, an SII DSC220C, and a Hewlett-Packard 4192A, respectively.

27.2.2 Preparation of the Polymer-ion-gel Actuator Using P(VDF/HFP) as the Polymer Network

An ion gel consisting of [C_2mim][NTf_2] and P(VDF/HFP) (17 wt%) was prepared using the following procedure: P(VDF/HFP) was dissolved in [C_2mim][NTf_2] at 120°C for 3 h. The heated solution was cast on a glass plate and cooled at room temperature to form the electrolyte sheet (thickness = 100 μm). An electrode layer containing carbon materials, [C_2mim][NTf_2], and P(VDF/HFP) as the binder polymer, was prepared using the following procedure. The mixtures were completely mixed and then pressed at 120 °C under 1 – 2 MPa for AC, OC, and VGCF and 4 MPa for MWCNT (thickness of the electrode sheet = 60 – 80 μm). In order to reduce the contact resistance at the electrolyte-electrode interface, an electrolyte sheet was sandwiched between two electrode sheets under a pressure in a thermoregulated oven at 130°C for 3 h.

The polymer ion-gel actuators were made by cutting the electrolyte-electrode composite into a 2 × 7 mm strip and sandwiching it between two cupper electrodes to fix one end (Fig.27.2).

27.2.3 Preparation of Ion-gel Film Using SMS as a Polymer Network

SMSs were dissolved in tetrahydrofuran (THF) as cosolvent, and the solutions were added to [C_2mim][NTf_2]. The mixed solution was cast on a Petri dish at room temperature by evaporating most of the THF cosolvent, and then dried under vacuum at 100 °C for 12 h. The obtained films were sandwiched between glass slides the thickness of the Teflon spacer used being 0.1 – 0.5 mm, and pressed at 130 °C.

27.3 Results and Discussion

27.3.1 Dependence on Carbon Materials

27.3.1.1 Characterization of the Carbon Materials

OC contains periodical pores (ca. 50 nm) in hard carbon that are prepared by pyrolysis of poly(furfuryl alcohol) resin. AB is not the main material of the electrode; rather, it is an auxiliary conductive material. This is because of the lower electric conductivity of AC and OC.

Figure 27.4 shows the SEM images of AC, OC, MWCNT, and VGCF. Both AC and OC have a granular-structure with small aspect ratios (length/diameter ratios nearly equal to 1), whereas MWCNT and VGCF both have rod-like-structures with large aspect ratios (>100).

Figure 27.5 shows the XRD patterns of AC, OC, AB, MWCNT, and VGCF. (002) peaks at $2\theta = 27°$ ($d = 3.3$ Å) can be attributed to the distance between graphene sheets. The strong (002) peaks in the patterns of AB, MWCNT, and VGCF indicate that these carbon materials have higher electric conductivity. However, the absence of the (002) peak in the patterns of AC and OC suggests that these carbon materials have lower electric conductivity.

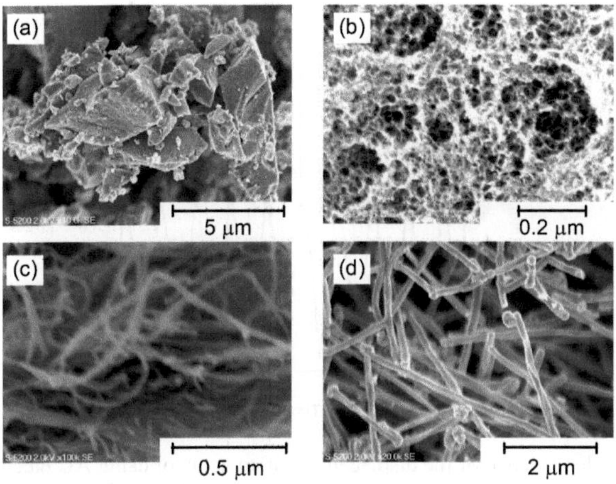

Fig. 27.4 SEM images of the carbon materials. (a) Activated carbon (AC); (b) 50 nm inverse-opal-carbon (OC); (c) multi-walled carbon nanotube (MWCNT); and (d) vapor growth carbon fiber (VGCF)

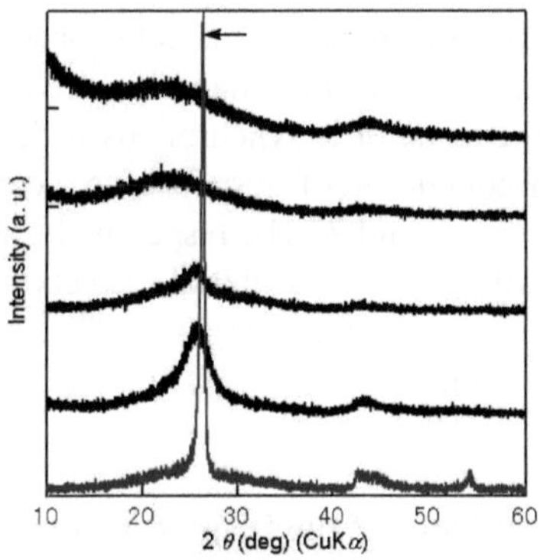

Fig. 27.5 XRD patterns of carbon materials. (a) AC; (b) OC; (c) AB; (d) MWCNT; and (e) VGCF. The peaks at $2\theta = 27°$ ($d = 3.3$ Å) are assigned to distance between graphene sheets

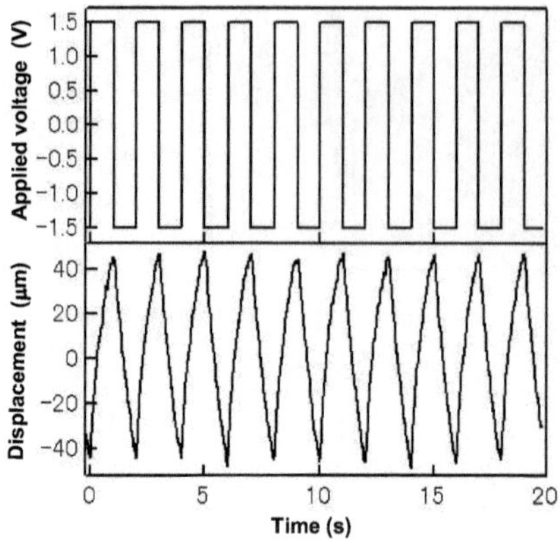

Fig. 27.6 Time dependence of the displacement of an actuator using AC electrodes, driven by square-wave form voltage (± 1.5 V) at a frequency of 0.5 Hz

27.3.1.2 Response Speed

Figure 27.6 shows the typical response of the actuator when we use AC electrodes that are driven by a square-wave form voltage at a frequency of 0.5 Hz. The displacement of the tip of the actuator is synchronous with switching the voltage polarity. The maximum frequencies of response to actuation by AC, OC, and MWCNT electrodes were 5, 5, and 25 Hz, respectively. These results indicate that the electric conductivity of the carbon materials affects the actuation response, because the MWCNT electrode, which possesses a well-developed graphite structure, has higher electric conductivity.

27.3.1.3 Scan Rate Dependence of the Displacement

Although there are scarcely any changes in the magnitudes of the displacements for MWCNT and VGCF actuators within the voltage scan rates used in this study (0.05 – 5 Vs^{-1}), the displacement of the AC and OC actuators is affected by the scan rates (Fig.27.7). When the scan rate (frequency) is less, the AC and OC actuators give rise to large displacements due to effective charging of the electric-double-layer, which results from the large surface area. As the scan rate increases, the displacement rapidly diminishes due to the IR drop of the actuators. On the other hand, for the MWCNT and VGCF actuators, which comprise electrodes with higher electric conductivity, fast charging and discharging seem to be possible. This results in displacement that is independent of the scan rate.

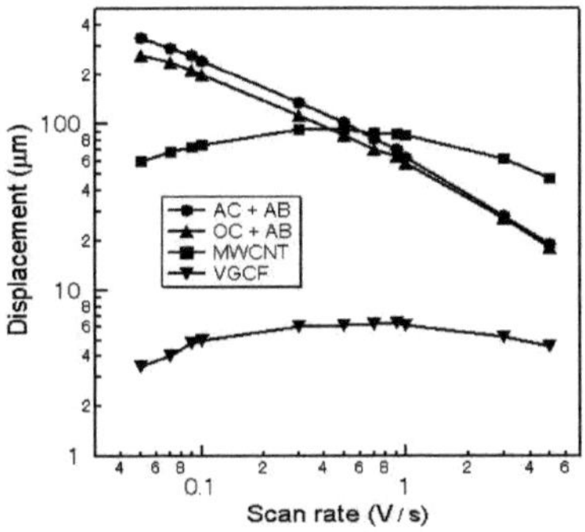

Fig. 27.7 Scan rate dependence of the displacement for the actuators using four kinds of carbon materials as electrodes

27.3.1.4 Cycle Life

Figure 27.8 shows the cycle property of polymer ion-gel-actuators using AC, OC, and MWCNT electrodes when a triangle-wave-form voltage of ±1.5 V was repeatedly applied at a scan rate of 500 mVs^{-1} (frequency = 0.083 Hz). The δ value shown in the vertical axis indicates the time (cycle number) dependence of the deterioration in the actuator performance (eq. 27.1).

$$\delta(\%) = d_n / d_1 \times 100, \qquad (27.1)$$

where d_n and d_1 are displacement at the nth cycle and the first cycle, respectively. The observed deterioration in the displacement with the cycle number can be attributed to an increase in the contact resistance between the carbon grains. The MWCNT actuator shows a lower degree of deterioration, compared with the AC and OC actuators. In order to make continuous conduction paths, AC and OC need to have many contacts because of their low aspect ratios. In contrast, rod-like MWCNT has a higher aspect ratio and thus easily forms the conduction path. Thus, large deterioration is observed in the AC and OC actuators.

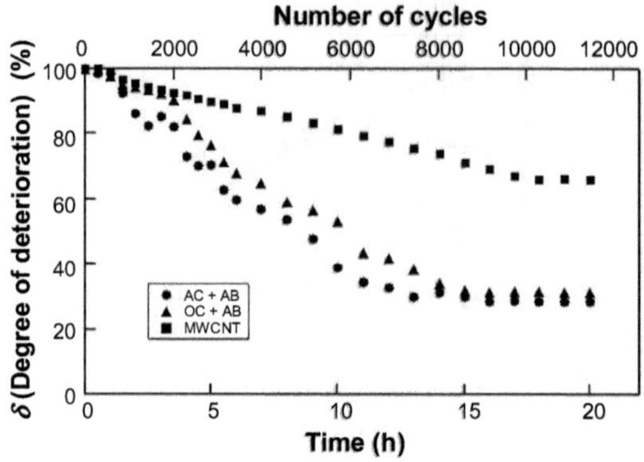

Fig. 27.8 Degree of deterioration of AC, OC, and MWCNT actuators driven by triangle wave form voltage (± 1.5 V) at a frequency of 0.083 Hz (scan rate = 500 mV/s). δ values are evaluated from eq. (27.1)

27.3.2 Characterization of SMSs and Application for the Actuator

27.3.2.1 Preparation and Characterization of the Ion-gels

Ion-gels were prepared by mixing [C_2mim][NTf_2] and SMS in several weight fraction ratios. Although the films with an ionic liquid weight fraction of below 92wt% seemed to be transparent and self-standing, as shown in Fig.27.9, SMS film (without ionic liquid, i.e., 0wt%) and 24wt% ionic liquid-mixed film were hard and fragile. The obtained composite film showed increasingly flexibility with the weight fraction ratio of ionic liquid.

Fig. 27.9 A photograph of the SMS/[C_2mim][NTf_2] gel film where weight fraction of [C_2mim][NTf_2] is 81 wt%

27.3.2.2 Observation of the Ion-gel Film Using Transmission Electron Microscopy

Figure 27.10 shows transmission electron microscopy (TEM) images of the obtained ion-gel, containing 87wt% of [C_2mim][NTf_2]. Approximately 20 – 30 nm of domain structures were observed as black spheres. Stained spheres are recognized as aggregated polystyrene, because polystyrene is easy to strain using ruthenium oxide as a dying agent. These images indicate that the polystyrene domain forms a micro-phase separation structure, which influences the physical cross-linked point of the ion gels.

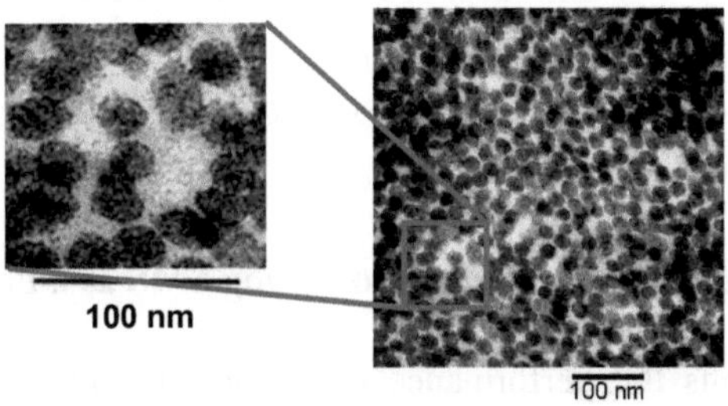

Fig. 27.10 TEM images of SMS/[C_2mim][NTf_2] composite ion-gel, where weight fraction of [C_2mim][NTf_2] is 87wt%. Sample film was strained by ruthenium oxide

27.3.2.3 Ionic Conductivity of the Ion-gel Composites

Figure 27.11 shows ionic conductivity for the SMS/[C_2mim][NTf_2] composite gel as a function of temperature. After increasing with the SMS weight fraction ratio, the ionic conductivity gradually decreased over all the temperature regions. However, the composites, which contained 87wt% and 92wt% ionic liquid, showed almost identical values of ionic conductivity for the ionic liquid. Therefore, we can successfully solidify the ionic liquid without losing the benefits associated with the use of ionic liquids.

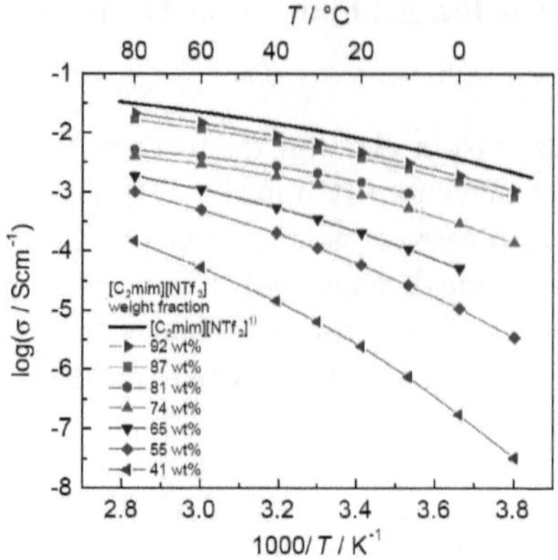

Fig. 27.11 Arrhenius plot of ionic conductivity for SMS/[C_2mim][NTf_2] composites

27.3.2.4 Performance of Ion-gel Actuator Using SMS as a Polymer Network

Figure 27.12 exhibits the performance of the ion-gel actuator using SMS and P(VDF/HFP) as polymer networks of ion-gel. The response speed of the actuator using SMS was inferior to that using P(VDF/HFP). This result was due to use of a different kind of binder polymer (P(VDF/HFP)) in the electrode for the SMS-used actuator. Moreover, the interface of the electrode and ion-gel resulted in large resistance. Namely, contact resistance induces degradation of response speed. Although the SMS-used actuator has a large contact resistance, displacement was superior to the P(VDF/HFP)-used actuator. This indicates that the actuator with higher ionic conductivity of ion-gel shows good performance in terms of maximum displacement.

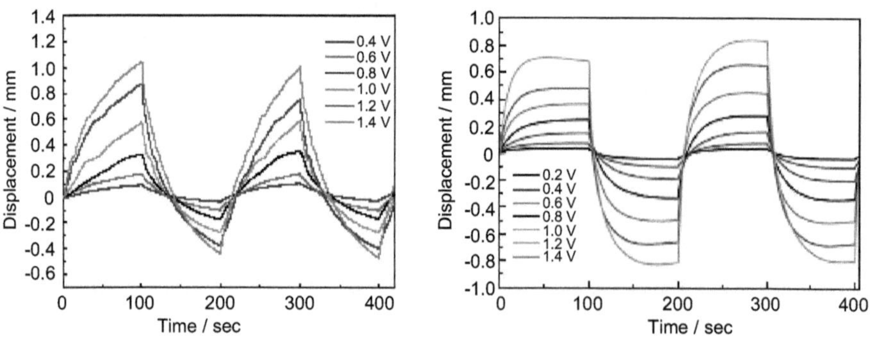

Fig. 27.12 Time dependence of the displacement of an actuator using SMS (left) and P(VDF/HFP) (right) as a polymer network of ion-gel, driven by square-wave form voltage at a frequency of 0.005 Hz

27.4 Conclusion

The polymer ion-gel-actuators using several carbon materials as electrodes have been prepared and demonstrated to be operable at low voltages under atmospheric conditions. The AC actuator exhibits the maximum displacement of ca. 0.35 mm at ±1.5 V (triangle wave form voltage; scan rate = 0.05 V/s). The difference in the surface area and electric conductivity of the carbon materials greatly affects the actuator performance.

Triblock copolymer (SMS) consisting of polystyrene and PMMA was successfully synthesized using the ATRP method. The ion-gel composites were prepared by mixing the obtained SMS and $[C_2mim][NTf_2]$. The polystyrene unit formed a micro-separated structure, as shown by observation of TEM images for ion-gel. Polystyrene domains serve as a physical cross-link point of ion-gel, because polystyrene and PMMA are insoluble and soluble, respectively, to ionic liquids such as $[C_2mim][NTf_2]$. Moreover, the results of DSC measurement support the hypothesis that the PMMA unit of SMS is selectively dissolved in ionic liquid. The ionic conductivity of ion-gels showed 10^{-2} Scm^{-1} at room temperature. In particular, the ion-gels with a higher weight content of ionic liquid had almost identical values to those of $[C_2mim][NTf_2]$. The ion-gel actuator using SMS as a network polymer of ion-gel exhibited improved performance in terms of displacement. It is expected that the ion-gels that control the molecular weight of PMMA and polystyrene and form good interface between the electrode and ion-gel also show better performance of maximum displacement and response speed.

Acknowledgments This research was supported by Grant-in-Aid for Scientific Research (No. 438-17040016 and No. 438-19016014) from the Japan Ministry of Education, Culture, Sports, Science and Technology (MEXT).

References

1. Bar-Chohen Y (2001) Electroactive Polymer (EAP) Actuators as Artificial Muscles. SPIE Press
2. Bar-Cohen Y (2001) MST news, International Newsletter on Microsystems and MEMS. Germany
3. Shahinpoor M (2003) Ionic polymer-conductor composites as biomimetic sensors, robotic actuators and artificial muscles-a review. Electrochim Acta 48: 2343
4. Smela E (2003) Conjugated polymer actuators for biomedical applications. Adv Mater 15: 481
5. Nanjo S, Watanabe M, Asai K, Yokoyama K, Yamamoto M (2004) Development of EAP actuators based on EDLCs using ion gels as electrolytes (I) – Proposal of new actuator driven at low voltage –. Abstract of 71st Annual Meeting of the Electrochemical Society of Japan: 262
6. Honda T, Kokubo H, Watanabe M (2006) Materials for Polymer-Ion-Gel-Actuator. Proceedings of 1st International Symposium on Next-Generation Actuators Leading Breakthroughs. MEXT Grant-in-Aid for Scientific Research on Priority Areas 438: 135
7. Kato Y, Kokubo H, Watanabe M (2007) Response of EAP actuators using ion-gel electrolytes. Abstract of 2nd International Congress on Ionic Liquids: 384

8. Laurer J H, Mulling J F, Khan S A, Spontak R J, Lin J S, Bukovnik R (1998) Thermoplastic Elastomer Gels. II. Effects of Composition and Temperature on Morphology and Gel Rheology. J Polym Sci B Polym Phys 36: 2513
9. Laurer J H, Khan S A, Spontak R J, Satkowski M M, Grothaus J T, Smith S D (1999) Morphology and Rheology of SIS and SEPS Triblok Copolymers in the Presence of a Midblock-Selective Solvent. Langmuir 15: 7947
10. Noda A, Hayamizu K, Watanabe M, Pulsed-Gradient Spin-Echo ^{1}H and ^{19}F NMR Ionic Diffusion Coefficient, Viscosity, and Ionic Conductivity of Non-Chloroaluminate Room-Temperature Ionic Liquids. J. Phys. Chem. B 105: 4603

Chapter 28
Compact MR Fluid Actuator for Human Friendly System

Junji Furusho[1], Takehito Kikuchi[1], Kikuko Otsuki[1], Hiroya Abe[2], Makio Naito[2] and Katsuhiro Hirata[1]

Abstract There is a strong demand for a human-machine-coexistent machine, e.g. a power-assist system or a computer-aided rehabilitation system, in the face of the super-aging society. However, the research and development of the actuator that considers safety for such a robot system is hardly achieved. In this study, we research and develop human-friendly compact MR fluid actuator (MRF actuator) with high safety. In this paper, at first, we describe the concept of the human-machine-coexistent system. Secondly, basic design of the MR actuator with multi-layered disks and narrow-gaps, and development of 5Nm and 40Nm MR actuators are explained. Finally, we discuss development of New MR fluids with nanopaticles.

28.1 Introduction

The MRF actuator consists of three parts; an input part (an actuation part), a reduction part and a transmission part. Any kind of conventional actuators and reduction devices can be used for the input part and the reduction part, respectively. The transmission part is called "MR fluid clutch" (or "MRF clutch" short for magneto-rheological fluid clutch). We can fabricate the MRF clutch with good-controllability of transmission torque by using MR fluid. In this system shown in Fig.28.1, the input part is controlled by very simple method, e.g. constant voltage, at very low speed. On the other hand, the transmission torque can be controlled with high-performance only using the MRF clutch. This system is human-compatible because of the safety based on the robustness of simple control system.

[1] Junji Furusho, Takehito Kikuchi, Kikuko Otsuki, and Katsuhiro Hirata

Graduate School of Engineering, Osaka University

[2] Hiroya Abe and Makio Naito

Joining and Welding Research Institute, Osaka University

28.2 Clutch-driven Actuator Using MR Fluid (MR Fluid Actuator)

The MRF actuator consists of an input part (actuation part), a reduction part (reduction gears) and a transmission part [1] (see Fig.28.1). The transmission part is called "MRF clutch". The MRF clutch is a clutch device using the MR fluid (MRF) as a working fluid. MR fluid is a fluid whose rheological properties can be controlled by applying an electric current [2]. The response of viscous change is very rapid (about several milliseconds) and repeatable in large range. Therefore, the devices using the MRFs have good-controllability, higher torque/inertia ratio and lower inertia than other conventional transmission devices, e.g. powder clutches and electromagnetic clutches. On the other hand, we can make up high performance MRF brake by fixing the input shaft of the MRF clutch [3].

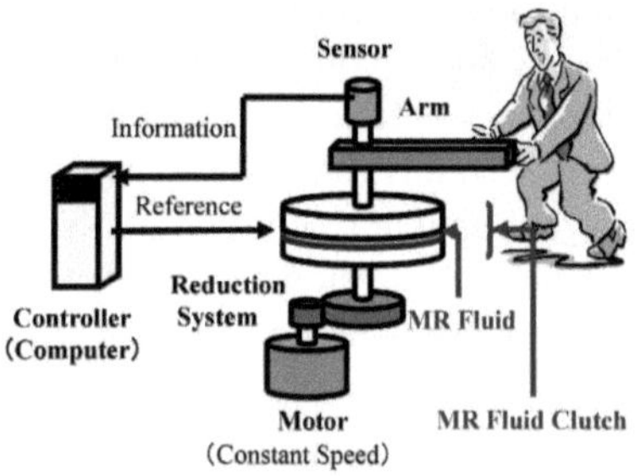

Fig. 28.1 Conceptual illustration of Human-Machine-Coexistent Robot Systems using MR actuators

Figures 28.2 show the basic structure of a conventional disk-type MR fluid clutch. MRF is filled around a disk. Magnetic flux flows through layers of the MRF and the disk shown in this figure. And, the shearing stress generated by a viscous change is transmitted by the disk as an output torque.

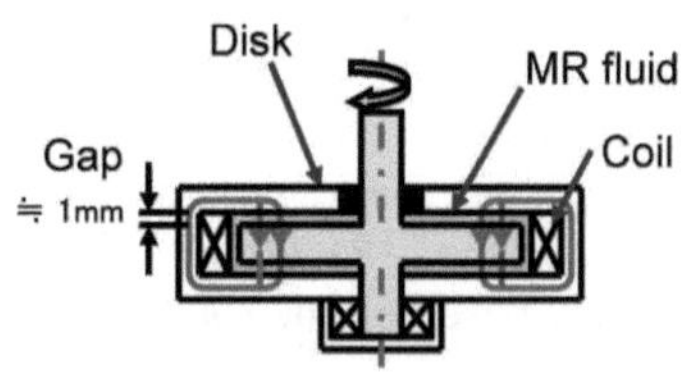

Fig. 28.2 Basic structure of a conventional disk-type MR fluid clutch

In this study, we mainly use the disk-type structure, because it is easy to be built up the multi-layer structure for the amplification of the output torque.

28.3 Basic Structure of Compact MR Fluid Clutch (CMRFC)

A conceptual drawing of the compact MRF clutch device is shown in Fig.28.3. A coil is rolled round the output shaft and it generates the magnetic flux shown by the dashed line in the drawing. Multi-layered disks are fixed on the input shaft and the output shaft, respectively, and the MRF is filled between these disks.

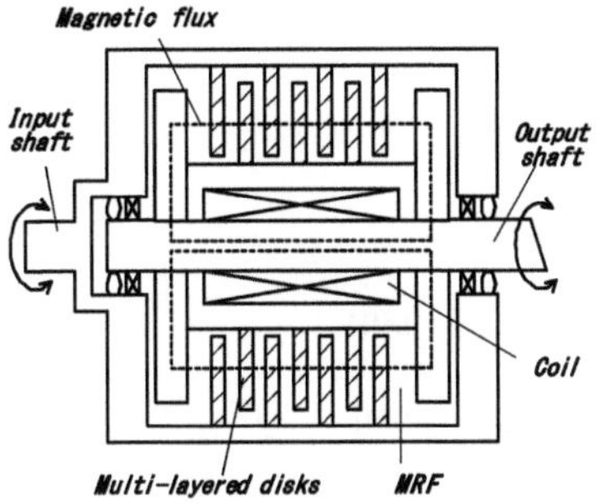

Fig. 28.3 Basic structure of a compact MR device with multi-layered disks

As you see in this figure, multi-layer structure is utilized for the amplification for an output torque. However, to make use of the viscous change of the MRF effectively, we have to impress a magnetic flux of about 0.5~1.0 tesla on the MRF layers. Because MRFs are nonmagnetic materials, its magnetic permeability is very small. Therefore, it is necessary to reduce the total gap of the MRF layers.

From this restriction of the gap-size, total gap of the MRF layers should be less than 0.5~1.0 mm. Therefore, we suggest to utilize narrow-gaps of 10~100μm. The gap-size, the diameter of disks, the number of the layered disks and the number of turns of the magnetic coil should be decided based on the result of magnetic analysis and processing (or assembling) accuracy of the multi-layered disks. At the same time, it is necessary to note not only the structure but also the material, especially MRF itself. Generally speaking, particles whose diameters are more than 1/10 of the distance of a gap are not ensured to be filled sufficiently in it. Because commercial MRFs consist of micron-size particles (1~10μm), there is a possibility of becoming the obstacle of steady performance. In this study, therefore, we also try to develop new MRF with nanoparticle.

28.4 5Nm-Class CMRFC

On the basis of the multi-layer structure mentioned above, we developed 5Nm-Class CMRFC. Figure 28.4 shows appearance and cross-sectional view of it. Table 28.1 shows specification data of the 5Nm-Class CMRFC.

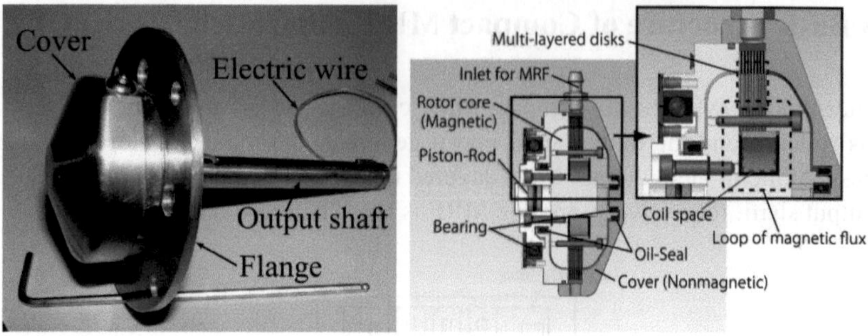

(a) Photo (b) Cross-section

Fig. 28.4 5Nm-Class CMRFC

Table 28.1 Specification of 5Nm-Class CMRFC

Total thickness [mm]	32
Outer diameter [mm]	52
Number of disks	9(input)+8(output)
Number of MRF layer	18
Turning number of coil	191
Idling torque [Nm]	0.15
Max. torque at 1A [Nm]	5.0~6.0
Weight [g]	237

Basic characteristics tests were conducted for this device. During tests, input part of the clutch was fixed and output part was rotated by a servo-motor with constant speed of 1 rad/s. Figure 28.5 (a) shows static torques of the device. White circles show experimental data and black squares show theoretical values. We can calculate estimation torque of the MRF devices with magnetostatic analysis. Figure 28.5 (b) shows a step response of it. A time constant of this device is about 20 milliseconds.

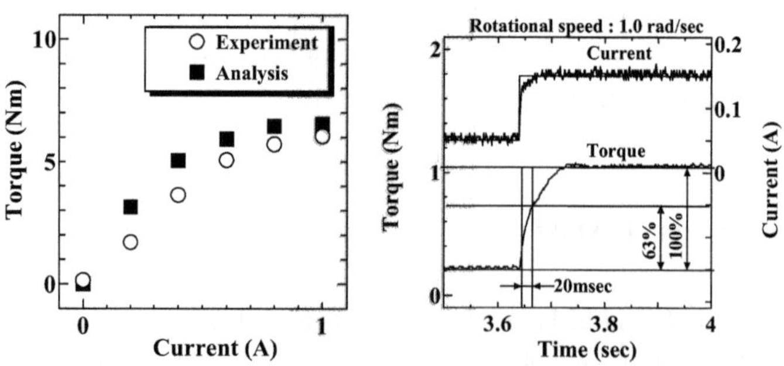

(a) Static torque (b) Step response

Fig. 28.5 Basic characteristics of 5Nm-Class CMRFC

28.5 40Nm-Class CMRFC

On the basis of the multi-layer structure mentioned above, we developed 40Nm-Class CMRFC. Figure 28.6 shows appearance and cross-sectional view of it. Table 28.2 shows specification data of the 40Nm-Class CMRFC.

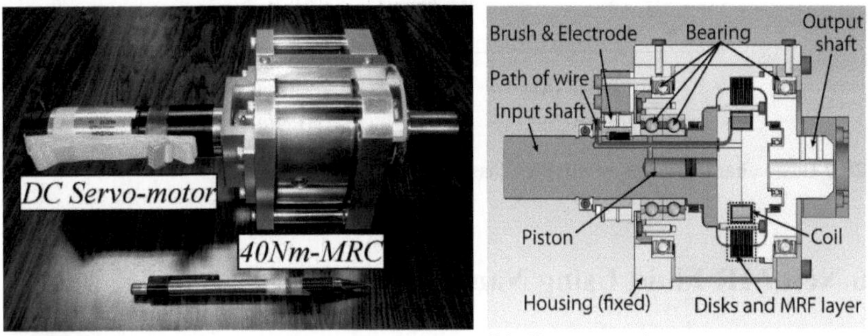

(a) Photo (b) Cross-section

Fig. 28.6 40Nm-Class CMRFC

Table 28.2 Specification of 40Nm-Class CMRFC

Total length [mm]	120
Outer width [mm]	96
Number of disks	18(input)+17(output)
Number of MRF layer	36
Turning number of coil	200
Idling torque [Nm]	1.0
Max. torque at 1A [Nm]	40
Weight [g]	2200

Basic characteristics tests were conducted for this device. During tests, input part of the clutch was fixed and output part was rotated by a servo-motor with constant speed of 1 rad/s. Figure 28.7 (a) shows static torques of the device. Black circles show experimental data and white squares show theoretical values. Figure 28.7 (b) shows a step response of it. A time constant of this device is about 36 milliseconds.

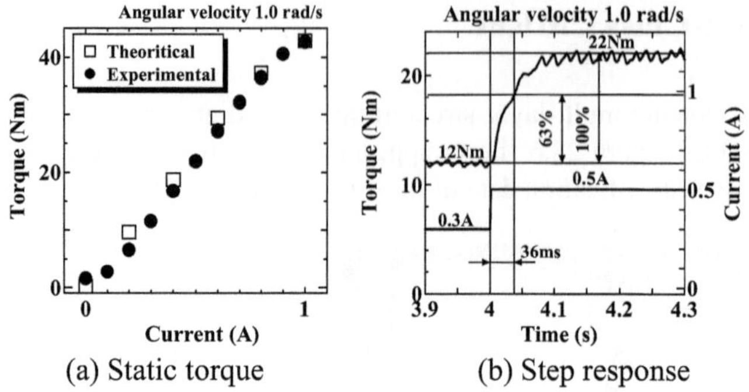

Fig. 28.7 Basic characteristics of 40Nm-Class CMRFC

28.6 New MR Fluid Using Nano-size Particles

The majority of existing MRF is composed of micron-sized Fe particles suspended in a nonmagnetic carrier fluid. The particles may lead to unwanted abrasion of the components in contact with the fluid. Also, they are susceptible to settling in the absence of frequent mixing due to predominant gravity forces. Nanoparticle dispersed fluid, nano-MR fluid, would be desirable.

In this study, Fe nanoparticles were synthesized by hydrogen arc plasma method, and their surfaces were modified to be hydrophobic for enabling higher solid loading into a carrier fluid. The MR performance was also examined.

Figure 28.8 shows a TEM image of the synthesized Fe nanoparticles. The particles had a spherical shape with a chain-like structure. The specific surface area was about 9m2/g of which corresponding diameter was about 90nm. The hydrogen arc plasma was invented in Japan [4], which possesses the advantage of higher productivity compared to other method. In this method, Fe bulk metal was melted and evaporated by the high temperature of the plasma, and the nanoparticles were formed through homogeneous nucleation. It is noted that the crystallinity of the present Fe nanoparticles was quite high, as shown in Fig.28.9.

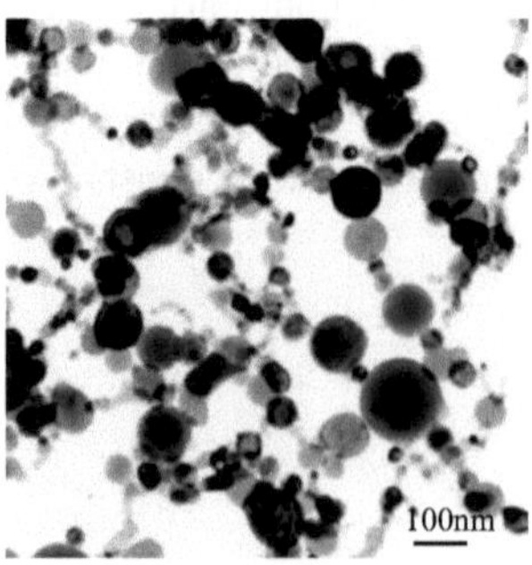

Fig. 28.8 TEM image of the present Fe nanoparticles

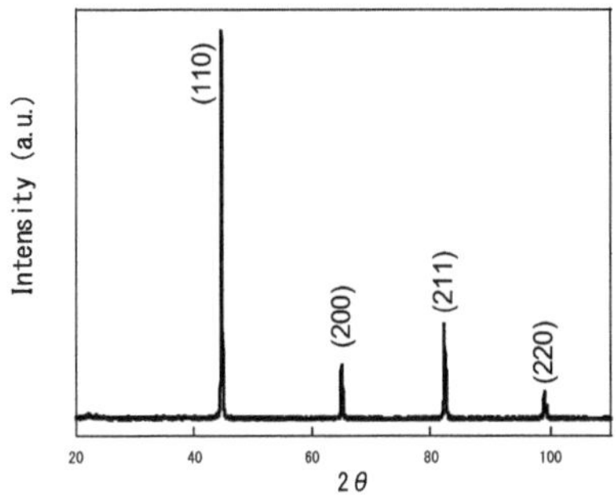

Fig. 28.9 XRD pattern of the present Fe nanoparticles

Figure 28.10 shows the photograph of the Fe nanoparticle-dispersed system. For this preparation, the surface modification must be needed for enhancing affinity to the carrier fluid of silicone oil. In as-received state, natural oxidation film with thickness of a few nanometers was formed on Fe nanoparticle, giving to hydrophilic surface. In our investigation, the surface was well changed to be hydrophobic after reaction with silan coupling agent. Up to now, the solid volume fraction of 10% was obtained. The hydrophobic surface is also preferable in term of oxidation resistance.

Fig. 28.10 Fe nanoparticle-dispersed system

28.7 Conclusion and Future Work

In this paper, we developed the MRF actuators with the multi-layered disks and narrow-gaps (several ten micrometers) for a human-machine-coexistent system. Static and step torque of these brakes were examined. This brake has good stability and rapidity of its torque characteristics. For the next step, we are going to do advanced experiments, for example, ramp, linearized torque, sine-waved torque and so on. Moreover, to achieve the new actuator that can be used in a practical use, further improvements are necessary for the MRF, design of the magnetic circuit and processing method and assembly method of basic structure of MRF clutch.

Acknowledgments This work was financially supported by a Grant-in-Aid for Scientific Research on Priority Areas, No.438 "Next-Generation Actuator Leading Breakthroughs".

References

1. Lin J, Asaoka H, Sakaguchi M et al (2000) Applying Study Connected with Development of New Actuator Using MR Fluid. Proceedings of the 2000 Japan-USA Flexible Automation Conference: 413-416
2. Carlson J D and Jolly M R (2000) MR fluid, foam and elastomer devices. Mechatronics 10: 555-569
3. Kikuchi T, Furusho J, Yamaguchi Y et al (2006) Design of the high-performance MR brake and its characteristics. Proceedings of the 10th International Conference on ER Fluids and MR Suspensions: 132
4. Cui Z, Zhang Z, Hao C et al (1998) Structures and properties of nano-particles prepared by hydrogen plasma method. Thin solid films 318: 76-82

Chapter 29
Development of Actuator Utilizing Hydrogen Storage Alloys

Masayuki MIZUMOTO[1], Takeshi OHGAI[1] and Akio KAGAWA[1]

Abstract The actuators utilizing hydrogen storage alloys (HSAs) have been developed. In order to convert a volume expansion of HSA accompanied by hydrogen absorption into a bending motion, the actuators have a bimorph structure which consists of sheet-shape HSA and non-HSA. Pd-Ni and V-Ti based HSAs, which have high pulverization resistance on hydrogen absorption - desorption cycles and enough ductility to form into sheet, are used for the actuators. The shape change behavior of the V-Ti alloy actuators was improved by the sputtered Pd layer which would act as a catalyst and a protective layer against oxidization. Pd-Ni alloy actuators exhibited cyclic bending motion accompanied by hydrogen absorption-desorption cycles. The shape change behavior of the actuators could be controlled by controlling the hydrogen pressure. By forming the sample shape into the "L" shape, the rotational motion could be achieved without modifying the basic bimorph structure of the actuator.

29.1 Introduction

Hydrogen storage alloys (HSAs) are the alloys which can absorb and desorb hydrogen reversibly. Nowadays some actuators utilizing this unique characteristic of the HSAs are investigated and some have been put into practical use. Most of these actuators utilize the pressure of hydrogen gas desorbed from the HSA [1]. In general, to control the motion of actuators utilizing the pressure of hydrogen gas discharged from HSAs, thermal and pressure control systems are required. These control systems result in the difficulty in making the actuators compact due to large and complex devices such as heater, temperature controller and pistons. On the other hand, it is known that the HSAs exhibit a large volume expansion as shown in Fig.29.1, about 10~25%, accompanied by the hydrogen absorption of the

[1] Masayuki MIZUMOTO, Takeshi OHGAI and Akio KAGAWA

Department of Materials Science and Engineering, Nagasaki University alloys. This volume expansion is much larger than that arising from the thermal expansion of metals (the thermal expansion coefficient is of the order of 10^{-5} /K). In addition, since the volume expansion arising from the hydrogen absorption of the alloys results from the phase transformation of metal into hydride, it is expected that the generation of powerful driving force can be obtained as an actuator. Honjo et al. have reported the properties of the actuator utilizing the volume expansion of $LaNi_5$ base HSAs [2, 3]. However, the response and the shape change behavior of the actuator were not sufficient because fine $LaNi_5$ powder was embedded in polymer matrix and the volume expansion of fine $LaNi_5$ powder would not utilized effectively. Thus it was found that sheet-shaped HSAs would be desirable for the application to the actuators which utilize the volume expansion of the HSAs. One of the present authors reported that vanadium-base and palladium-base HSAs had some excellent properties such as high pulverization resistance on hydrogen absorption - desorption cycles, easy activation, enough ductility to form into foils, in comparison with intermetallic compound alloys such as $LaNi_5$ alloy [4, 5]. In addition, hydrogen absorption-desorption plateau pressure (plateau pressure in the pressure - composition isotherm) of the alloys would be set around 1 atm at room temperature by controlling the alloy composition as shown in Fig.29.2. This means that the alloys can absorb and desorb hydrogen without complex thermal control devices, resulting in a possibility of the development of simple, compact and powerful actuator. From these facts, it is expected that palladium-base and vanadium-base HSAs are quite suitable for the micro-power-actuator.

To develop the actuator utilizing the palladium-base and vanadium-base HSAs, the actuators which have bimorph structure consisting of Pd-Ni alloy or V-Ti alloy and Cu-plating were fabricated. The shape change behavior of the actuators was investigated and the motion control technique for the actuators was developed.

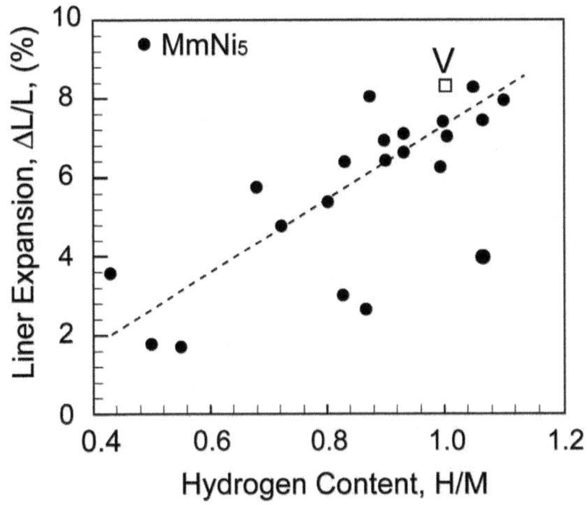

Fig. 29.1 Relationship between linear expansion of HSAs and hydrogen content

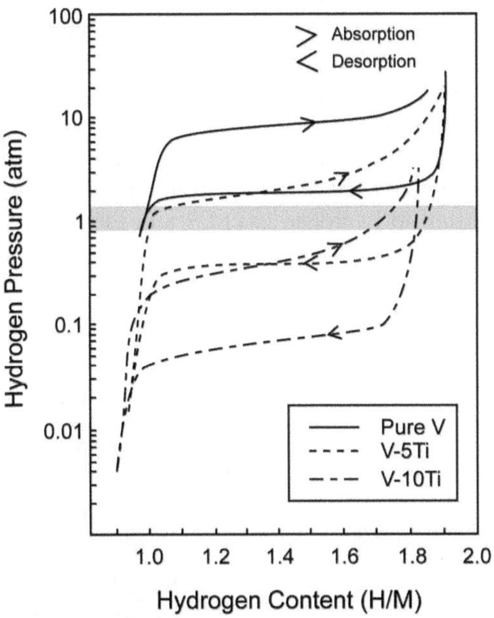

Fig. 29.2 Effect of titanium concentration on the hydrogen absorption-desorption plateau pressure of V-Ti alloys at 293K

29.2 Fabrication Process of Bimorph Structure Utilizing HSA

The V-3, 5at%Ti alloys and Pd-11at%Ni alloys were prepared by arc-melting pure materials (vanadium > 99.9wt%, titanium > 99.99wt%, palladium > 99.9wt% and nickel > 99.9wt%) under an argon gas atmosphere. The alloys were remelted several times and were annealed for 15 hours at 1073K in vacuum for homogenization. After repeated rolling and annealing for stress relieving, the alloys were rolled into thin plate with 20μm or 40μm thickness. The alloys were cut into rectangular sheet samples of 1.25~5mm x 50mm. It is considered that the actuator should have a bimorph structure consisting of HSA and non-HSA to convert the volume expansion accompanied by hydrogen absorption into the bending motion of the actuator. Thus, pure copper was plated on one side of the rectangular sheet sample by electro-plating. The thickness of Cu-plating was equal to that of rectangular sheet sample. After plating, the actuator was held at 1073K for 2 hours in vacuum to join the interface between the alloy and Cu-plating by diffusion bonding. In the case of the actuator utilizing V-Ti alloys, the surface of the V-Ti alloy side of the actuator was coated with sputtered film of pure palladium with about 1.0×10^2 nm thickness to avoid the oxidation of the surface of the V-Ti alloy. In addition, the sputtered pure palladium layer would be expected to act as a catalyst to convert the hydrogen molecules into hydrogen atoms, resulting in an improvement of the response of the actuator.

The actuator was set in the apparatus shown in Fig.29.3 for in-situ observation of the shape change behavior and was heat-treated at 673K for 3 hours in vacuo followed by an exposure to hydrogen gas of 3 atm for activation. Shape change behavior of the actuator was observed by CCD camera through a viewport. To measure the displacement of the top end of the actuator precisely, KEYENCE laser displacement sensor, LK-G155 was used to monitor the horizontal displacement of the top end of the actuator.

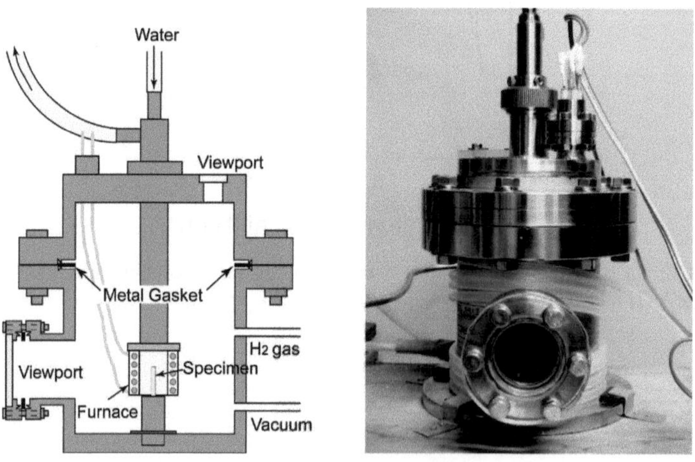

Fig. 29.3 Shape change measurement apparatus

29.3 Effect of the Width of Bimorph Structure Specimen on Its Bending Behavior

It is easily expected that the width of bimorph structure would influence the bending motion of the actuator due to the deformation of the actuator in width direction. In order to investigate the effect of the width of bimorph structure on the bending behavior, the actuator width was varied from 1.25 to 5mm. Fig.29.4 shows the appearance of the actuator with various widths after hydrogen absorption. In this figure, left side of the actuator is Cu-plating and right side is V-Ti alloy. Each actuator bent clearly to the Cu-plating side. This indicates that the volume expansion of V-Ti alloy accompanied by hydrogen absorption was converted to a bending motion. The actuator with 5mm wide exhibits the largest deformation as shown in Fig.29.4 (c). However the deformation of the actuator in the width direction was also large, resulting in the generation of the cracks and spalling of Cu-plating. On the other hand, when the actuator width was lower than 2.5mm, no cracks and spalling of Cu-plating was observed. Thus, it is suggested that the actuator width should be lower than about 2.5mm.

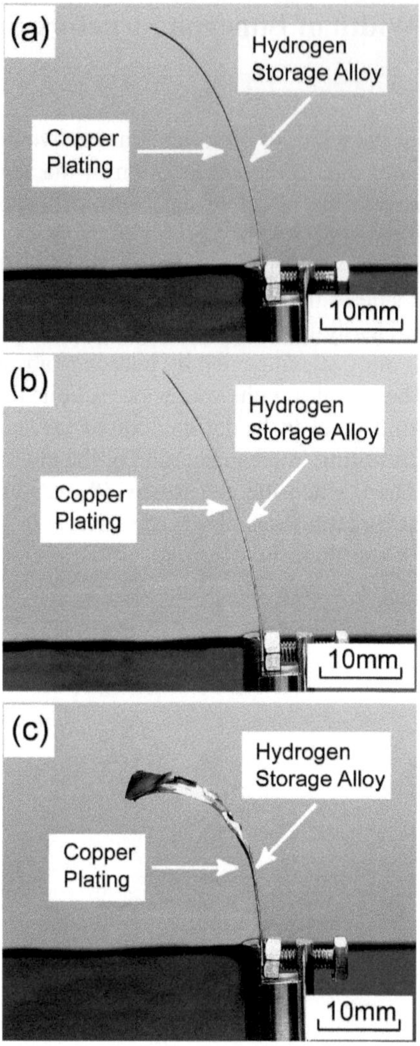

Fig. 29.4 Effect of the specimen width (*W*) on the shape change behavior of the bimorph structure specimens. (a) *W*=1.25 mm, (b) *W*=2.5 mm, (c) *W*=5 mm

29.4 Effect of Sputtered Pure Pd Layer on the Bending Behavior of V-Ti Alloy Actuators

V-base alloys are easily oxidized. To avoid the oxidation of V-Ti alloy, the surface of the V-Ti alloy side of the actuator was coated with sputtered film of pure palladium. In order to investigate the effect of the sputtered pure palladium layer on the response and the bending behavior of V-Ti alloy actuator, the bending behavior of V-5at% Ti alloy actuator with or without the sputtered layer was investigated The results are shown in Fig.29.5. On hydrogenation, both V-5at%Ti alloy actuators started to bend to the Cu-plating side and the bending speed of both actuators was almost constant during hydrogen absorption. However, the actuator without the sputtered layer seemed to deform irregularly. This irregular bending behavior was attributable to incomplete removal of the surface scale on the activation process. On the other hand, the actuator with the sputtered layer deformed uniformly and smoothly. The actuator with the sputtered layer exhibited higher maximum displacement of the top end of the actuator which was 45% larger than that of the actuator without the sputtered layer. These improvements in the bending behavior were attributable to the sputtered pure Pd layer which would act as a catalyst and a protective layer against oxidization of the V-Ti alloy surface. When the inside of the apparatus was evacuated up to 10^{-5} Torr, the displacement recovered slightly but could not recover to the original shape. The following equation gives a bending strain, ε as a function of distance from a neutral plane, η and the radius of curvature at the mid point, ρ.

$$\varepsilon = \frac{\eta}{\rho} \tag{29.1}$$

Calculated bending strain was 1.4×10^{-3} and that was much higher than a yield strain of pure vanadium, about 8.0×10^{-4}. Thus, it is considered that this irrecovered deformation of the actuator would be attributable to the plastic deformation introduced in the actuator during the hydrogen absorption. This problem would be solved by controlling the amount of hydrogen absorbed into HSAs.

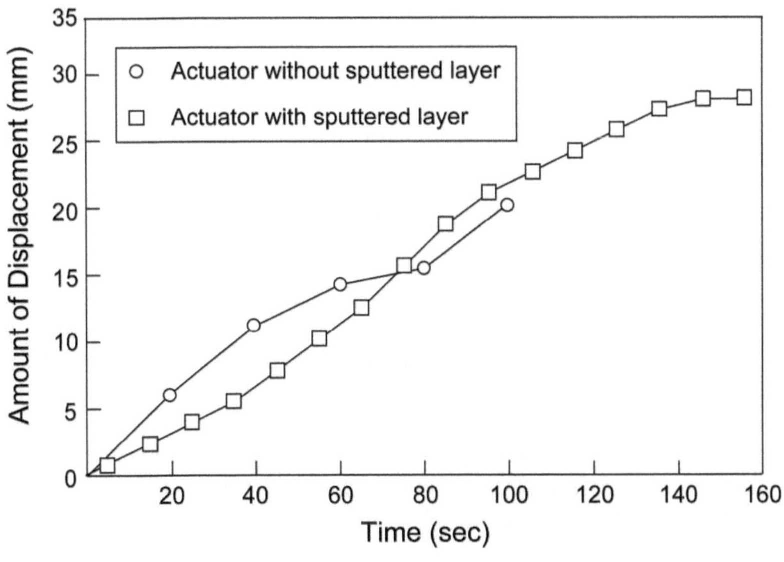

Fig. 29.5 Bending behavior of a V-5at%Ti alloy actuators with or without sputtered pure Pd layer on hydrogen absorption

29.5 Bending Behavior of Pd-Ni Alloy Actuators

Palladium-base alloys have a good resistance to pulverization on hydrogenation, similar to vanadium alloys, and the alloy surface was easily activated for hydrogen absorption. By alloying with nickel, the plateau pressures for hydrogen absorption and desorption can be controlled and Pd-11at%Ni alloy has the plateau pressures of nearly 1 atm[6]. In addition, the actuator showed a recover of displacement in vacuo after the hydrogen-absorbed actuator was held in air for 24 hours. Figure 29.6 shows a change in the displacement of the top end of the Pd-11at%Ni alloy actuator on repeated hydrogen absorption – desorption cycles. In the experiments, hydrogen absorption was stopped on the displacement of 7.5mm, where the displacement remained in the region of the elastic deformation. In the first cycle, the bending displacement of 7.5mm required about 6 minutes after hydrogen introduction. The displacement completely recovered in 45 minutes after evacuation up to 10^{-5} Torr. In comparison with absorption process, the bending speed was greatly reduced. Present authors have reported that a response time of bending motion for Pd-11at%Ni alloy actuator decreased with increasing the number of hydrogen absorption-desorption cycles[6]. This delay of response is considered to be caused by Up-hill diffusion [7, 8], i.e., hydrogen diffuses toward a higher concentration portion of the HSA. It is known in the earlier work that a reduction in thickness of the alloy is effective to reduce effects of Up-hill diffusion and thereby result in an increase of bending speed of the actuator. Therefore, the thickness of the HSA and the Cu-plating were changed from 40μm to 20μm. Figure 29.7 shows a variation in displacement of the top end of the actuator with time on hydrogen absorption in Pd-11at%Ni alloy. The actuator with HAS thickness and Cu-plating thickness of 40μm shows a displacement of 20mm in ten minutes, while the actuator with HAS thickness and Cu-plating thickness of 20μm shows a displacement of 23mm in one minute, showing 10 times faster bending speed. The following well-known equation gives a diffusion time, *t*, as a function of diffusion coefficient, *D* and diffusion distance, *L*.

$$t \propto \frac{L^2}{D} \tag{29.2}$$

Since a diffusion coefficient is to be equal in both the case at room temperature, the diffusion time for the HSA with 40μm thickness should be 4 times longer than that for the HAS with 20μm thickness. However, the experiment reveals more than 10 times faster bending motion in the latter actuator. This may be caused by a reduction of the effect of Up-hill diffusion which disturbs hydrogen diffusion in a distorted host lattice. By thinning the HSA, the gradient of hydrogen concentration in the direction of sample thickness should be lowered and the distortion of the lattice may be reduced, and hence an obstruction for hydrogen diffusion is weakened.

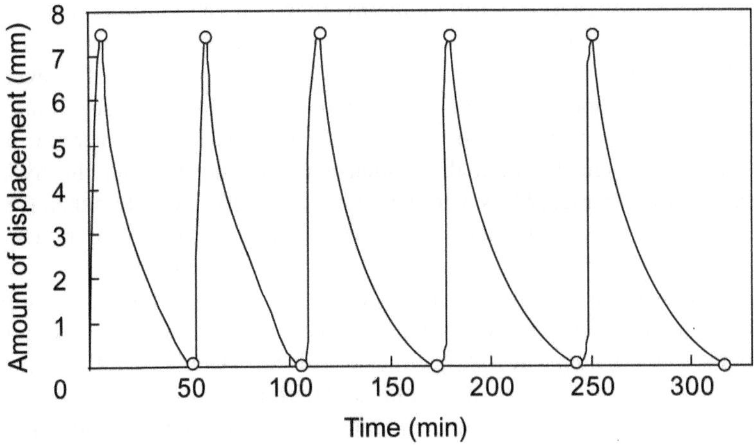

Fig. 29.6 Relationship between displacement and time on cyclic hydrogenation of Pd-11at%Ni alloy actuator

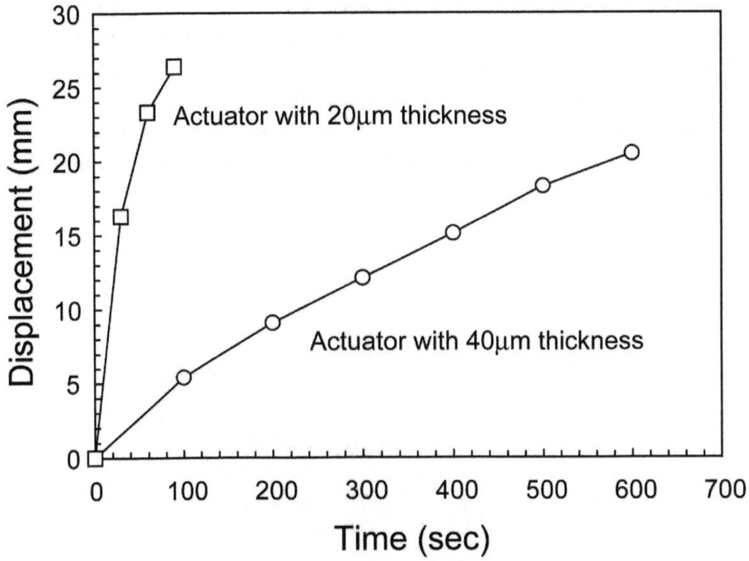

Fig. 29.7 Effect of actuator thickness on bending behavior of Pd-11at%Ni alloy actuator

29.6 Control of Bending Behavior of the Actuator

In order to develop the motion control technique for the actuator utilizing HSA, the bending behavior of Pd-11at%Ni alloy actuator at different hydrogen pressure was observed, since it is expected that the amount of hydrogen absorbed into the

HSA can be controlled by controlling the hydrogen pressure. In this work, hydrogen pressure from 0.4 to 1.6 atm was applied by 0.2 atm increments. Figure 29.8 shows the bending behavior of Pd-11at%Ni alloy actuator observed at every 0.2 atm. The actuator stopped bending after about 600sec from the hydrogen introduction. The displacement of the top end of the actuator seemed to increase irregularly. The relationship between the displacement increments and the hydrogen pressure is shown in Fig.29.9. When the hydrogen pressure was less than 0.8 atm, the displacement increments increased rapidly with increasing the hydrogen pressure and the maximum displacement increment was observed at 0.8 atm. When the hydrogen pressure was higher than 0.8 atm, the displacement increments decreased gradually. This variation in the displacement increments is considered to correspond to the amount of hydrogen absorbed into HSA. Present authors have reported that Pd-11at%Ni alloy has a plateau pressure around 0.8 atm and a small hysteresis on hydrogen absorption and desorption by alloying with nickel[6]. From this fact, it is expected that Pd-11at%Ni alloy can absorb most of total amount of hydrogen around 0.8 atm. In addition, the volume expansion of HSAs accompanied by hydrogen absorption would depend on the amount of absorbed hydrogen. Thus, it is considered that the Pd-11atNi alloy actuator exhibited large displacement increments near the plateau pressure, around 0.8 atm. On the other hand, at pressures over 0.8 atm, the increments in the amount of absorbed hydrogen would decrease, resulting in the decrease in the rate of hydrogen absorption and the displacement increment. From these results, it is suggested that shape change behavior of actuators utilizing HASs can be controlled by controlling hydrogen pressure.

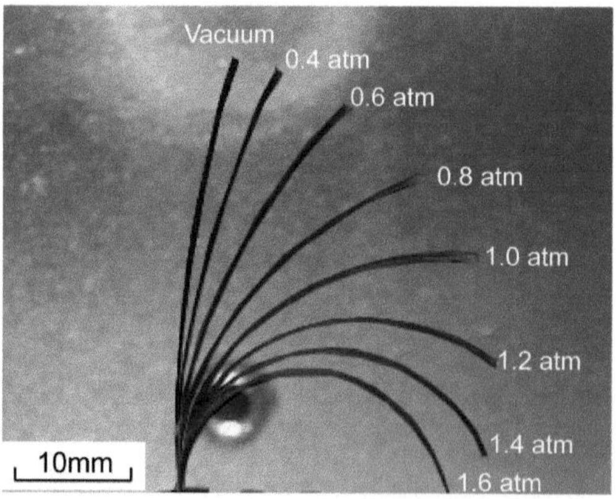

Fig. 29.8 Bending behavior of Pd-11at%Ni alloy actuator observed at every 0.2 atm

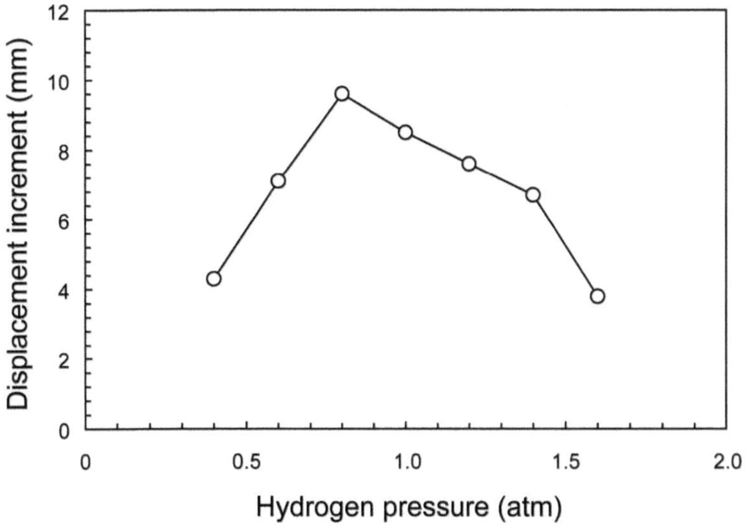

Fig. 29.9 The relationship between the displacement increments and the hydrogen pressure in Pd-11at%Ni alloy actuator

29.7 Rotational Motion of Actuators

Figure 29.10 shows a schematic illustration of Pd-11at%Ni alloy actuator which exhibits rotational motion. In order to convert the volume expansion of HSA into rotational motion of the actuator, the bimorph structure was flexed to a right angle along the longitudinal direction as shown in the figure. At the flexed corner of the actuator, the bending motion of the wide side is expected to be counterbalanced by the deformation of narrow side, which would suppress the bending motion along the longitudinal direction. On the other hand, the deformation in the width direction of the structure would not be constrained, resulting in the generation of torsional deformation, i.e., rotational motion. Figure 29.11 shows the appearance of the Pd-11at%Ni alloy actuator before and after hydrogen absorption showing a rotational motion. When the inside of the apparatus was evacuated up to 10^{-5} Torr, the shape of the actuator recovered to the original shape. This result suggests that various motions of the actuators such as bending and rotational motions can be exhibited by fabricating the bimorph structure into the appropriate shape.

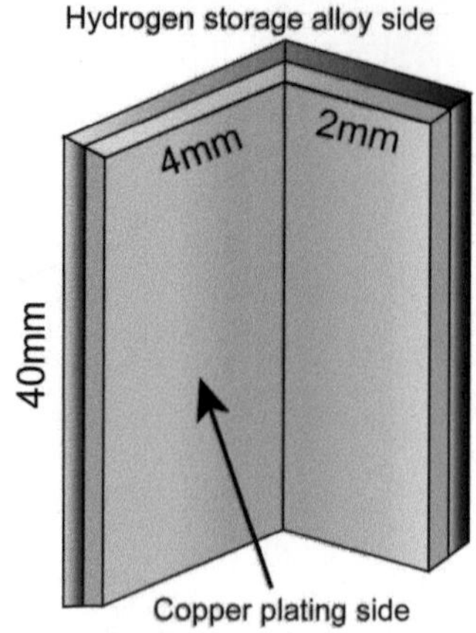

Fig. 29.10 schematic illustration of Pd-11at%Ni alloy actuator which exhibits rotational motion

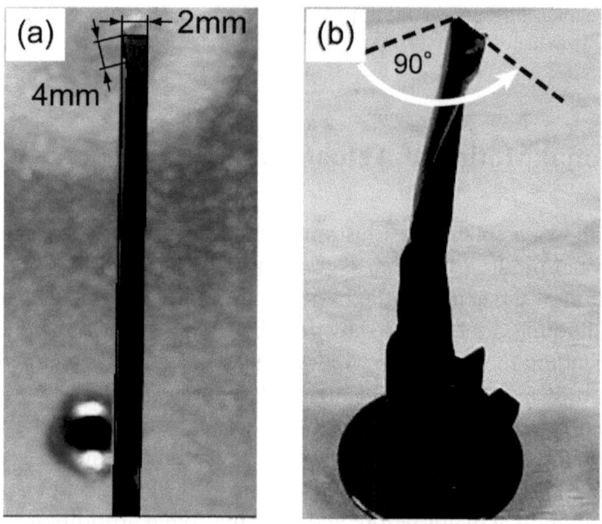

Fig. 29.11 Appearance of the Pd-11at%Ni alloy actuator which would exhibit rotational motion. a) before and b) after hydrogen absorption

29.8 Summary

The actuators utilizing the volume expansion of HSAs accompanied by the hydrogen absorption can be developed. Pd-Ni alloys and V-Ti alloys are suitable HSAs for the actuators because they have some excellent properties such as high pulverization resistance on hydrogen absorption - desorption cycles, easy activation, and enough ductility to form into foils. The actuators have a bimorph structure which consists of sheet-shape HSA and non-HSA. In order to convert the volume expansion of HSA into the motion of the actuator efficiently, it is necessary to join the interface between HSA and non-HSA by diffusion bonding. The actuators utilizing HSAs show a good response and exhibit a large displacement in bending motion. The shape change behavior of the actuators can be controlled by controlling the hydrogen pressure. By fabricating the bimorph structure into the appropriate shape, the rotational motion can be obtained without modifying the basic structure of the actuator.

Acknowledgments This research was partially supported by the Ministry of Education, Culture, Sports, Science and Technology, Grant-in-Aid for Priority areas, 19016019, 2007.

References

1. Tamura H et al (2003) Hydrogen Storage Alloys. NTS Inc., Tokyo
2. Honjo T, Yabe H, Tsubuteishi S, Uchida H, Nishi Y (2003) Chemical Composition Dependence of the Properties of a LaNix (x=3.8 to 6.5) Hydrogen Storage alloy Film Actuator. J. Japan Institute of Metals vol. 67 no. 4:145-148
3. Honjo T, Uchida H, Matsumura Y, Nishi Y (2004) Improvement of Response Time of Thin Film Unimorph Structure using Hydrogen Storage alloy by Platinum Surface Treatment. J. Japan Institute of Metals, vol.68, no.2:58-61
4. Kagawa A (1995) Absorption of Hydrogen by Vanadium-Titanium Alloys. Reports of the Faculty of Engineering, Nagasaki University, vol.25, no.47:223-239
5. Kagawa A, Ono E, Kusakabe T, Sakamoto Y (1991) Absorption of Hydrogen by Vanadium-rich V-Ti-based Alloys. J. Less-Common Met., vol.172-174:64-70
6. Mizumoto M, Ohgai T, Kagawa A (2009) Bending behavior of Cu-plated Pd-Ni alloys ribbon driven by hydrogenation. J. Alloys Compd., vol.482:416-419
7. Lewis FA, Baranowski B, Kandasamy K (1987) Uphill diffusion effects induced by self-stresses during hydrogen diffusion through metallic membranes. J. Less –Common Met., vol.134:27-31
8. Baranowski B (1989) Stress-induced diffusion in hydrogen permeation through $Pd_{81}Pt_{19}$ membranes. J. Less –Common Met., vol.154:329-353

Chapter 30
Development of New Actuators for Special Environment

Toshiro HIGUCHI[1] and Toshiyuki UENO[2]

Abstract Here, we describe new actuators for special temperature environment (extremely high and low). One is thermally driven actuator based on self-excited oscillation. The actuator consists of a metal disk with weigh and rolled bimetal which the bending deformation causes torque for the oscillation. The self-excited oscillation is easily available using ceramic heater in electrical furnace and can be generated not only bimetal but also single metal ribbon, which the results indicate that the actuator can be useful in extremely high temperature. In temperature range between -273 and 200°C, a magnetostrictive actuator using Iron-Gallium alloy (Galfenol) is suitable. Galfenol is an iron-based magnetostrictive material with magnetostriction of more than 200 ppm, Young's modulus of 70GPa and relative permeability of 70. The advantages of the actuator are simple, rigid, easily miniaturized and low voltage driving. Here, a vibrator and positioning device using machined Galfenol demonstrate strong potential of the material for micro applications. The displacement of the actuator was also verified under 500°C which the limitation is determined by Curie temperature of the material.

30.1 Introduction

Actuator endurable and functional in extremely high and low temperature environments (-200 to 1000 °C) is essential for the advancement of science and technology. However, conventional actuators such as electromagnetic (motor) and piezoelectric (PZT) types are not available due to the low Curie temperature, low lubrication, melting of the adhesive and isolation, and large thermal-structural deformation. Our solution in the development is to utilize thermal expansion or

[1] Toshiro HIGUCHI

Department of precision machinery engineering, University of Tokyo

[2] Toshiyuki UENO

School of electrical and computer engineering, Kanazawa University magnetostrictive material of high Curie temperature. So far, we have investigated micro actuator photothermally driven by Laser pulse [1], however, found that it is not practical because of low energy conversion and difficulty of driving condition and positioning alignment (Laser) in micro scale [2]. Here, we propose thermal driven actuator based on self-excited oscillation using bending of bimetal, simple with a few components, and describes the experimental results to discuss the possibility for high temperature usage. Micro magnetostrictive (Fe-Ga alloy) actuator advantages of low voltage driving, high mechanical strength and wide temperature operation is also discussed.

30.2 Thermally Driven Actuator

30.2.1 Configuration and Principle

Thermal actuator has two configurations, one is rolling and the other is oscillation type [3]. The rolling type is a ring of bimetal strip. (two metals layers with different thermal expansion coefficient bonded together), which the outer layer has larger coefficient. When it is placed and heated on a plate of high temperature (more than 50°C higher than the environment), it rotates and rolls to one direction. The oscillation type is a metal disk with a weight (the center of mass is located below the center) bound with a bimetal (the outer layer has larger coefficient). The disk rolls periodically right and left and does not continue the oscillation because of damping by friction and air, however it on the heated plate (> 50°C) does continue the oscillation as shown in Fig.30.2. That motion is self excited oscillation. The principle how the torque for the rotation is generated by the heated plate is explained as follows (see Fig.30.3). Normally, the thermal conduction resulting temperature distribution in bimetal becomes asymmetric due to imperfect shape and contact condition when the actuator is put on the plate. If the temperature of left side is higher than the right and the thermal deformation (bending inside) of the left is larger than the right, the summation of the deformation yields the torque which makes the ring to rotate to right direction. The torque is also generated continually during the rotation because the temperature just behind the contacting (back side to moving direction) is higher than that of the forward (the front side before the contact is cooled by heat dissipation). Figure 30.3 right shows the thermal imaging of the rolling bimetal (going to right). It was observed that the temperature of back side (after contact) to the moving direction is higher and we felt thrust force to right direction at the time.

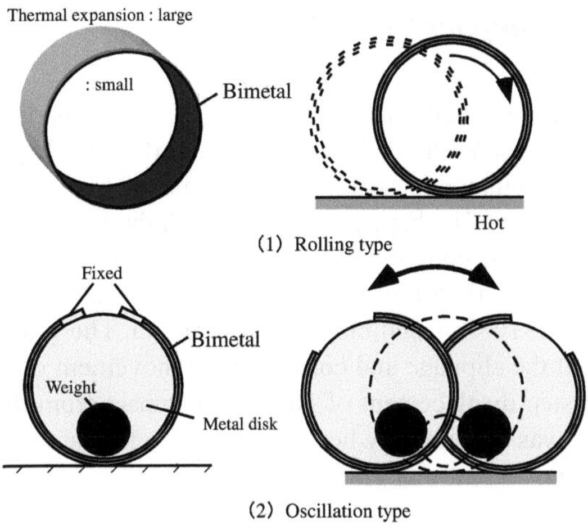

Fig. 30.1 Configuration of actuator, rolling (top) and oscillation type (bottom)

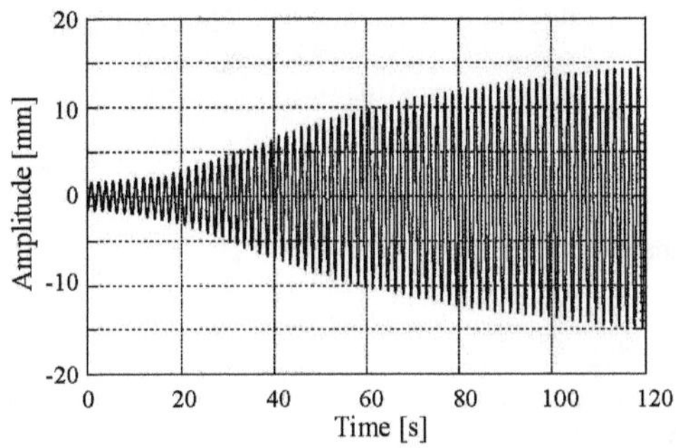

Fig. 30.2 Self-excited oscillation of disk with bimetal on heated plate (70°C)

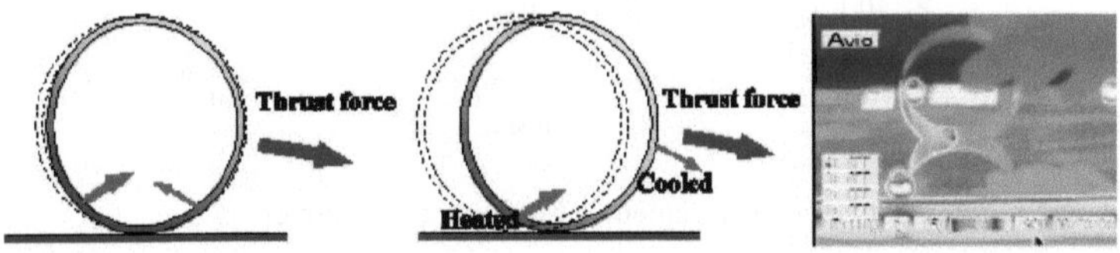

Fig. 30.3 Principle of generating torque (left) and thermal imaging of rolling bimetal. (right). The asymmetric temperature distribution seems to generate torque for one directional rotation

30.2.2 Thermal Impact Drive Actuator

Thermal impact drive actuator as shown in Fig.30.4 is constructed using the principle [4]. The actuator is the disk with weight and bimetal, placed on L shaped mover. When the actuator is heated on the plate of high temperature, the disk oscillating hit the bars of L shape in a cycle, which the impact force generates step displacement. Prototype consists of the disk of brass (40 mm diameter and 20 mm width) with bimetal of 0.15 mm thickness consists of 22Ni-4Cr-Fe and 36 Ni-Fe (bending ratio 14.5×10^{-6}/K), and mover of Aluminum. The gear of rack and pinion is used to prevent the slippage and constraint the movement of the disk. As shown in Fig.30.5, the step displacement of 10 μm per impact from the oscillation of 15 mm amplitude was obtained when the plate temperature of 140 °C. The continuous step movement was verified.

Fig. 30.4 Thermal impact drive actuator

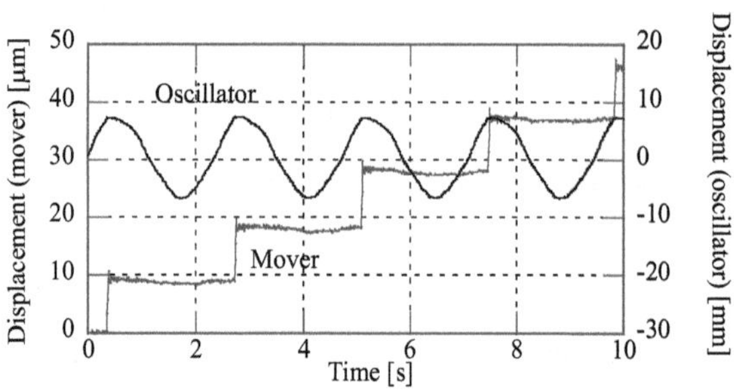

Fig. 30.5 The relation between displacements of oscillator and mover

30.2.3 Temperature Characteristics

The thermal actuator based on the self excited-oscillation is considered endurable for high temperature usage (<1000°C) in electric furnace because of the following reasons.

1. Simple with a few components which the thermal deformation does not influence on the movement.
2. Heat source is readily available by ceramic heating device inside chamber. The electrical energy for the drive and control is supplied via heat-resistive wire.
3. The movement based on impact is less affect by the variation of sleeve and thermal deformation.

The oscillation in high temperature is investigated by experimental setup shown in Fig.30.6. A disk with bimetal is placed and heated on a ceramic heater (Joule heating using Nichrome wire) fixed in the chamber of electrical furnace and its movement is observed through quartz glass window. (The vibration would be possibly measured by Laser displacement sensor). The temperate of the heater monitored by thermal coupling is controlled by current from temperature controller. So far we have verified that the oscillation is maintained under 300°C with heater temperature of 400°C. In this case, iron disk with bimetal was used as oscillator. Above the temperature, the Nichrome wire was cut by melting.

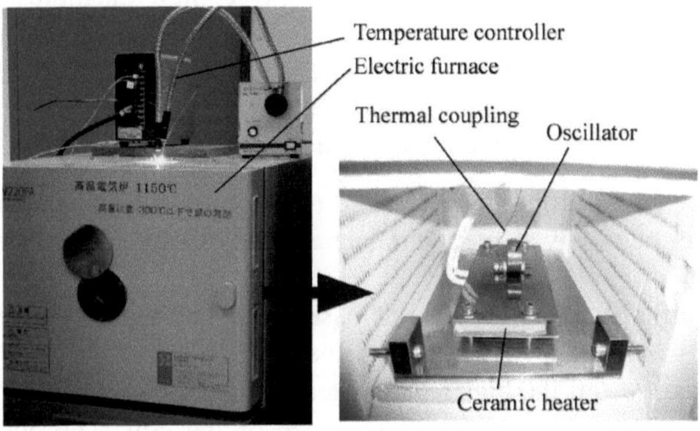

Fig. 30.6 Experimental setup for high temperature environment. The self-excited oscillation was verified under 300°C

We found that the oscillation is excited using not only bimetal but also single metal. Table 30.1 compares the lowest temperature of plate and disk when the oscillation was observed. Disks with a strip of stain-less and brass could also generate the oscillation even they solely do not bend with temperature rise. The reason is considered that the temperature gradient in thickness direction varying with the time and movement might cause bending deformation like bimetal. The result indicates that the oscillator can be constructed of single metal such as SUS304 endurable in high temperature usage, and is not affected by structural deformation. (all components expand with same ratio). We are trying to understand how the heat conduction (resulting thermal distribution and deformation) and oscillatory motion are related in changing time, and appropriate shape and materials for the design of high temperature actuator.

Table 30.1 Comparison of lowest temperatures of plate and disk with different materials where self-excited oscillation was observed

Material	Bimetal (Ni-Cr-Fe+Ni-Fe)	Stainless (SUS304)	Brass (C5191)
Temperature of plate (°C)	50	100	130
Temperature of disk (°C)	60	120	160

30.3 Micro Magnetostrictive Actuator

We are investigating the potential of micro magnetostrictive actuators using iron-gallium alloys (Galfenol). Galfenol is an iron-based magnetostrictive material developed by Clark [5, 6]. This material has a high magnetostriction which exceeds 200 ppm, a zero-field Young's modulus of ~70 GPa, and a relative permeability of 60-200. The strain is not as large as that of PZTs or Terfenol-D, but the material has the advantages of machinability [7] and ductile properties. Microactuators made using Galfenol will have the following benefits:

1. Good mechanical properties. Because of its good mechanical properties, the actuator can be of any shape, miniaturized, and durable under tensile and bending forces.
2. Simple configuration and ease of assembly. The minimal components are Galfenol, a magnetic yoke, and a drive coil. A pre-stress mechanism is not necessary in view of the high tensile strength of Galfenol (> 400 MPa).
3. Low drive voltage due to low impedance of the coil. The actuators can be driven by voltages of a few volts with a small power supply.
4. Wide temperature operation range. Low temperature operation in liquid nitrogen has been demonstrated, and high temperature operation is limited by Curie temperature.

30.3.1 Micro Magnetostrictive Vibrator [8]

The magnetostrictive vibrator consists of a pin of iron-gallium alloy ($Fe_{81.6}Ga_{18.4}$, Galfenol), a drive coil, and housing as shown in Fig.30.7. The overall dimensions are 2 mm diameter and 6 mm long (fixed on a brass fixture in the picture of Fig.30.7 right). The coil is made of 0.04 mm diameter polyimide wire with 270 turns to give a resistance of 17 Ω. The end caps attached to the pin and housing for the flux path are made of Permalloy. A magnetic field is applied by a closed magnetic circuit along the axial direction of the Galfenol pin resulting in axial displacement. The Galfenol pins, 1 mm diameter by 5 mm long, were fabricated from a 6.35 mm diameter by 50 mm long rod prepared by the Free Stand Zone Melting technique. The rod was divided into 1 mm square pieces, 5 mm long by wire EDM. These pieces were then milled into a cylindrical shape by an ultra precision cutting machine.

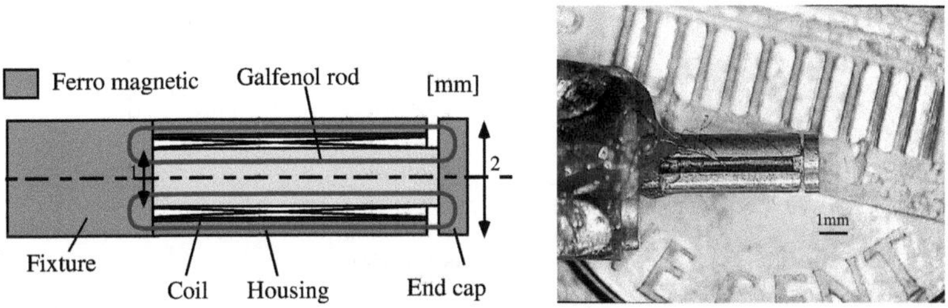

Fig. 30.7 Vibrator using a Galfenol rod of 1mm diameter. Fixture and housing are welded to the rod

Figure 30.8 shows strain-field curves. The vibrator exhibited the maximum displacement of 1.2 μm (240 ppm) and a steeper rise in the strain with increasing field (higher piezomagnetic constant). The frequency responses of the displacement against the input current and impedance were measured as shown in Fig.30.9. The first resonance around 70 kHz and a high bandwidth of more than 30 kHz were observed in both vibrators. Because of the low drive voltage the impedance is constant until the cut-off frequency is approached, above which it increases due to inductance effects. The frequency response with respect to drive voltage is considered the multiplication of top and bottom in Fig.30.9 in this case, mostly dominated by the inductive reactance.

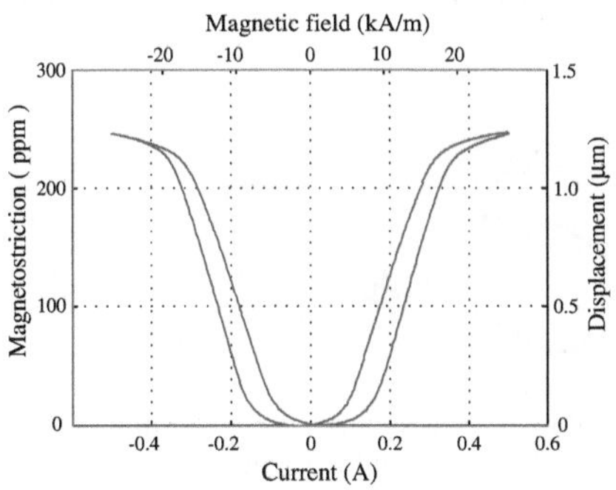

Fig. 30.8 Strain-current (magnetic field) butterfly curves of Galfenol vibrator taken for a frequency of 100 Hz

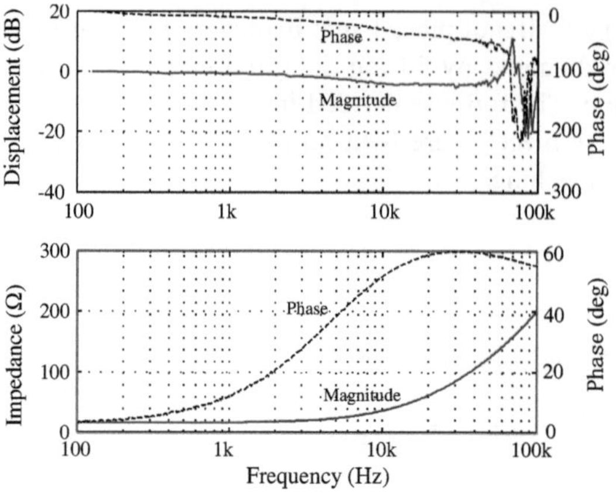

Fig. 30.9 Frequency response of the displacement divided by current (top) and impedance (bottom)

The displacement of the Galfenol vibrator is still not as large as can be obtained with a PZT stack (>1000 ppm) or Terfenol-D (>1500ppm), but its advantages over conventional ones are in its high robustness against tensile and bending stresses. In addition, the low drive voltage (5 V for a strain of 200 ppm) and the inductive nature of the coil are of great advantage for power supply and drive circuit design. For example, it could be used as a speaker as shown in Fig.30.10. Here, the vibrator was attached to an iron plate via a permanent magnet and was driven by a portable music player (iPod nano music player, Apple Inc.). Even with a current in the milliamp (mA) range, the iron plate could generate a clear sound from the vibrations generated by the Galfenol vibrator. This study demonstrates that Galfenol has high applicability in micro-actuator applications for generation of a high frequency vibration using simple, low cost designs.

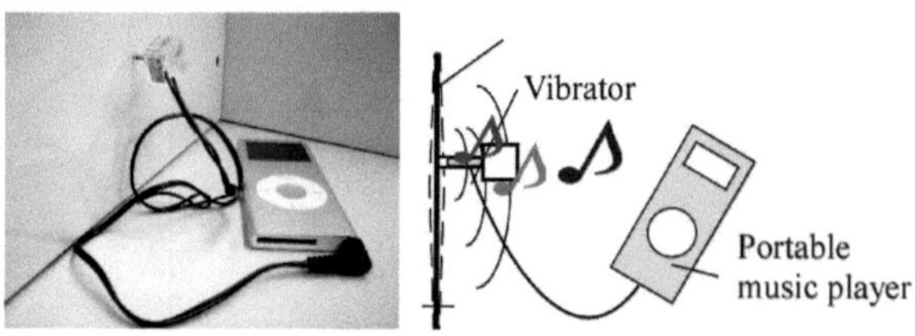

Fig. 30.10 The Galfenol vibrator used as a speaker. The iron plate which was vibrated by the attached micro actuator generated clear audible sound

30.3.2 Micro Positioning Device [9]

The vibrator is also used as positioning device as shown in Fig.30.11. The actuator is composed of a Galfenol rod, coil, yoke and frictional rod on which a mover

(1.8 g) slides. The mover, two brass bars with half circle hole is clamped on the rod by elastic force of two springs adjusted by two bolts. As same, the displacement of the mover is precisely controlled by the current of feed back control. In addition, the mover is driven with long stroke by saw current wave form of slowly increased and rapidly decreased based on the smooth impact drive mechanism as shown in Fig.30.12. When the current is slowly increased, the mover travels distance as same as the extension of the Galfenol pin, because the mover is fixed on the rod by the friction. On the other hand, when the current is rapidly decreased, there is slippage on the mover because it cannot follow the quick movement of the pin. In a cycle, the mover movement is equivalent to the slippage distance at the shrinkage of the pin. By repeating these operations, the mover can be driven with a long stroke. To reverse the moving direction, the form of the current is modified, such that it is rapidly increased and slowly decreased. Figure 30.13 shows the responses by the saw current of 0.15 A at the driving frequencies of 2 kHz. We could verify the motion as described, where the movement is caused by the slippage at rapid displacement of the rod. The velocities were 0.63 in forward and 1.0 mm/s in back motion respectively.

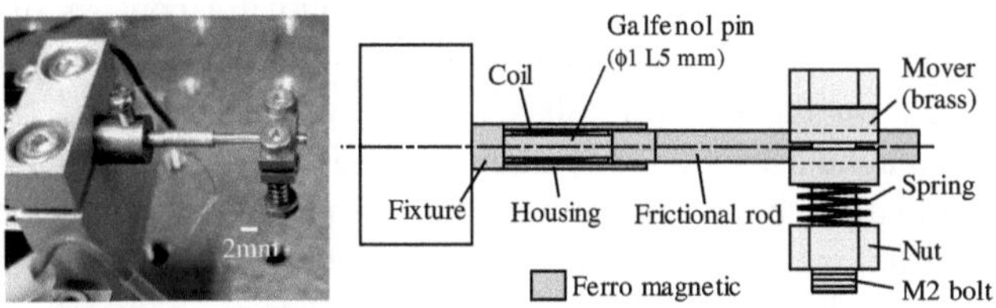

Fig. 30.11 Configuration of micro positioning device

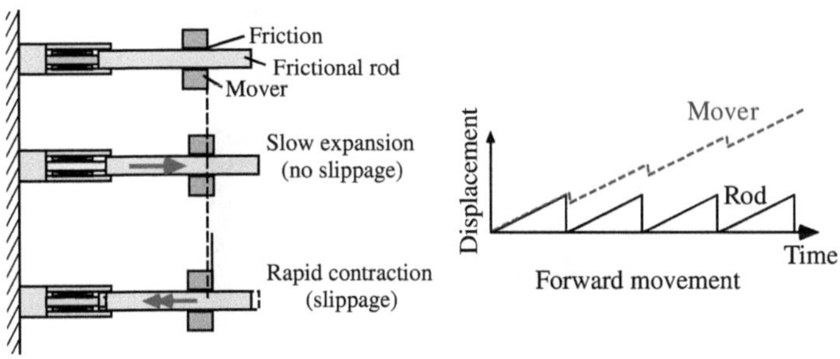

Fig. 30.12 Principle of movement

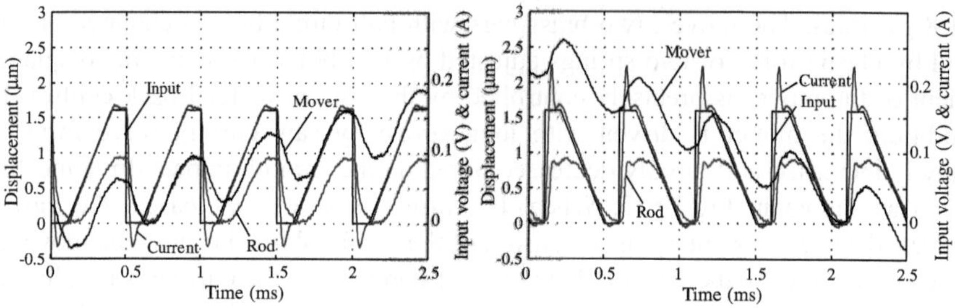

Fig. 30.13 Time response at saw current of 2 kHz (left: forward and right: back)

30.3.3 Temperature Characteristics

Low temperature operation in liquid nitrogen has been demonstrated [10], and high temperature operation is limited by Curie temperature of Galfenol (around 700 °C). Fig.30.14 shows an experimental apparatus to measure the displacement of a Fe-Ga actuator under high temperature. The actuator consists of a Galfenol rod of 8mm diameter and 50mm length, iron yoke and one-layer coil wound on a ceramic cylinder on the rod was placed in a electrical furnace. As shown in Fig.30.15, the magnetostriction was maintained almost 100% under 100°C and decreased gradually above the temperature. In addition, the actuator was sustainable in the measurements under heat cycle of temperature difference around 500°C. Therefore, micro magnetostrictive actuator of simple configuration, is expected to have wide temperature operation ranges from -250 to 500 °C inherited from high Curie temperature and high mechanical strength of the materials, and high compatibility with structural and magnetic materials (Fe) with close thermal expansion ratio of 11 ppm/K.

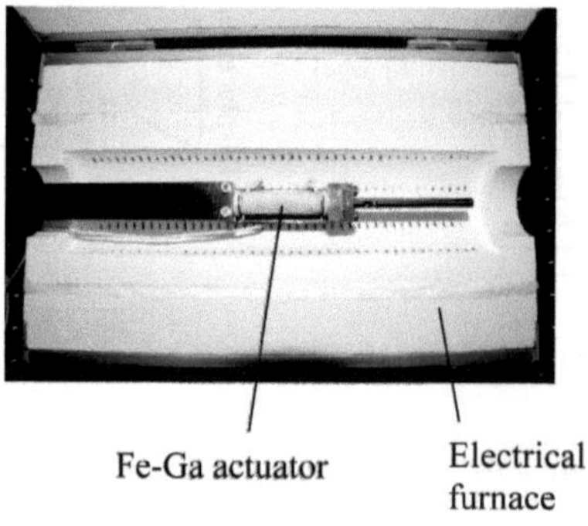

Fig. 30.14 Experimental setup to measure the temperature characteristics of Fe-Ga Actuator

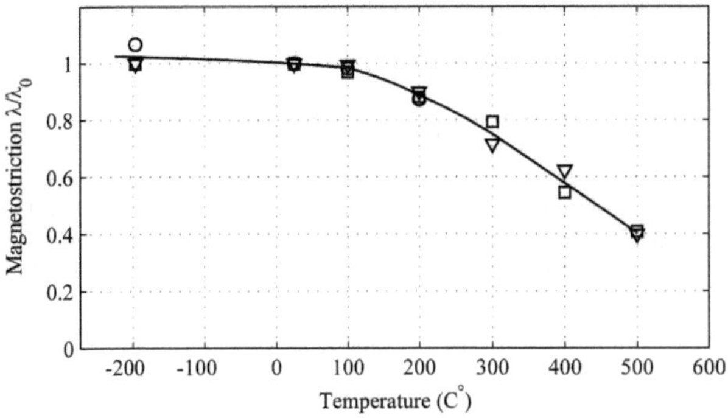

Fig. 30.15 Magnetostriction dependent on temperature

30.4 Conclusion

Two types of actuator for special temperature environment were described in this paper. In the future, the fundamental factors for the self-excited oscillation will be examined experimentally and theoretically to understand the principle and design criteria. Thermal impact driven actuator with oscillator made of single metal will be also fabricated for high temperature usage. Novel micro magnetostrictive actuators using Fe-Ga will be fabricated and investigated the temperature operation range and heat-resistivity for practical usage in hash environment.

Acknowledgments This study is financially supported by grants-in-aid for scientific research of the Ministry of Education, Culture, Sports, Science and Technology of Japan on priority areas, "Next-Generation Actuators Leading Breakthroughs".

References

1. Ohmichi O, Yamagata Y and Higuchi T (1997) Micro impact drive mechanisms using optically excited thermal expansion. J. MicroElecMech Sys. 6, 3 :200-207
2. Ueno T and Higuchi T (2006) Development of New Actuator Using Thermal Deformation. in Proc. of 1st Int. Symp. Next generation Actuator Leading Breakthroughs 149–152
3. Harada Y, Suehara T, Ueno T and Higuchi T (2006) Development of an Actuator using Thermal Deformation. Proc. 2006 spring conf. of the Jpn. Soc. Precision Eng. 803-804
4. Suehara T, Harada Y, Ueno T and Higuchi T (2006) Development of a Self-propelled System using Thermal Deformation. Proc. 2006 spring conf. of the Jpn. Soc. Precision . Eng. 805-806
5. Clark EA, Wun-Fogle M, Restorff BJ (2000) Magnetostrictive Properties of Body-Centered Cubic Fe-Ga and Fe-Ga-Al Alloy. IEEE, Trans. Mag. 37 :3238-3240
6. Clark EA, Wun-Fogle M, Restorff BJ (2002) Magnetostrictive property of Galfenol alloys under compressive stress. Materials Transaction 43 :881-886

7. Ueno T, Summers E, Higuchi T (2007) Machining of Iron-Gallium for Microactuator. Sensors and Actuators A 137 :134-140
8. Ueno T, Summers E, Wun-Fogle M and Higuchi T (2007) Micro Magnetostrictive Vibrator using Iron-Gallium Alloy. J. Mag. Soc. Jpn. 31, 4 :372-375
9. Ueno T and Higuchi T. Smooth Impact Drive Mechanism using Iron-Gallium Alloy. 13th Int. Symp. on Apl. Electromagnetics and Mechanics, will be published.
10. Ghodsi M, Ueno T, Teshima H, Hirano H, Higuchi T and Summers E (2007) Zero-Power Positioning Actuator for Cryogenic Environments by Combining Magnetostrictive Bimetal and HTS. Sensors and Actuators A 135, 2 :787-791

Chapter 31
Applications of Electrostatic Actuators within Special Environments

Akio YAMAMOTO[1]

Abstract Typical electromagnetic motors are not preferred in some special environments due to their incompatibilities with the special characteristics of the environments. Electrostatic motors have great potentials for the use within those special environments, since they are non-magnetic and structurally simple. This chapter introduces some trials to utilize electrostatic motors in those special environments. The motor mainly focused on is a film-based electrostatic linear motor that has a large force generation capability such as thrust force of several tens of newtons. The chapter illustrates the successful operations of the motor in some special environments such as high vacuum and strong magnetic field. For strong magnetic field environment, an example of application studies using the electrostatic linear motor is also described, which is biomechanical modeling performed within a magnetic resonance (MR) scanner. It is shown that the electrostatic motor is one of the promising MR compatible actuators.

31.1 Introduction

Selection of actuators is one of the most important issues in mechatronic system design, since characteristics of actuators are directly reflected to kinematic performances of the developed systems. At present, for the most of the mechatronic systems, electromagnetic actuators are selected as their driving power sources. Electromagnetic actuators are, in fact, far superior to other actuators from the viewpoint of system design; they offer a wide variety in terms of size, the design methodology has been clearly established, and peripheral equipments including motor drivers and controllers can be easily found. They are definitely the best candidates in actuator selections for the most of the mechatronic systems.

Despite the numerous advantages of the electromagnetic motors, there are still needs for other actuators that are based on different actuation principles. Some of

[1] Akio YAMAMOTO
Department of Precision Engineering, School of Engineering, The University of Tokyo

those needs can be found in systems designed for special environments, where the use of typical electromagnetic actuators is limited. One example of such special environments is a strong magnetic field. Strong or precise magnetic fields are utilized in various scientific equipments or in advanced manufacturing systems. If one tries to design a mechatronic system that works with those systems, selection of actuators can be difficult, since typical electromagnetic actuators are not compatible with those magnetic fields; electromagnetic actuators can disturb the precise magnetic field or, adversely, their operations can be disturbed by the strong magnetic field. For those environments where electromagnetic actuators are not preferred, different actuation principles should be explored.

This chapter focuses on electrostatic actuators as alternative actuators for those special environments and introduces trial studies that have been made by the author's research group. First, the chapter discusses the advantage of electrostatic actuators in special environments. Then, examples of studies, which applied electrostatic actuation technologies within special environments, will follow. The environments dealt in this chapter include cryogenic environment using liquid nitrogen, high vacuum, and strong magnetic field. Among various electrostatic actuators, the paper especially focuses on a film-based high-power electrostatic motor, a unique electrostatic motor characterized by its superior force generation capability.

31.2 Advantages of Electrostatic Actuators in Special Environments

Advantages to utilize electrostatic actuators within special environments can be discussed from the two different viewpoints. From the viewpoint of the application side, the actuators' characteristics, such as non-magnetic and structural simplicity, would be regarded as the advantages of the actuators. Since the largest disadvantage of the electromagnetic motors in special environments is the incompatibility with strong magnetic fields, being non-magnetic is one of the important characteristics that alternative actuators should possess. The simple structures would be also fascinating characteristics for special environments applications. Systems working within special environments sometimes have severe space limitations. Mechatronic devices working with those systems are not allowed to occupy large space, and therefore, simple and compact structures of electrostatic actuators will be preferred. In addition, electrostatic actuators do not need functional materials such as piezoelectric elements or permanent magnets, which suffer from the Curie temperature limitations. Therefore, electrostatic motors would easily cope with extreme temperatures.

On the other hand, from the viewpoint of developers of electrostatic actuators, special environments seem fascinating as a target application area. One of the problems regarding electrostatic motors is weak output force in normal condition.

The weak force comes from the limitation imposed on electrostatic field in atmosphere. Typically, electrostatic field is limited by electric breakdown. In a typical atmospheric condition, breakdown occurs at 3 MV/m if the device is about the size of several centimeters. At this maximum field strength, the electrostatic force per unit area is calculated as 40 N/m^2. This value is far smaller than typical magnetic force, which can easily reach up to 4×10^5 N/m^2. To increase the output force of electrostatic motors, the limitation of the field strength needs to be overreached, and it can be possible in some special environments. For example, according to Paschen's law, breakdown voltage increases in high vacuum, or in micro scale. Another example is liquid nitrogen to create cryogenic environment, where breakdown voltage is much higher than in atmosphere. Of course, not all the special environments are good for electrostatic motors, but some special environments surely have a great potential to realize high performance electrostatic actuations.

31.3 Electrostatic Actuation in Liquid Nitrogen

Liquid nitrogen is widely used medium for cryogenic applications. Developing actuation technology for liquid nitrogen can contribute to various studies performed in cold environments. One of such applications can be found in biological studies. In biological studies, liquid nitrogen is used for, e.g., cryopreservation of biological cells. Actuation technology in liquid nitrogen can highly automate cryopreservation or thawing processes, which could lead to, for example, a development of μTAS devices including cryopreservation processes. Since liquid nitrogen has a large dielectric strength of over 80 MV/m [1, 2], which is about 30 times larger than the atmosphere, it seems promising environment for electrostatic actuators.

To confirm the potential of electrostatic actuations in liquid nitrogen, an electrostatic device was tested for particle transportation [3]. The prototype device is shown in Fig.31.1. The device was set at the bottom of a liquid nitrogen container to transport frozen particles. The particles used in the experiments were frozen droplets of Mannitol solution, which is medium used in biological experiments. Figure 31.2 shows a droplet motion in liquid nitrogen when a three-phase pulse voltage was applied. During the transportation, no electrical discharge and electrochemical reaction were visually observed. The motion of the droplet was found to be synchronous and stepwise, meaning that positioning of a droplet can be easily performed on the device. The result confirmed the effectiveness of electrostatic actuation in the environment, especially for cryogenic applications in biology.

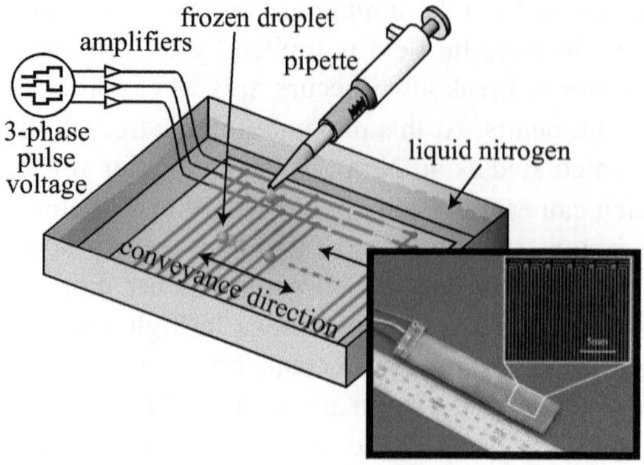

Fig. 31.1 Particle transportation setup for liquid nitrogen environment

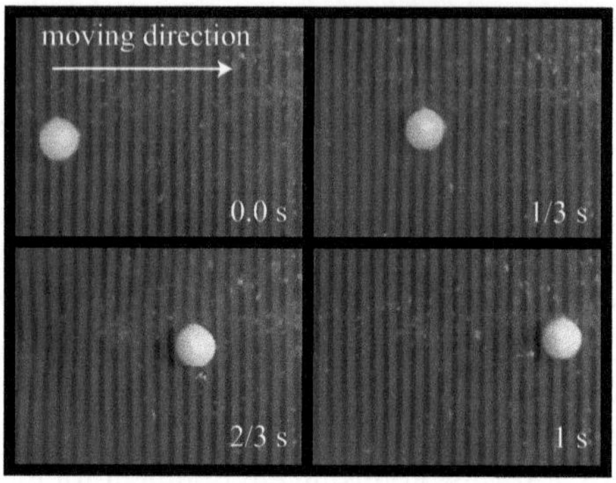

Fig. 31.2 A frozen particle of Mannitol and its motion by the electrostatic transportation device in liquid nitrogen (700 $V_{0\text{-}p}$, 5 Hz)

31.4 High-Power Electrostatic Motor and Its Application to Special Environments

31.4.1 High-Power Electrostatic Motor

In recent actuator research, electrostatic actuators have been studied mainly in micro region as MEMS actuators. However, some pioneering studies have revealed the great potential of electrostatic force in larger scale [4, 5]. For example, it was reported that film-type motors developed by Egawa, et al. [4] or by Niino, et al. [5] generated thrust force of several newtons, or more, with characteristic length of several centimeters. These achievements proved that electrostatic actuators can be utilized to drive ordinary mechatronic devices.

We have been trying to apply such high-power motors to special environments. The motor we have mainly focused on has been a film-based linear motor called Dual Excitation Multiphase Electrostatic Drive [5]. The motor has a characteristic length of several centimeters and generates thrust force of several tens of newtons. The basic structure of the motor is shown in Fig.31.3. In its simplest configuration, the motor consists of a pair of thin plastic films. Each film contains three-phase parallel and skewed [6] electrodes that are aligned at regular intervals (typically 200 μm). The motor is driven by high-voltage three-phase signals such as 1 to 2 $kV_{0\text{-}p}$ to generate practical thrust force. Since such high voltages cause electric discharges of the air, the motor is normally immersed in dielectric liquid, such as Flourinert™ (3M) or silicone oil.

The motor operation is synchronous; the displacement or speed of the slider can be controlled in an open loop control. If combined with a position sensor to detect the relative electrode position, closed-loop servo control can be achieved with a simple feedback controller. Considering such characteristics that can hardly be found in other alternative actuators, introducing this motor into special environments would bring considerable benefits.

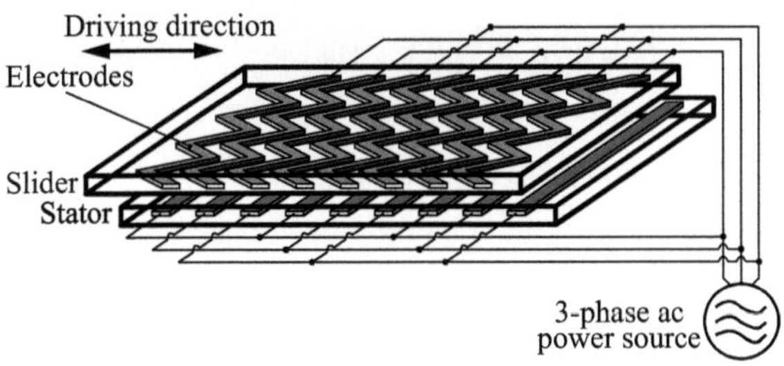

Fig. 31.3 Schematic illustration of a high-power electrostatic motor

31.4.2 Performance Evaluation in High Vacuum

The motor is typically operated in dielectric liquid to realize high field strength, which in turn leads to larger thrust force. According to Paschen's law, the breakdown voltage of air becomes larger as the pressure decreases. Therefore, in considerably high vacuum, the motor is expected to have a large force generation capability due to the increased breakdown voltage, even without dielectric liquid.

The motor was tested in vacuum using a prototype shown in Fig.31.4 [7]. Since lubrication is one of the major problems in vacuum, the motor is equipped with a linear guide to avoid the direct friction between the slider and stator. At a chamber pressure of 10^{-3}Pa, the prototype was driven in an open loop by applying a three phase ac voltage with amplitude of 1.5 $kV_{0\text{-}p}$ and a frequency of 1 Hz. The resultant motion is shown in Fig.31.5. The driving signal was switched every 3 seconds to alter the driving direction. No stepping-out was observed for the operation. The motor was tested continuously for 15 hours and no noticeable temperature raise was observed.

Our experiments also verified that the prototype motor generated the same thrust force as it did in Fluorinert [7]. The prototype was also tested in terms of positioning servo control. Using a built-in position sensor [8], a closed-loop positioning control was demonstrated [7].

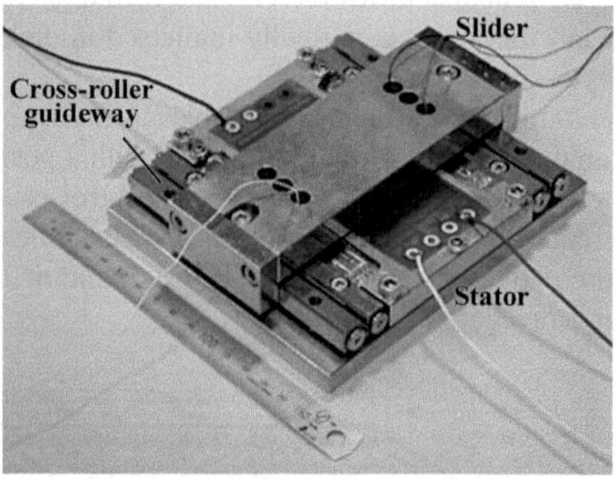

Fig. 31.4 Prototype motor for driving test in vacuum

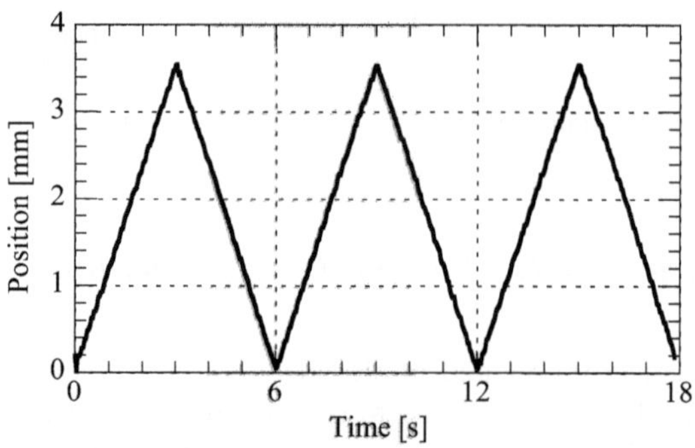

Fig. 31.5 Slider displacement in open loop driving at 10^{-3}Pa

31.5 Applications of High-Power Electrostatic Motors for MRI Related Studies

31.5.1 Needs for Non-Magnetic Actuators in MRI Related Studies

Magnetic resonance imaging (MRI) is now widely used as a standard tool not only in medical diagnosis, but also in several scientific studies that include biomechanics and neuroscience. In biomechanics studies, for example, an MRI system can be used with actuators in such a manner that reactions of bodies against physical motion stimuli provided by an actuator are measured by a magnetic resonance (MR) scanner to analyze mechanical properties of body parts. In such applications, actuators that can simultaneously work with MR scanners are imperative. However, typical electromagnetic motors are not compatible with the strong and precise magnetic field of MR scanners [9]. Many studies trying to apply non-magnetic actuators to this environment are in progress [9-11], which include hydraulic drives [10] and ultrasonic motors [11], but most of the non-magnetic actuators have difficulties in terms of servo control. Thus, there is still a considerable need for a new non-magnetic actuator, especially for one with a good control performance. The high-power electrostatic motor introduced in the previous section is a synchronous actuator and would possibly satisfy those demands.

31.5.2 Evaluations of the High-Power Electrostatic Motor in MR Environment

A prototype electrostatic motor was fabricated using non-magnetic materials and evaluated in an MR scanner [12], as shown in Fig.31.6. Measurement of thrust force verified that the force generation capability is not degraded by the strong magnetic field of the MR scanner. To investigate the impact of the motor operation on MR imaging, signal-to-noise ratios (SNR) of MRI images were analyzed. Images were taken while the motor was operating near the edge of the MR scanner's gantry. The measured result shown in Fig.31.7 indicates that the impact of the electrostatic motor to the imaging is low.

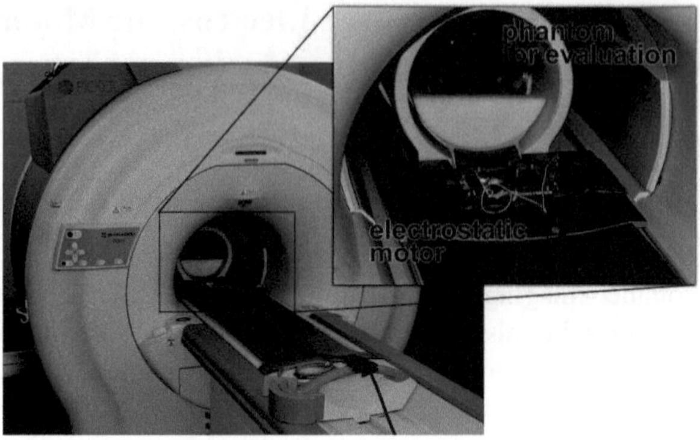

Fig. 31.6 Electrostatic motor working in an MR scanner of 1.5 T

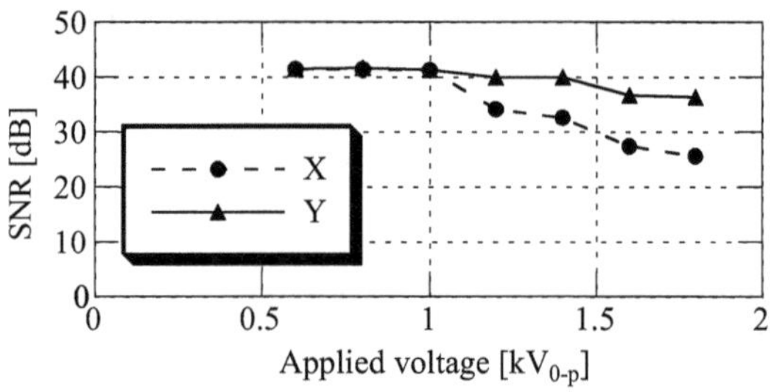

Fig. 31.7 Relationship between SNR of MR images and applied voltages to the electrostatic motor. The motor was positioned on the bed, 35 cm away from the RF-coil, and was driven parallel (X) or perpendicular (Y) to the magnetic field. The corresponding thrust force of the motor is about 4 N for 1 kV, and about 9 N for 1.6 kV

31.5.3 Application to Biomechanical Modeling

The electrostatic motor was applied for biomechanical studies [13-15]. Deformation of biological samples was visualized using motion-triggered cine-MRI sequence, which can visualize dynamic motion of samples from iterative motion. In this sequence, the motor repeats the same iterative motions, during which an MR scanner collects data that will later be reconstructed into one set of sequential images. The motion of the electrostatic motor was controlled in an open loop. Since the use of typical sensors is also difficult within MR environments, the capability of open loop motion control is beneficial.

Figure 31.8 shows the experimental setup to measure muscles on a human upper arm [14]. The electrostatic motor was situated on the bed of an MR scanner to push the upper arm iteratively. The motor motion was synchronized with trigger pulses based on which the MR scanner performed the cine imaging sequence. The motor repeated reciprocating motion of 12-mm stroke for 256 times, with an approaching speed of 24 mm/s and retrieving of 12 mm/s. The finally obtained images are shown in Fig.31.9. The mesh patterns observed in the images are magnetic “tags”, which enables tracking of the deformation for further analysis.

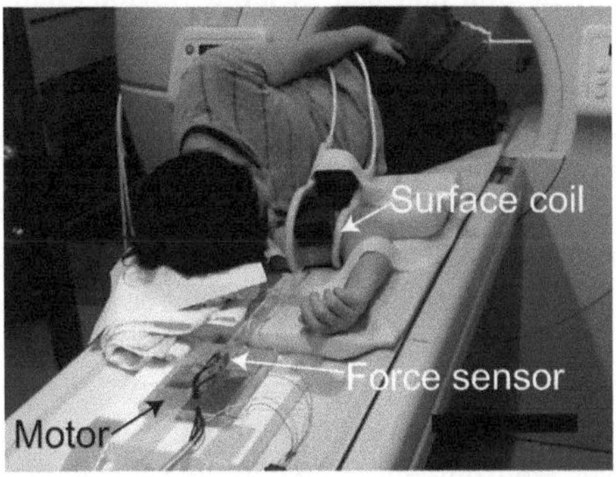

Fig. 31.8 Experimental setup to measure mechanical properties of muscles on a human upper arm. The scanner’s bed was withdrawn from the gantry for photographing; the real measurement was performed inside the gantry

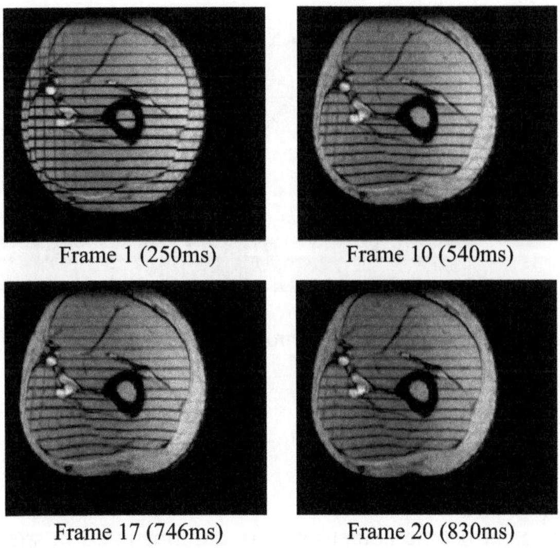

Fig. 31.9 Cine MR images of human upper arm deformation

31.5.4 Applications for Haptic Interfaces

Another interesting application within MR environments can be found in neuroscience. To investigate sensorimotor activities in a human brain, MR-compatible haptic devices are demanded. A haptic device is a mechatronic device used for human-computer interaction and can generate various haptic stimuli to a human operator. Control performance of actuators is important in haptic devices, and therefore, it would be a promising application for the electrostatic motor.

The prototype device shown in Fig.31.10 was fabricated using two electrostatic motors, whose maximum thrust force is about 18 N [16]. The device is equipped with an MR-compatible 2-DOF optical force sensor to measure interaction force from a human operator. Haptic rendering was performed by admittance control scheme. In admittance control [17], a controller reads interaction force and calculates the position where the stick of the device should be positioned. In the prototype, the positioning is done in an open loop control. The prototype device has not yet been tested within MR environment. In our future work, it will be improved and brought into MR environments to be applied for neuroscience studies on sensorimotor functions.

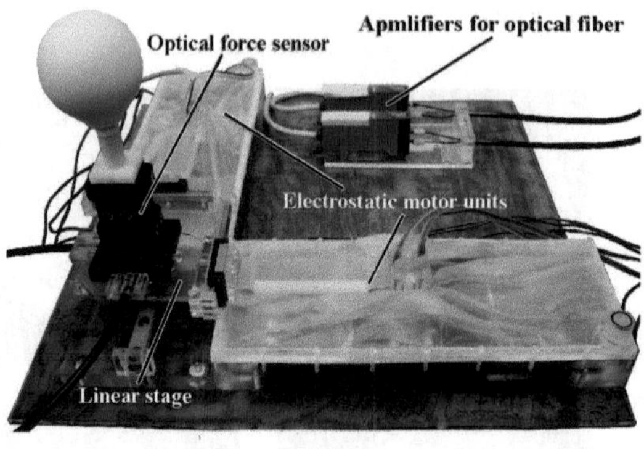

Fig. 31.10 Prototype haptic device using electrostatic motors and an optical force sensor

31.6 Conclusions

In this chapter, electrostatic actuators were demonstrated in special environments that include cryogenic, high vacuum, and strong magnetic fields, which are harsh for typical electromagnetic motors. Special focus was placed on a high-power motor called Dual Excitation Multiphase Electrostatic Drive.

Although the described results are still fundamental, the advantages of using electrostatic motors in special environments have been confirmed. Especially strong magnetic fields seem one of the promising application areas. Since, in addition to MRI, many other systems utilize strong magnetic fields, various interesting applications could be found for electrostatic motors, which we would like to pursue in our future work.

Acknowledgments This work was partly supported by a Grant-in-Aid for Scientific Research on Priority Areas, No. 16078203, from MEXT of Japan. The biomechanical measurement described in 1.5.3 was performed in cooperation with Computational Biomechanics Unit of RIKEN in Japan. The optical force sensor described in 1.5.4 was developed in Robotic Systems Lab (LSRO) of Ecole Polytechnique Fédérale de Lausanne (EPFL) in Switzerland.

References

1. Swan DW, Lewis TJ (1960) Influence of Electrode Surface Conditions on Electrical Strength of Liquified Gasses. J Electrochem Soc, 107(3):180-185
2. Mathes KN (1967) Dielectric Properties of Cryogenic Liquids. IEEE Trans Electr Insul, El-2(1):24-32
3. Yamamoto A, Nakajima T, Kudoh K, Higuchi T (2006) Direct Electrostatic Transportation of Frozen Droplets in Liquid Nitrogen for Single Cryopreserved Cell Processing. Proc IEEE MEMS 2006:382-385
4. Egawa S and Higuchi T (1990) Multi-Layered Electrostatic Film Actuator, Proc IEEE MEMS 1990:166-171
5. Niino T, Higuchi T, Egawa S (1995) Dual Excitation Multiphase Electrostatic Drive. Proc IEEE IAS 1995:1318-1325
6. Yamamoto A, Niino T, Higuchi T (2006) Modeling and Identification of an Electrostatic Motor. Precis Eng, 30(1):104-113
7. Yamamoto A, Yasui H, Nishijima T, Higuchi T (2003) Electrostatic Linear Servo Motor with Built-in Position Sensor for Vacuum Environment. Proc IEEE ISIE 2003, 2:928-933
8. Nishijima T, Yamamoto A, Yasui H, Higuchi T (2006) A Built-in Displacement Sensor for an Electrostatic Film Motor. Meas Sci Technol, 17(10):2676-2682
9. Gassert R, Yamamoto A, Chapuis D, Dovat L, Bleuler J, Burdet E (2006) Actuation methods for applications in MR environments. Concepts Magn Reson, 29B(4):191-209
10. Moser R, Gassert R, Burdet E, Sache L, Woodtli HR, Erni R, Maeder W, Bleuler H (2003) An MR compatible robot technology. Proc IEEE ICRA 2003:670-675
11. Flueckiger M, Bullo M, Chapuis D, Gassert R, Perriard Y (2005) fMRI compatible haptic interface actuated with traveling wave ultrasonic motor. Proc IEEE IAS 2005:58-63
12. Yamamoto A, Ichiyanagi K, Higuchi T, Imamizu H, Gassert R, Ingold M, Sache L, Blueler H (2005) Evaluation of MR-compatibility of Electrostatic Linear Motor. Proc IEEE ICRA 2005:3669-3674
13. Yamamoto A, Rajendra M, Hirano Y, Kataoka H, Yokota H, Himeno R, Higuchi T (2007) Motion Generation in MRI Using an Electrostatic Linear Motor for Visualizing Internal Deformation of Soft Objects by Tagged Cine-MRI. Proc IEEE ISIE 2007:2741-2746
14. Mayoran R, Yamamoto A, Kataoka H, Oda T, Yokota H, Hirano Y, Himeno R, Higuchi T (2007) Application of an Electrostatic Film Motor in MRI-related Biomechanical Measurement. Proc JSME 2007 annual congress, 5:257-258
15. Rajendra M, Yamamoto A, Oda T, Kataoka H, Yokota H, Himeno R, Higuchi T (2008) Motion Generation in MR Environment Using Electrostatic Film Motor for Motion Triggered cine-MRI. IEEE/ASME Trans Mechatron, 13(3):278-285
16. Hara M, Matthey G, Yamamoto A, Chapuis D, Gassert R, Bleuler H, Higuchi T (2009) Development of a 2-DOF Electrostatic Haptic Joystick toward MRI/fMRI related Studies. Proc IEEE ICRA 2009:1479-1484
17. Yokokohji Y, Hollis R, Kanade T (1996) What you can see is what you can feel – development of a visual/haptic interface to virtual environment, Proc IEEE VRAIS 1996:46-53

applications could be found for electrostatic motors, which we would like to pursue in our future works.

Acknowledgments This work was partly supported by a Grant-in-Aid for Scientific Research [illegible] No. 20078[illegible] from MEXT [illegible]

References

[illegible]

Chapter 32
Micro Actuator System for Narrow Spaces Under Specific Environment

Takefumi KANDA[1]

Abstract This chapter describes micro actuator systems for narrow spaces under specific environments. In the fields of scientific or medical instruments, there is a great demand for micro sensors and actuators. We have fabricated some micro sensors and actuators which can be used under such specific environment. Two types of micro actuator systems for narrow spaces have been fabricated and evaluated. Those are for precise tool control unit and attitude control unit. In the precise tool control unit, a micro servo motor using the micro ultrasonic motor and the magnetic encoder have been designed and fabricated. In the attitude control, flexible displacement sensor made of piezoelectric polymer and shape memory alloy actuator driven by optical source.

32.1 Introduction

Many micro systems for narrow spaces have been studied for negotiating small pipes for inspections or human bodies for endoscopes [1]. For example, examinations and surgeries using endoscopes require a precision actuator system to control the configuration of the tools (cameras, manipulators and other instruments) and to drive them. In other cases, instruments for measurement may require specific environments, such as a strong magnetic field. Under these conditions, precise control of the instruments is also demanded.

The goal of this work is to develop a micro actuator system for narrow spaces under such specific environments. This system requires micro sensors and actuators. In this work, micro sensors and actuators for precise tool control and attitude control are integrated in an actuator system for narrow spaces. A micro ultrasonic motor, micro magnetic encoder, flexible displacement sensor, and shape memory alloy actuator were developed for the system.

[1] Takefumi KANDA

Graduate School of Natural Science and Technology, Okayama University

32.2 Structure

The micro actuator system for narrow spaces mainly consists of two parts: the precise tool control unit and the attitude control unit.

For the precise tool control unit, a micro servo motor was developed. A micro ultrasonic motor and micro magnetic encoder are used to achieve precise control of tools. The ultrasonic motor is driven by vibration without magnetic devices. Therefore, using a magnetic sensor for the sensing is possible. The motor, including the magnetic encoder, has a diameter of 2.5 or 3.0 mm.

For the attitude control unit, an actuator is used to control the attitude of a stick- or cylindrical-shaped body. In addition, the attitude of the body is detected by a sensor that has a flexible structure. To achieve these features, we used a shape memory alloy actuator driven by an optical source and a flexible displacement sensor using piezoelectric polymer. The optical waveguide in the actuator and the flexible displacement sensor were deposited using a paste injection system.

32.2.1 Micro Ultrasonic Motor and Sensor

The micro servo motor consists of a micro ultrasonic motor and a micro encoder. Each device uses a piezoelectric vibrator and a micro magnetic resistive sensor.

32.2.1.1 Micro Ultrasonic Motor Configuration

The micro ultrasonic motor is shown in Fig.32.1. This motor is the cylindrical type micro ultrasonic motor and uses a cylindrical piezoelectric vibrator [2–4]. The motor consists of a rotor, bearing, piezoelectric vibrator, and glass case. The rotor is joined to the output shaft and is made of stainless used steel. The bearing is made of poly (tetrafluoroethylene) (PTFE). The vibrator and the bearing are supported by a glass case. The glass case has a diameter of 1.8 mm and a height of 5.8 mm.

To generate traveling waves, four divided electrodes are located on the outer surface of the piezoelectric vibrator. On the inside of the vibrator, an electrode is also deposited. These electrodes are made of non-electro-plating nickel. The four electrodes on the outer surface and the electrode on the inner surface of the cylinder-type piezoelectric vibrator are used to oscillate the vibrator. The rotation direction is switched by changing the phase difference of the applied voltage sources between the outer electrodes [5].

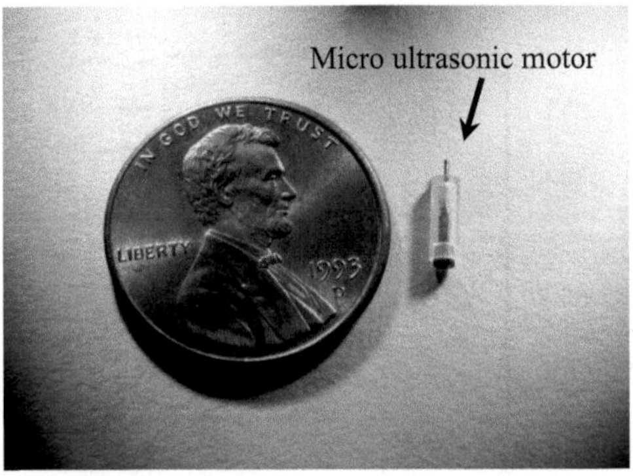

Fig. 32.1 Photo of micro ultrasonic motor

32.2.1.2 Evaluation of Micro Ultrasonic Motor Performance

To obtain large output torque, the relationship between the revolving velocity of the rotor, pre-load values, and applied voltage was measured experimentally. The surface velocity of the rotor shaft, which was joined with the rotor, was measured. A laser surface velocity meter was used to measure the surface velocity, and the revolution speed was estimated. The relationship is displayed in Fig.32.2. When the pre-load and the applied voltage were 1.5 mN and 29 $V_{p\text{-}p}$, respectively, the revolving velocity was 10,000 rpm. However, as shown in Fig. 32.2, the revolving velocity reached its peak when the applied voltage values were 20 and 29 $V_{p\text{-}p}$. In addition, the maximum revolving velocity was 1.1×10^4 rpm.

To evaluate the output torque of the motor, the starting performance of the motor was measured experimentally. The revolving velocity was measured with the laser surface velocity meter. From the experimental results, the revolving speed and starting torque were evaluated using the calculated inertia and measured values for the relationship between the revolving velocity and time when the pre-load values were 1.0 or 2.0 mN.

The relationship between the estimated starting torque and the applied voltage is shown in Fig.32.3. The driving frequency was set at 314 kHz. The maximum starting torque was estimated to be 5.5 μNm when the driving voltage and the pre-load were 40 $V_{p\text{-}p}$ and 1.0 mN, respectively. As can be seen, the starting torque values peaked when the applied voltage values were over 30 $V_{p\text{-}p}$.

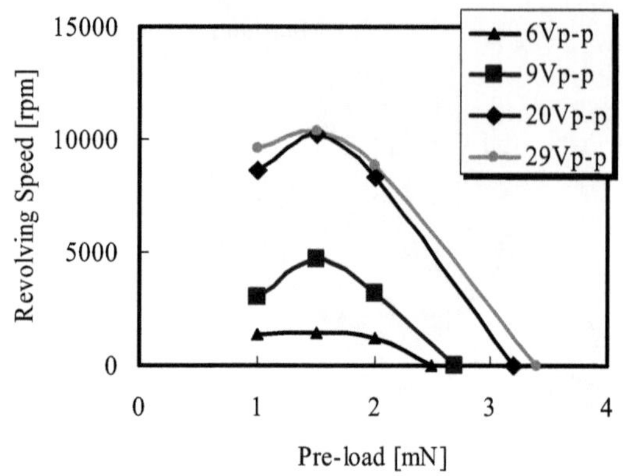

Fig. 32.2 Relationship between revolving velocity and pre-load

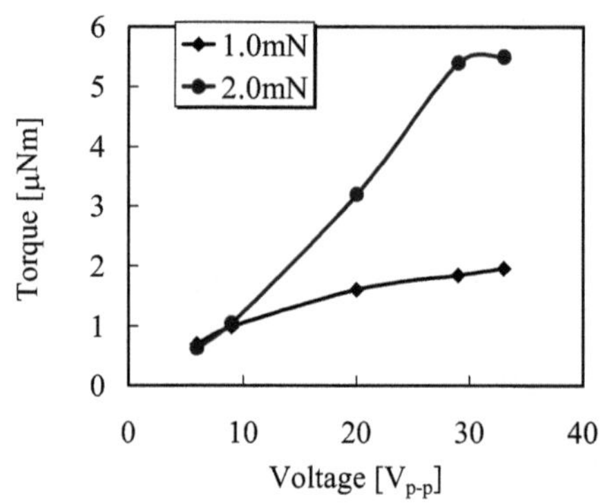

Fig. 32.3 Relationship between estimated starting torque and applied voltage

32.2.1.3 Micro Encoder and Servo Motor

The micro encoder for detecting the rotating condition, which is generated by the micro ultrasonic motor, was created by using a micro magnetic resistive sensor.

A schematic of the micro encoder is shown in Fig.32.4. The shaft driven by the micro motor is connected with the magnetic drum, which has a magnetic slit pattern. The magnetic resistive sensor detects the magnetic pattern on the drum.
The minimum magnetic pattern pitch is 40 μm from the resolution of the sensor. Therefore, when the magnetic drum has a diameter of 2.3 mm, the angular resolution is estimated to be 2°. In our evaluation, there were 10 magnetic patterns and the pattern pitch was 0.43 mm.

The output 2-phase signal values from the magnetic resistive sensor are plotted in Fig.32.5. The detected signal was converted to a pulse wave using a pick-up circuit. As shown in Fig.32.5, the sensor detected the magnetic pattern on the drum.

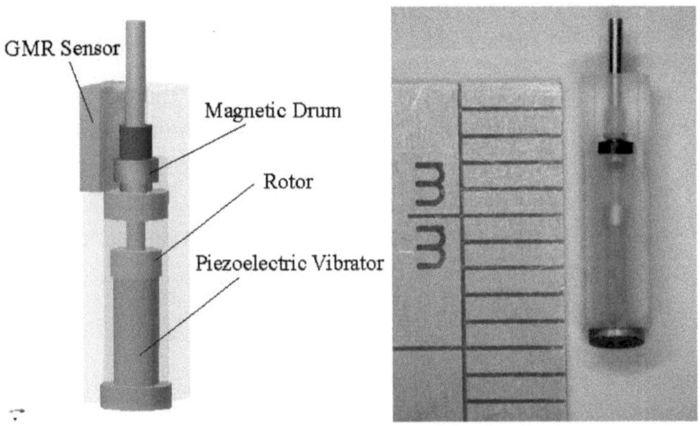

Fig. 32.4 Schematic and photograph of micro servo motor using micro magnetic resistive sensor

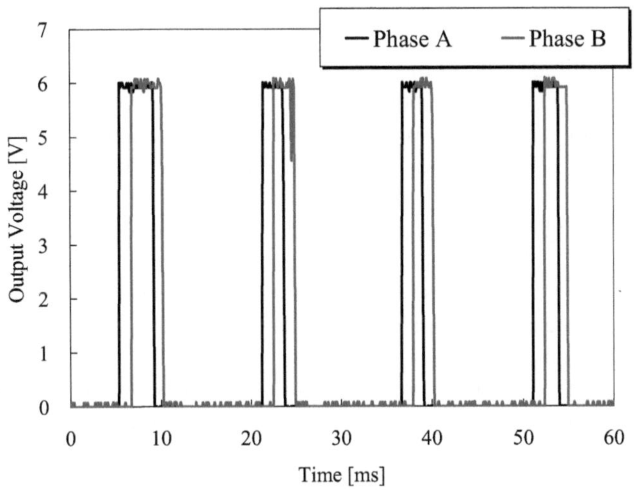

Fig. 32.5 Output signal from magnetic resistive sensor using magnetic drum and pick-up circuit

32.3 Attitude Control Unit

32.3.1 Structure and Evaluation Results

For the attitude of a stick- or cylindrical-shaped body, the unit consists of a shape memory alloy (SMA) actuator and piezoelectric polymer sensor. The basic structure is shown in Fig.32.6. To achieve a bending motion, the SMA actuator is driven optically. An optical waveguide is connected to that actuator. The sensor and optical waveguide are made of polymer paste material and patterned by a paste injection system.

32.3.1.1 Flexible Displacement Sensor

Displacement sensors that have flexible structures have been receiving increasing attention. This is mainly because the need for control of soft actuators is increasing. Many types of sensors having a flexible structure for soft actuators have been reported [1, 6–9]. In this work, the actuator generates large displacement, as discussed in the next section. Hence, the system needs a sensor that has a flexible structure for attitude control of the actuator.

A paste type piezoelectric polymer, poly (vinylidene fluoride-trifluoroethylene) copolymer (P(VDF/TrFE)), was used to achieve a flexible displacement sensor. This polymer is different from many other types of piezoelectric polymers, which need an extension process to obtain a piezoelectric effect. This polymer can be formed by casting or spin coating.

A schematic of the paste injection process is shown in Fig. 32.7. In this work, the paste injector nozzle was controlled by a 3°-of-freedom positioning machine. The patterned piezoelectric polymer and conductive paste film was deposited by the paste injection system [10].

A close-up of the patterned sensor is shown in Fig.32.8. The sensor detects the bending of the plate. The sensor is 20 mm in length, 5mm in width, and 30 μm in piezoelectric-film thickness.

The sensing performance of the fabricated sensor was evaluated by measuring the output voltage of the sensor. A bending deformation was given by a pneumatic actuator, and tip displacement was measured by a laser displacement sensor. The pick-up voltage was amplified by a charge amplifier circuit. The step response of the fabricated sensor is plotted in Fig.32.9. The sensor detected the motion of the plate but the stability needs to be improved.

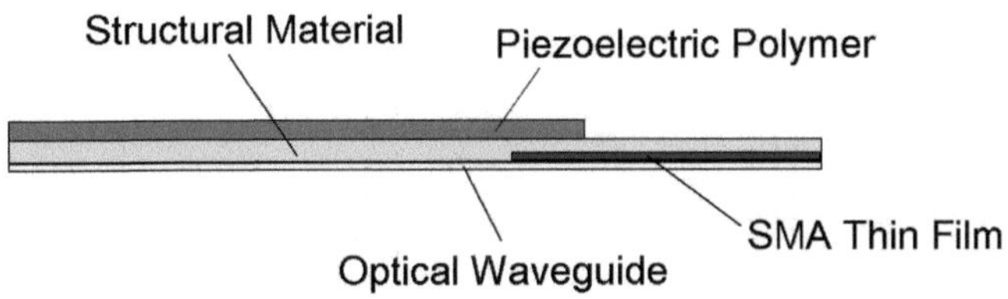

Fig. 32.6 Configuration of attitude control unit

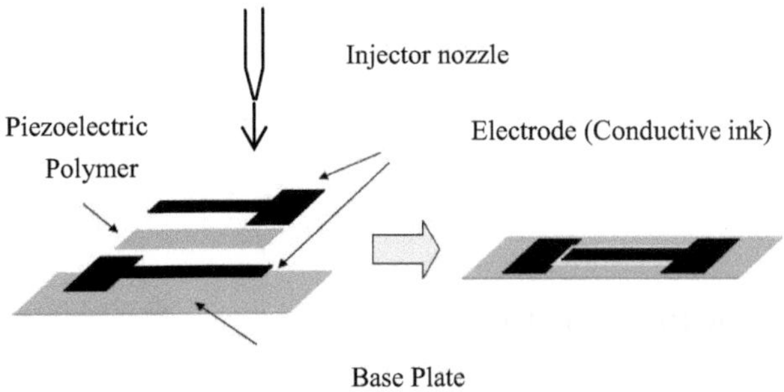

Fig. 32.7 Paste injection patterning process

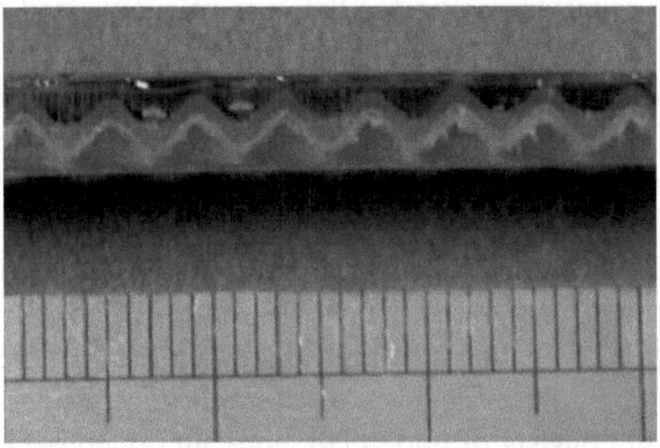

Fig. 32.8 Close-up of flexible displacement sensor using piezoelectric polymer

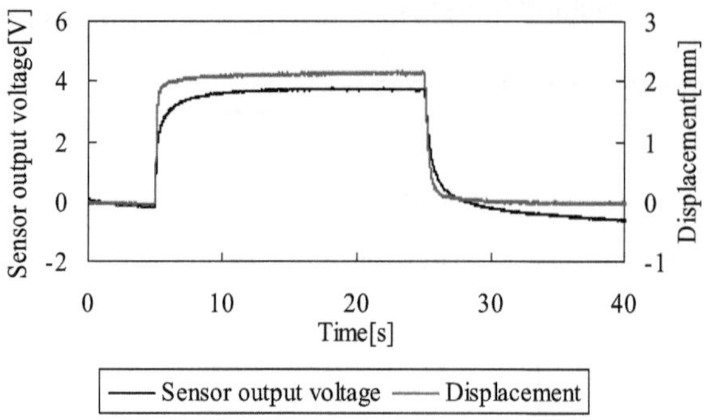

Fig. 32.9 Relationship between output voltage of sensor detected by charge amplifier circuit and tip displacement

32.3.1.2 SMA Actuator Driven by Optical Source

To control the attitude of a stick-type or cylindrical-shaped body, an SMA actuator was used to generate displacement. The SMA can be driven by heating. Electric residence heating by an electric current is mainly used. However, the electric residence heating requires wiring for the current supply and this wiring also becomes heated.

In this work, heating of the SMA was achieved by an infrared radiation source. By use of an optical waveguide, the optical source can drive the SMA actuator without any electrical wiring.

A schematic of the actuator is shown in Fig.32.10. An SMA sheet was attached to a Cu substrate. An optical fiber was connected to the optical waveguide element, which was made of polymer films.

The fabricated actuator is shown in Figs. 32.11 and 32.12. The length of the driving section is 45 mm. The thickness of the optical waveguide is 0.77 mm. The waveguide element consists of underclad, core, and overclad layers. Each layer is made of polymer optic material.

The fabricated sensor was driven by the infrared radiation source in an experiment. The measured detected force is plotted in Fig.32.13. The wavelength of the optical source was 830 nm. When the tip displacement was 4.2 mm, the generated force was 35mN. In this experimental result, the attenuation through the optical waveguide was large.

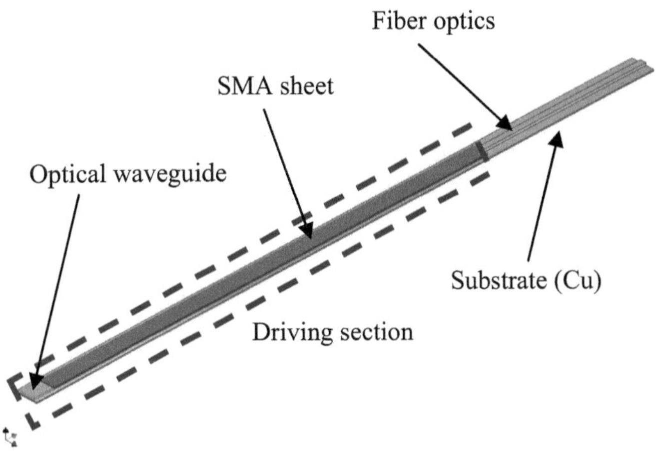

Fig. 32.10 Schematic of attitude control actuator made of SMA and optical waveguide

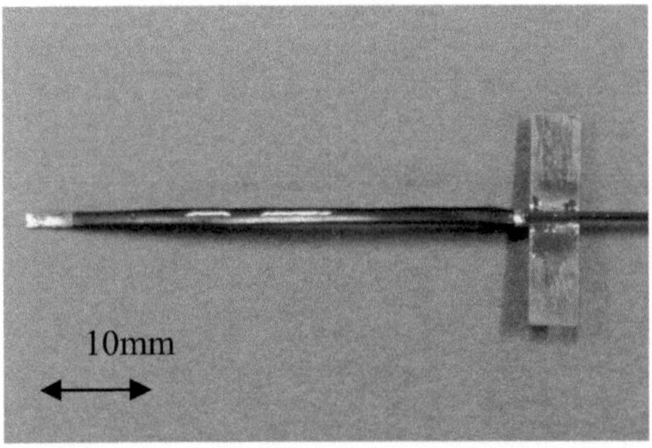

Fig. 32.11 Photograph of fabricated actuator

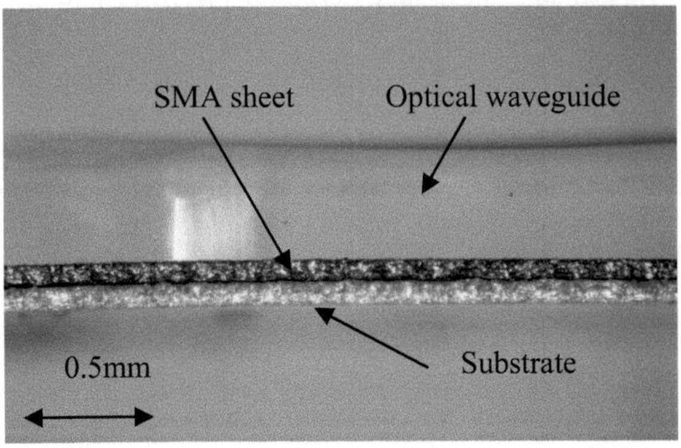

Fig. 32.12 Close-up of fabricated attitude control actuator

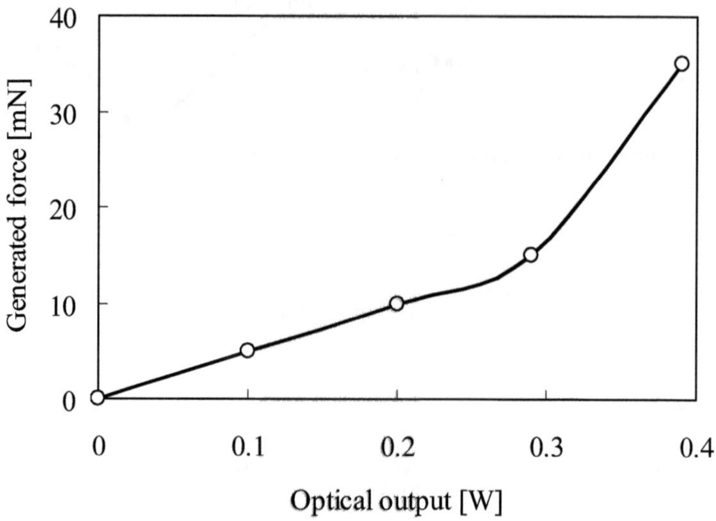

Fig. 32.13 Relationship between generated force at tip of actuator and optical output of optical source

32.4 Conclusion

Two components for achieving micro actuator systems for narrow spaces have been fabricated and evaluated.

For a precise tool control unit, a micro servo motor using a micro ultrasonic motor and magnetic encoder has been designed and fabricated. For attitude control, a flexible displacement sensor and SMA actuator driven by an optical source were evaluated.

The performance of some micro actuator systems consisting of these components have been under testing in some experimental conditions.

Acknowledgments This research was supported by the Ministry of Education, Culture, Sports, Science, and Technology through a Grant-in-Aid for Scientific Research on Priority Areas, No. 438, "Next Generation Actuators Leading Breakthroughs".

References

1. Wakimoto S, Suzumori K, Kanda T (2005) Development of Intelligent McKibben Actuator. International Conference on Intelligent Robots and System 2005 (IROS 2005): 2271-2276
2. Kanda T., Makino A, Ono T, Suzumori K, Morita T, Kurosawa M K (2006) A Micro Ultrasonic Motor using a Micro-machined Cylindrical bulk PZT transducer. Sensors and Actuators A 127: 131-138
3. Kanda T, Makino A, Oomori Y, Suzumori K (2006) A Cylindrical Micro-Ultrasonic Motor Using Micromachined Bulk Piezoelectric Vibrator with Glass Case. Jpn. J. Appl. Phys. 45: 4764-4769
4. Kanda T, Oomori Y, Kobayashi A, Suzumori K (2006) Cylindrical Piezoelectric Vibrators For Micro Ultrasonic Motors. 10th International Conference on New Actuators: 592-595
5. Morita T, Kurosawa M, Higuchi T (1999) Cylindrical Shaped Ultrasonic Motor Utilizing Bulk Lead Zirconate Titanate (PZT). Jpn. J. Appl. Phys. 38: 3327-3350
6. Shinohara T, Dohota S, Matsushita H (2004) Development of a soft actuator with a built-in flexible displacement sensor. 9th International Conference on New Actuators: 383-386
7. Just E, Bingger P, Woias P (2004) Piezo-polymer-composite actuators a new chance for applications. 9th International Conference on New Actuators (ACTUATOR 2004), Bremen Germany :521-524,.
8. van der Smagt P, Groen F, Schulten K (1996) Analysis and control of a Rubbertuator arm. Biological Cybernetics: 75, 433-440
9. Leivo E, Wilenius T, Kinos T, Vuoristo P, Mantyla T (2004) Properties of thermally sprayed fluoropolymer PVDF, ECTFE, PFA and FEP coatings. Progress in Organic Coatings 49: 69-73
10. Yamamoto Y, Kure K, Iwai T, Kanda T, Suzumori K (2007) Flexible Displacement Sensor using Piezoelectric polymer for Intelligent FMA. International Conference on Intelligent Robots and System 2007 (IROS 2007): 765-770

References

[illegible]

Chapter 33
Development of Ultrasonic Micro Motor with a Coil Type Stator

Yuji FURUKAWA[1], Tomoyuki KAGA[2], Kenji MAKITA[3], Tetsuro WADA[3] and Akihei NAKAJIMA[3]

33.1 Introduction

The motor is one of the most important actuators and its miniaturization is strongly desired in the market in order to make electro – mechanical systems as small as possible.

Ultrasonic motors are commercially applied to a focussing device of camera and etc. and there exists further demands to miniaturize them so as to adopt to mobile phone, automobile actuators, and micro surgery devices like IVUS. However, the conventional ultrasonic motors are based on a surface propagation and transmission of vibration wave, inevitably, they need many parts and there is a certain limit to miniaturize.

Taking these points into account, the present paper proposes a new concept [1, 2] in which a coil type stator contacts on a cylindrical rotor directly, hence, the structure becomes very simple and can realize miniaturization.

Four different types of ideas and prototypes are proposed, manufactured and tested their fundamental performances.

[1] Yuji FURUKAWA

President ofthe Polytechnic University Japan

[2] Tomoyuki KAGA

Current Graduate Course Students, Tokyo University of Agriculture and Technology

[3] Kenji MAKITA, Tetsuro WADA, Akihei NAKAJIMA

Former Graduate Course Students, Tokyo University of Agriculture and Technology

33.2 Driving Principle of Ultrasonic Micro Motor with a Coil Type Stator

Fig.33.1 (a) shows schematic diagram of the developed ultrasonic micro motor. The prototype consists of an external vibrator, waveguide, stator and rotor. Waveguide made of stainless steel thin wire receives ultrasonic vibration from the vibrator that is attached to one end of it (hereinafter, the waveguide is described as an arm part of coil) and propagates the vibration to the coil type stator.

At the micro contact points between the outer surface of stator and the inner surface of rotor, rotational force acts toward helical direction, consequently, both a radical and axial driving force acts on the rotor as shown in Fig.33.1 (b). The details of mechanism are as follows. The micro frictional force works opposite to the forward direction of progressive wave at the contact points always. In addition, the frictional force is influenced by the axial direction of the coil as shown in Fig.33.1 (c). Therefore, the rotor receives both a rotary and aaxial force at the same time.

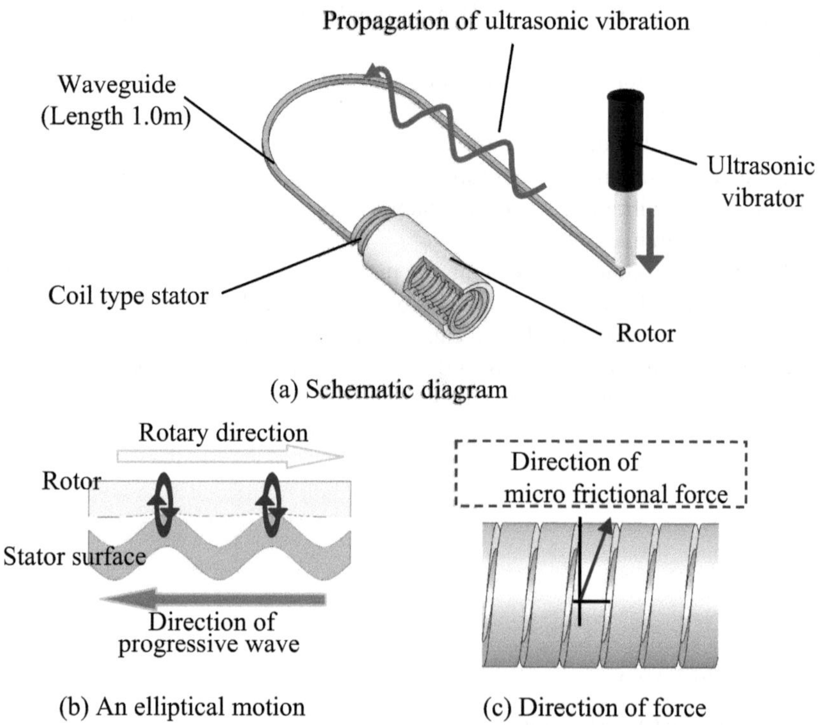

Fig. 33.1 Driving principle of the ultrasonic micro motor

33.3 Development of Ultrasonic Micro Motor with Coil-Type Stator

33.3.1 Structure of Ultrasonic Micro Motor

Fig.33.2 shows the structure of fabricated motor where a Langevin type ultrasonic vibrator is applied. The one end of the arm is fixed on a XYZθ Y stage which is set under the vibrator and pressed to the tip of vibrator. The progressive flexural wave propagates along the waveguide to the coil type stator. The cylindrical rotor is inserted into the rotor. A steel ball is bonded to the end of coil type stator to prevent the rotor throwing away axially. The size of motor is 0.8mm in diameter and 4.0mm in length. The photograph in right of Fig.33.2 is a cross sectional view of rotor and stator.

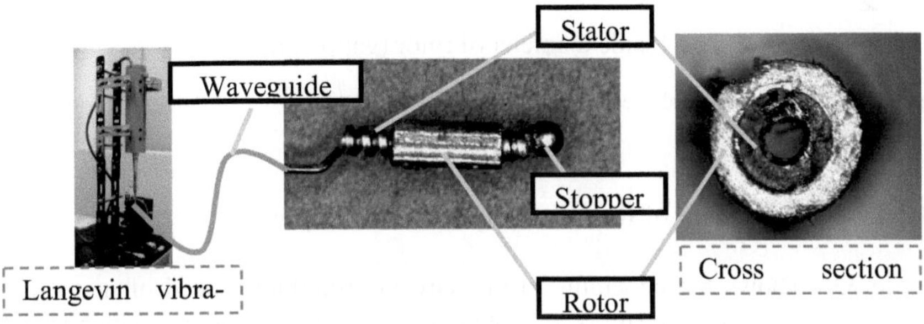

Fig. 33.2 Structure of micro motor

33.3.2 Driving Performance

33.3.2.1 Effect of Preload on Rotational Speed

The preload applied, between rotor and stator affects much on its friction driving performance. So, the gap between rotor and stator was set as shown in Table 33.1 , where the outside diameter of coil was kept at 534 μm while the inside diameters of rotor were varied. Under these combinations, the rotational speed was measured by a digital tachometer (SANWA Co. Ltd,SE-100). As can be seen from Fig.33.3, the tighter the gap is, the faster the rotational speed becomes, that may come from the higher frictional force with tighter gap.

Table 33.1 Inside diameter of rotors

Rotor No.1	1	2	3	4	5
Inside diameter [μm]	507.5	509.5	511	512.5	515
Gap [μm]	-1.5	+0.5	+2	+3.5	+6

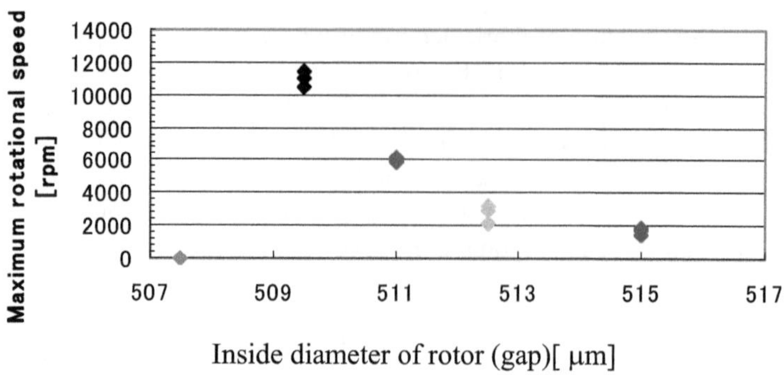

Fig. 33.3 Result of rotational speed in each gap

33.3.2.2 Measurement of Rotational Speed and Torque

Figure 33.4 shows the apparatus to measure the rotational speed of motor precisely, in which, a laser beam is irradiated onto a small teeth of gear which is connected to the rotor so as to reflect it intermittently, and this reflected light is caught by a laser displacement sensor whose data are stored in a memory of oscilloscope. The rotational speed is calculated from a series of pulse output. On the other hand, the torque was calculated from a rotational angular acceleration and a required rise time when starting up the motor at rest.

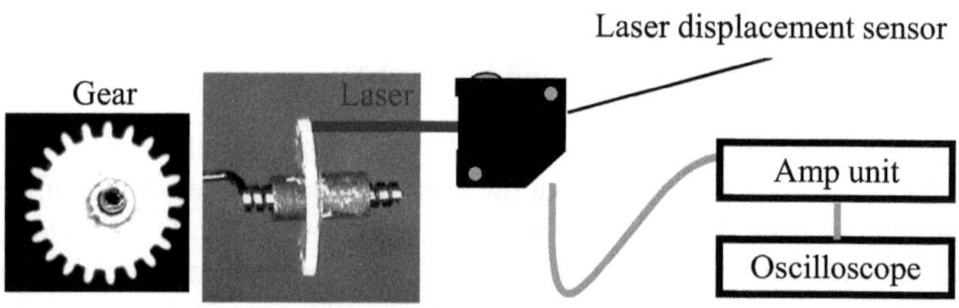

Fig. 33.4 Measurement system of rotational speed

The result of rotational speed and torque is shown in Fig.33.5 where the rotor No.1 specification of Table 33.1 was adopted. Look the transitional period of speed with time from at rest to steady state and the curve suggests the dynamics almost follows to a first order time lag system. On the other hand, the speed vs.

torque, namely N-T curve, can be calculated taking an approximated curve of N-t and inertia of motor, and the dashed line is almost inverse proportional. The deviation may be caused by an uncertainty of the frictional behavior between rotor and stator.

The starting torque was so small as 0.2μNm and seems not enough for a practical use and some better method to give optimum preload has to be developed.

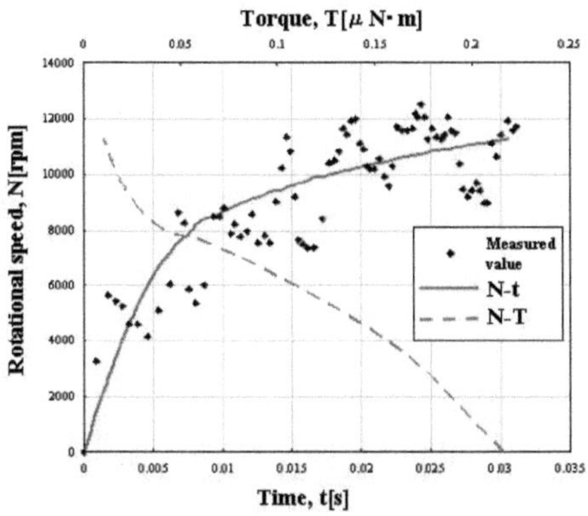

Fig. 33.5 Result of measured performance

33.3.3 Development of Ultrasonic Micro Motor with Outer Case

The rotor must be covered with outer case when practical use is considered. Fig.33.6 and Table 33.2 show the photograph of prototype and its specification. Stainless steel was used for rotor and stator and polyimid resin (Ti polymer) was adopted for the case taking smoother rotation and smaller wear at the same time. By this design, the rotor is supported both radically and axially.

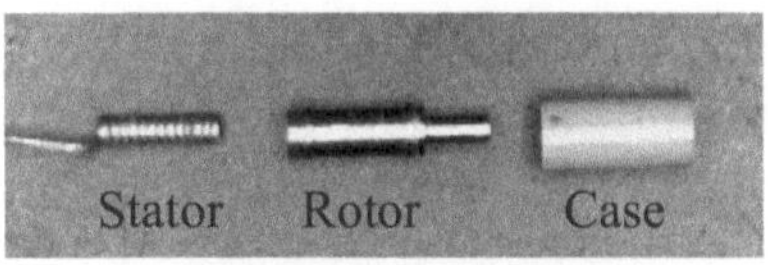

(a) Component of motor

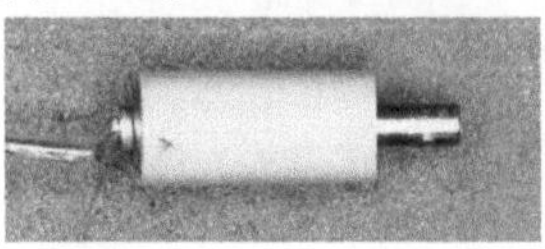

(b) Prototype of motor

Fig. 33.6 Photograph of prototype

Table 33.2 Specification of prototype

Wire size[mm]	0.15
Outer diameter of stator[mm]	0.55
Pitch of coil [mm]	0.18
Material of stator	SUS304
Outer diameter of casing[mm]	1.3
Length of motor [mm]	3.8
Length of waveguide [m]	1.0

In spite of these design, the prototype rotates rather in-stably as shown in Fig.33.7 that may be cased from an unsteady contact between rotor and stator as in the case of 1st prototype.

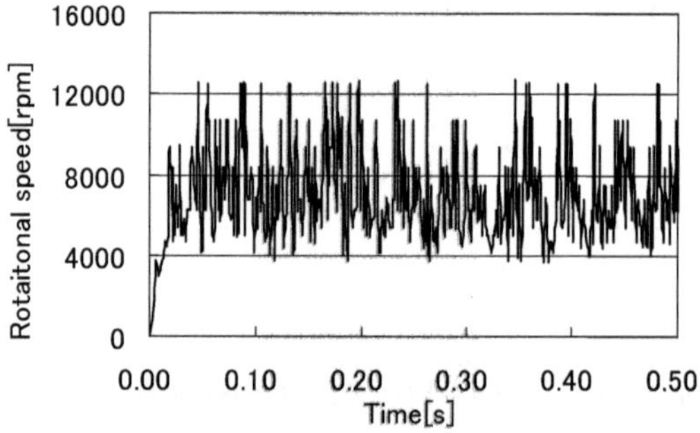

Fig. 33.7 Rotational speed of motor with case

This second prototype was miniaturized by the help of wrist watch production of Seiko Instruments Inc. and the world thinnest motor with 0.95mm diameter was realized as shown in Fig.33.8, the specification of which is in Table 33.3.

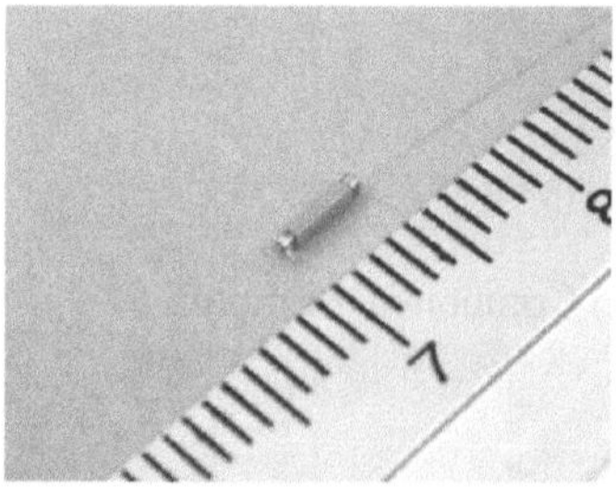

Fig. 33.8 Miniature motor (ϕ0.95mm)

Table 33.3 Specification of motor

Wire size[mm]	0.052
Outer diameter of stator [mm]	0.288
Material of stator[mm]	SPRON510
Rotational speed [rpm]	4000
Starting torque [μNm]	0.2

33.4 Development of Ultrasonic Micro Motor with Foil Type Stator

33.4.1 Principle of Ultrasonic Micro Motor with Foil Type Stator

In order to make a motor smaller, the coil of stator should be made of thin foil and Fig.33.9 (a) shows the schematic diagram of developed ultrasonic micro motor with foil type stator. Ultrasonic wave is propagated along the waveguide to the foil type stator instead of coil. As can be seen in Fig.33.9 (b), the vibration draws an elliptical motion at any point of surface, however, it saturates and decays with thinner foil such as less than 50μm taking the amplitude of vibration into account. The rotor rotates in the same direction of propagation of the progressive wave because the frictional force which is induced by expansion of stator by the ultrasonic wave becomes larger than the frictional force which is induced by the elliptical locus of ultrasonic wave.

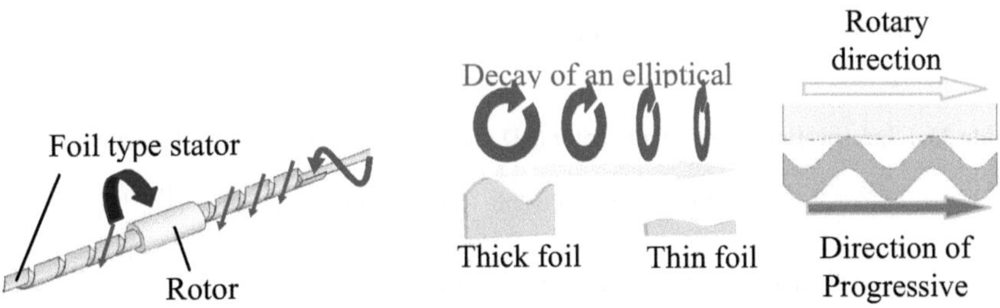

(a) Schematic of diagram (b) Decay of an elliptical motion

Fig. 33.9 Principle of ultrasonic micro motor with foil type stator

33.4.2 Prototype of Ultrasonic Micro Motor with Foil Type Stator

A prototype of ultrasonic micro motor with 2.7 mm outer diameter and 9.1 mm length has been developed as shown in Fig.33.10. It consists of a rotor, a stator (foil made coil), a case and a stopper, each size of which is shown in Table 33.4. The arm has 1.0 m length. The rotor is so placed around the stator as to fit smoothely between them adjusting the outer diameter of stator with the inner diameter of rotor. The rotor is covered with a cylindrical case and supported axially and radially by fitting one end of the case with the stopper. The foil and case are made of stainless steel and TI polymer respectively in order to give enough heat and wear resistance.

When an ultrasonic vibration is applied to the arm, a progressive wave propagates along with a waveguide and reaches to the foil type stator, and the rotor is driven at small contacts between the rotor and stator according to the principle described in Fig.33.1.

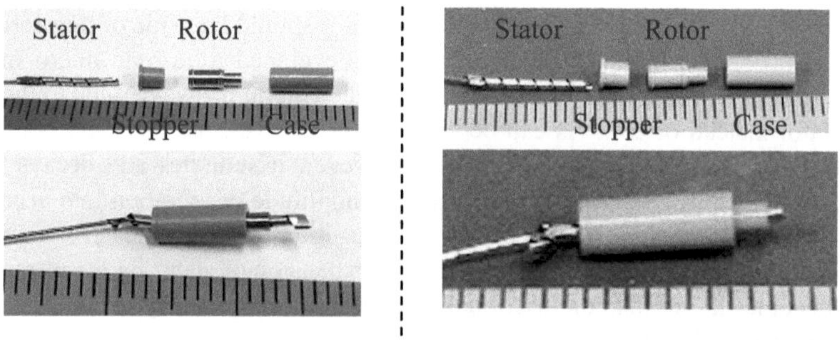

(a) Motor with rotor of stainless steel (b) Motor with rotor of Ti polymer

Fig. 33.10 Prototype of ultrasonic micro motor with foil type stator

Table 33.4 Specification of prototype motor (I.D.: inner diameter, O.D.: outer diameter)

[mm]

	I.D.	O.D.	Length	Thick	Width	Pitch
Rotor	0.9	2.2	6			
Stator	0.7	0.88	12	0.05	1.2	0.3
Casing	2.2	2.7	7			

33.4.3 Performance of Ultrasonic Micro Motor with Foil Type Stator

33.4.3.1 Direction of Rotor Rotation

In order to check the effect of rotor/stator arrangement on a rotational direction of motor, two different combinations were prepared, where the rotor is set outside of stator or inside. However, both of them rotated to the same direction of propagation of progressive wave and was opposite to the former ultrasonic motor described in Section 33.3 and 33.4.

33.4.3.2 Measurement of Rotational Speed and Torque

Table 33.5 compares the rotational speed of coil type and foil type stator. The former cross section is rectangular of 0.15mm×0.15mm, and the latter is thin foil rectangular of 0.05mm×1.2mm. For the same input of ultrasonic vibration, the foil coil could generate higher speed than the coil type stator, however, the motor with TI polymer-made rotor fluctuated between 2500-5300 rpm and that of stainless rotor 2500-3500rpm. The reason why the rotational speed does not stabilize seems be affected by rather unstable micro frictional conditions between rotor and stator surface, that must result in unstable frictional force, hence unstable rotational speed. This was verified from the fact that the tendency became more predominant with larger gap between rotor and stator.

Table 33.5 also shows the measured result of starting torque. In the case of foil type motor, it was found that a stainless-made rotor resulted in a higher starting torque than that of TI polymer-made. And the torque was 35μNm three times larger than that with foil type stator of same size. As the foil type stator is very flexible structurally both in radial and axial direction, the contact force, hence the friction force decreases and that might result in a lower torque. If the foil can be rigidly fixed to the case while keeping the transmission of ultrasonic vibration, it may be possible to miniaturize the motor with the foil coil.

Table 33.5 Measurement result of rotational speed and starting torque

Stator type	Material	Rotational speed	Starting torque
Foil type	Stainless steel	2500-3500rpm	12.4μNm
	Ti polymer	2500-5300rpm	10.2μNm
Coil type	Stainless steel	2500rpm	35.0μNm

33.4.3.3 Measurement of Wear of Rotor

It is necessary to inspect the amount of wear because it affects much on the driving force of motor. The inner diameter of rotor was measured three times after a certain interval of rotation as shown in Table 33.6. The TI polymer-made worn 7μm after 10 minutes drive, on the other hand, that of stainless steel reached to 40μm. It is required to develop better wear resistant material than TI polymer.

Table 33.6 Measurement result of wear of rotor

Material	0 minute	1 minute	10 minute
Stainless steel	0 μm	3 μm	40μm
Ti polymer	0 μm	2 μm	7 μm

33.5 Development of Direct Drive Ultrasonic Micro Motor

33.5.1 Development of Ultrasonic Micro Motor with Shortened Waveguide

In order to avoid attenuation of ultrasonic vibration in the waveguide, the PZT vibrator was attached to the coil type stator as shown in Fig.33.11, and the Langevin vibrator was omitted. Fig.33.12 shows the photograph of prototype that consist of a coil type stator, a rotor, a waveguide (Length: 2.0mm) and an ultrasonic PZT vibrator. An ultrasonic wave is exited at the PZT vibrator when supplied alternative voltage, and it propagates to the coil type stator directly. Table 33.7 shows its specification.

Table 33.8 shows the measured performance of prototype when the voltage changed every 10V. It was confirmed that design principle of Fig.33.11 can rotate the rotor, however, both of rotational speed and torque were not stable, so the vibration mode of PZT beam, the propagation of vibration, and the loss of transmission at the contact, etc. must be carefully adjusted.

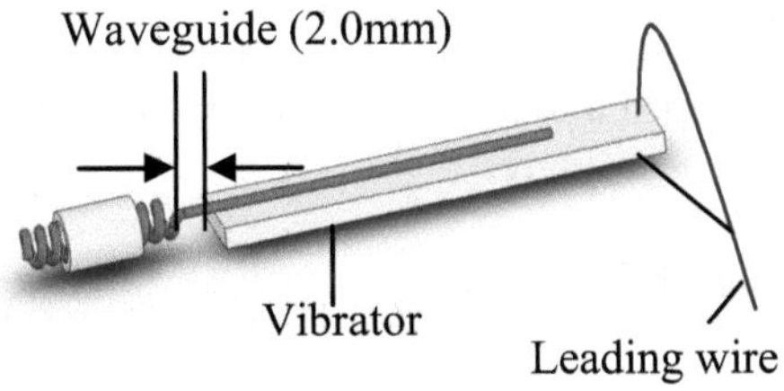

Fig. 33.11 The schematic of prototype

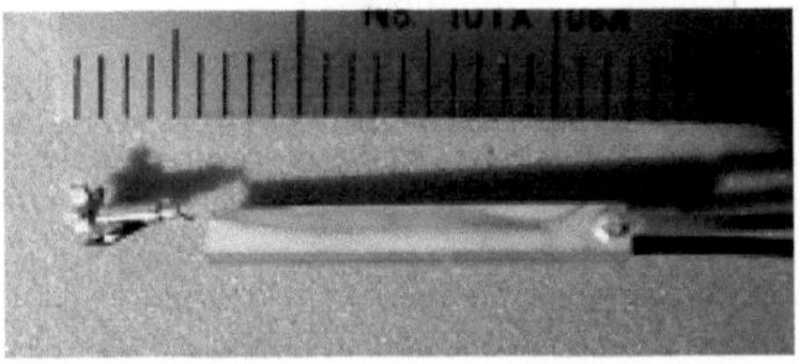

Fig. 33.12 The photograph of prototype

Table 33.7 Specification of prototype

Vibrator				Waveguide	Stator	Rotor
Frequency	Length	Thickness	Width	Length	O.D.	O.D.
100kHz	16.2mm	1.01mm	2.05mm	2.0mm	0.55mm	1.3mm

Table 33.8 Measured results of rotational speed and starting torque

Voltage[$V_{p\text{-}p}$]	10	20	30	40	50	60	70
Rotational Speed[rpm]	0	550	650	700	750	800	820
Starting torque[μ Nm]	0	0.25	0.11	0.35	0.48	0.16	0.47

33.5.2 Development of Ultrasonic Micro Motor with PZT Made Stator

It will bring more miniaturized ultrasonic micro motor if the stator itself can possess both the functions of vibrator and driver. So, the stator itself was made of PZT as shown in Fig.33.13, where the coil type PZT stator generates and propagates an ultrasonic vibration once excited by alternative voltage. For this purpose, a tube PZT was ground to a coil with a specified pitch by a specially designed grinding machine.

Figure 33.14 and Table 33.9 show the photograph and the specification of prototype respectively. This could rotate, although still unstable, when the alternative voltage of $90V_{p\text{-}p}$ is applied. It is necessary to verify the vibration mode of PZT coil and how the vibration is converted to the rotational force. However, the proposed principle can surely open a new way to develop an ultimately miniaturized ultrasonic micro motor in a near future.

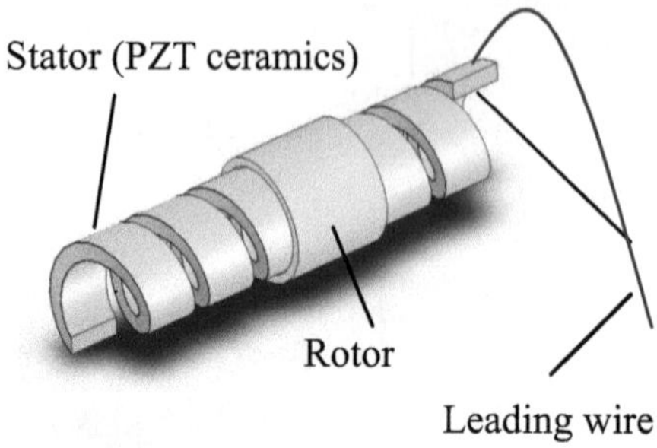

Fig. 33.13 The schematic of prototype

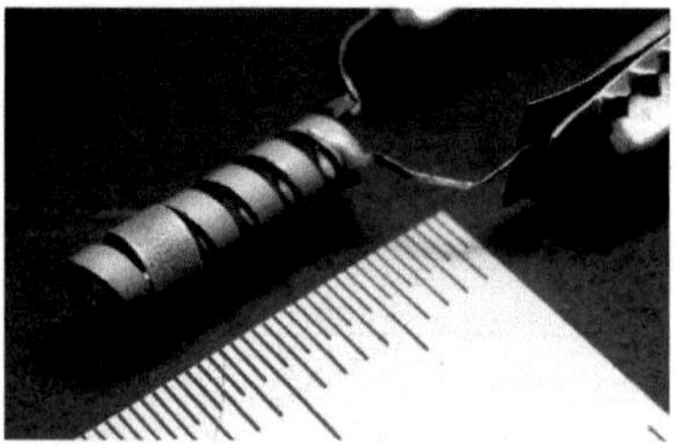

Fig. 33.14 The photograph of prototype

Table 33.9 Specification of prototype

Stator (Material: PZT ceramics)					Rotor
I.D.	O,D,	Length	Pitch	Board width	O.D.
2.0mm	3.0mm	15mm	2.0mm	1.2mm	3.2mm

33.6 Conclusion

In the present paper, new ideas of ultrasonic micro motors were proposed, their prototypes were prepared, and their performance was investigated.

First, the coil type stator could result in a simplified design and could realize less than 1mm diameter motor.

Secondly, the coil was replaced by a foil, and the possibility to miniaturize the motor size more was verified.

Thirdly, PZT made micro vibrator was attached to the stator directly and the waveguide could be omitted. This can miniaturize the whole system more.

Fourthly, the coil type stator itself was made of PZT, consequently, the motor system consists of only stator and rotor, which gave the simplest structure of ultrasonic micro motor.

However, there still remain many technological barriers before the proposals can be put in practical use. They are, to make the driving principle of the motor with coil type stator and foil type stator clear, to simulate the driving mechanism using finite element method, to miniaturize, to increase torque, and to minimize wear of rotor.

References

1. Moriya T, Furukawa Y, Akano Y, and Nakajima A (2005) Experimental Study on a Miniature Ultrasonic Motor Using a Coiled Stator : IECEI Technical Report, No. US2005-29, July, 2005 (in Japanese).
2. Nakajima A (2006) Development of a New Traveling Wave Ultrasonic Micromotor : MS thesis, Tokyo University of Agriculture and Technology.

Chapter 34
Self-Running Non-Contact Ultrasonically Levitated Stage

Daisuke Koyama[1] and Kentaro Nakamura[1]

Abstract A noncontact self-running ultrasonically levitated stage is discussed. First, a slider for a self-running standing wave-type, ultrasonically levitated, thin linear stage is proposed. The slider has a simple configuration and consists of an aluminum vibrating plate and a piezoelectric zirconate titanate (PZT) element. The slider can be levitated and moved using acoustic radiation force and acoustic streaming at 68 kHz and 69 kHz with the asymmetric vibration distribution of the slider's vibrating plate. A noncontact self-running ultrasonically levitated two-dimensional (2D) stage was developed. The proposed 2D stage was fabricated from a rectangular aluminum plate and four cantilever-type vibrating plates were integrated. The flexural lattice vibration mode on each vibrating plate assists the stage movement, and noncontact movement of the stage in two dimensions could be achieved by controlling the driving frequency.

34.1 Introduction

Noncontact sliding tables are employed in the fields of nanotechnology and precision machining and can resolve several problems occurred on contact-type actuators such as aging phenomena by abrasion, dust and noise generation, and oscillation of sliders. Especially, ultrasonically levitated actuators utilizing high intensity ultrasound have several advantages compared with air bearing and magnetic systems such as a simple configuration, low-profile and silence in operation. The authors have been investigating ultrasonically levitated sliding tables based on Near-Field Acoustic Levitation (NFAL) [1-3]. NFAL is a phenomenon in which the light plane object on a vibrating plate can be levitated by the acoustic radiation force from a vibrating plate [4, 5]. The acoustic radiation force is a static force produced by sound waves that is capable of levitating an object in the *near-field*

[1] Daisuke Koyama and Kentaro Nakamura

Precision and Intelligence Laboratory, Tokyo Institute of Technology compared with the wavelength of sound in air when both radiation force and the gravity force of the object are balanced. These noncontact type ultrasonic actuators utilize the traveling waves propagating along the stator beams or the slider itself in their movement mechanism. By generating the traveling wave, the acoustic streaming can be induced along the air gap, and the thrust force to the slider can be generated through the viscosity force of the air. It is well-known that acoustic streaming can be also induced in the field of asymmetric acousitc standing wave fields. First, the standing-wave-type ultrasonically levitated slider for a self-running sliding stage was investigated, and the noncontact straight back-and-forth motion of the slider was aimed. The motivation of this linear slider was the use for a compact-size noncontact 2D stage without guide rails which the conventional ultrasonically levitated table requires.

34.2 A Self-Running Standing Wave-Type Bidirectional Slider for the Ultrasonically Levitated Thin Linear Stage

The generation of acoustic streaming is necessary for most noncontact ultrasonic motors. The goal was to realize a standing wave-type ultrasonic noncontact slider for a linear stage with a simple and thin structure. The configuration of the slider is shown in Fig.34.1. The standing wave-type levitated slider consists of an aluminum plate ($30.0\times10.0\times1.0$ mm^3) and a PZT plate ($10.0\times10.0\times1.0$ mm^3). The polarization of the PZT plate is in the thickness direction, and the two components were bonded using epoxy and the mass of the slider was 1.62 *g* (52.9 N/m^2). By applying the input voltage to the PZT, flexural vibrations can be generated along the aluminum plate at frequencies ranging from 20 to 100 kHz (Fig.34.1(b)). The slider on the flat substrate can be levitated in the vertical direction (z-direction) by the acoustic radiation force radiated from the plate of the slider when the slider gravity and the radiation force are balanced. When the vibration distribution along the plate is asymmetric in the length direction (x-direction), the acoustic field in the air gap between the flat substrate and the levitated slider will also be asymmetric, and acoustic streaming will be induced along the air gap in the length direction. The slider can be moved in the same direction as the acoustic streaming due to the acoustic viscosity force [6-8]. The vibration mode of the slider can be changed by controlling the driving frequency, and the moving direction can be also altered with a single-phase drive. The size of the slider was determined through the finite element analysis (FEA) to obtain the large asymmetric vibration distribution for high thrust and levitation performance.

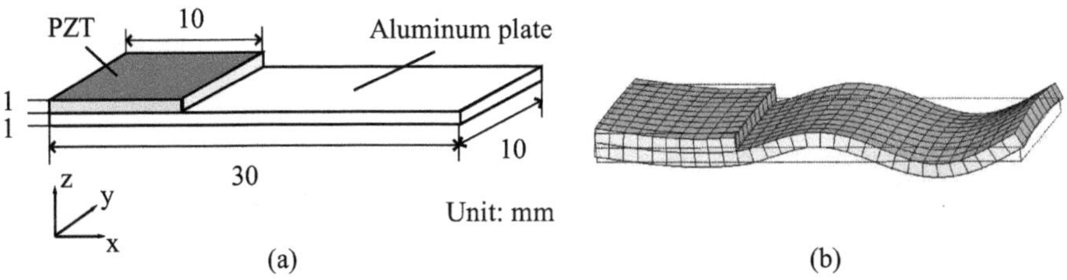

Fig. 34.1 Slider for a standing wave-type, self-running, ultrasonically-levitated sliding table: (a) configuration of the slider, and (b) flexural vibration mode at 35 kHz predicted by FEA

34.3 Estimation of the Slider Moving Direction by FEA

To change the direction of the slider, the direction of the acoustic streaming was estimated by calculating the sound pressure distribution in the air gap between the slider and the flat substrate. The slider is statically inclined due to the asymmetric vibration mode in the length direction when the slider is levitated by the acoustic radiation force. Considering the inclination of the slider, the sound pressure distribution in the inclined air gap was calculated by the FEA. The simulation model of the FEA is shown in Fig.34.2. The slider is surrounded by an air block. The displacement constraint condition was applied on the upper surface of the air block, implying that the upper surface was assumed to be the flat rigid substrate. The sound pressure distribution in the air gap was calculated via piezoelectric-structure-acoustic interaction analysis of the FEA. The levitation distances of both ends of the slider are 0.1 and 1 mm. The driving force of acoustic streaming was calculated from the computed results of the sound pressure distributions. The driving force of the acoustic streaming F could be expressed as [9]:

$$F = -\overline{(v\nabla)v} - \overline{v \operatorname{div} v}, \tag{34.1}$$

where v is particle velocity. The particle velocity v can be expressed as $v = j/\omega\rho \cdot \mathrm{grad}P$, where ω is the angular frequency, ρ is the density of the fluid, and P is the sound pressure, and can be calculated from the computed sound pressure distribution. The calculated sound pressure distributions and the driving force of acoustic streaming in the inclined air gap are expressed as gray scales and vector lines, as shown in Fig.34.3. The prominent driving frequencies with high sound pressures were 73 and 71 kHz. In both cases, the sound pressure on the left side is larger than that on the right side because the air gap is smaller on the left side. In Eq. 1, note that the driving force is derived from the gradient of the sound pressure distribution, and the generation of the acoustic streaming originates with the

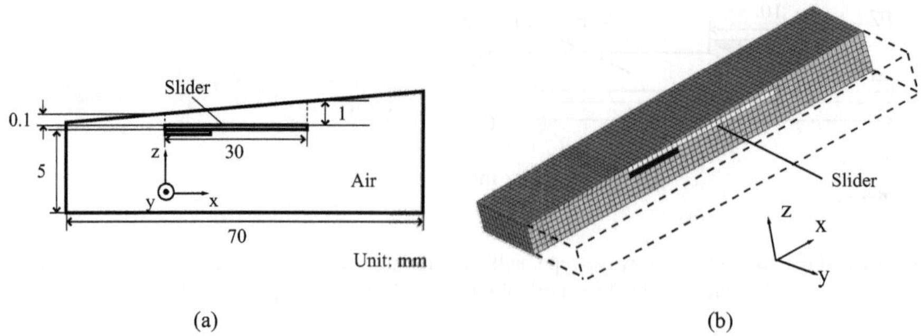

Fig. 34.2 Simulation model of the FEA. The slider is assumed to be statically inclined by the acoustic radiation force

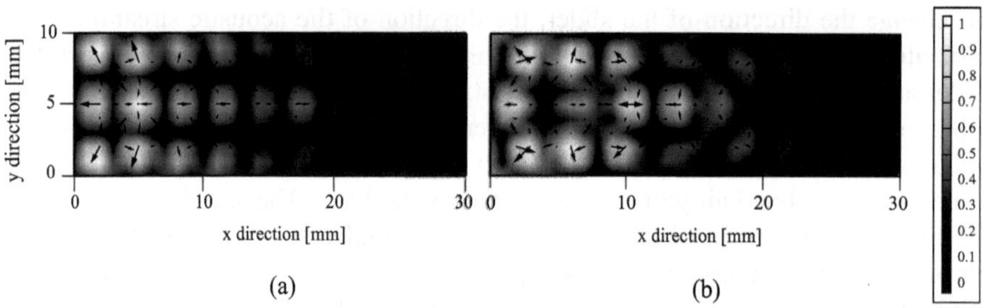

Fig. 34.3 Computed sound pressure distributions and the driving force of the acoustic streaming in the inclined air gap at (a) 73 kHz and (b) 71 kHz

asymmetric gradient of the sound pressure distribution. The driving forces were calculated at all nodes of the air gap (30.0×10.0 mm^2). We also defined the summation of the x-direction components of the diving forces at all nodes as the total driving force of the slider. Since the largest positive and negative driving force peaks appeared at 71 and 73 kHz, acoustic streaming is predicted to be generated in the positive and negative directions along the x-axis at the corresponding frequencies.

34.4 Levitation and Propulsion Characteristics of the Slider

The levitation and thrust characteristics of the slider were investigated. A slider driven using an input voltage of 80 V_{pp} can be levitated on the flat substrate at driving frequencies of 35, 68, and 69 kHz, and could be moved bi-directionally at 68 and 69 kHz. The vibration amplitude distributions of the slider at 68 and 69 kHz measured by a laser Doppler velocimeter are shown in Fig.34.4. The results at y = 0 mm are expressed as curved lines. The calculated results obtained by FEA in the previous section are also shown. These experimental and calculated results are in good agreement, although the corresponding frequencies between the experimental and calculated results are reversed, i.e., the experimental results at 68 and 69 kHz correspond to the calculated results at 73 and 71 kHz, respectively. The vibration amplitude of the right side of the plate (10 to 30 mm in the x-direction) was larger than that on the left side (0 to 10 mm in the x-direction), and the asymmetric vibration distribution in the length direction of the slider could be realized. The vibration modes of the slider plate at 68 and 69 kHz exhibited a lattice-shape vibration mode (which we term the *vibration mode A*) and a complex combination of the flexural and the lattice modes (which we term the *vibration mode B*).

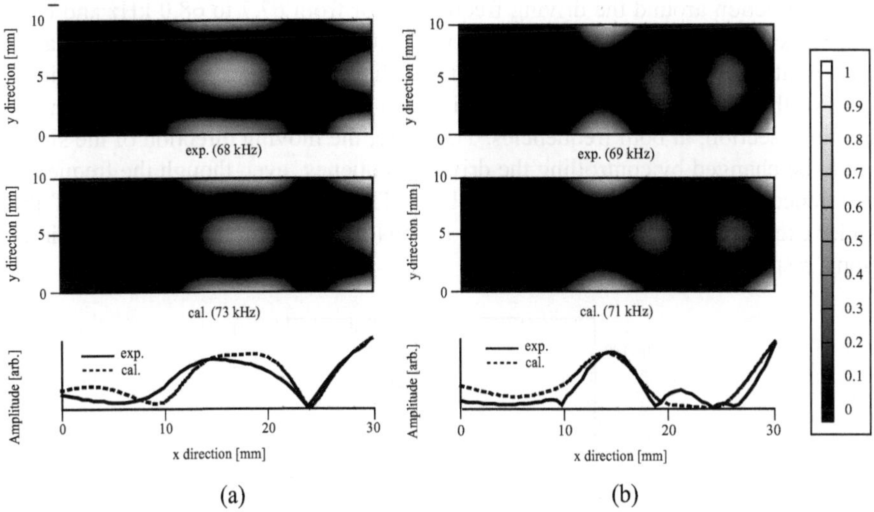

Fig. 34.4 Vibration amplitude distribution of the slider at (a) 68 kHz, and (b) 69 kHz (from top, experimental, calculated result, and plots at y = 0 mm)

According to the NFAL theory [10, 11], when the vibration mode of the sound source is the in-phase piston vibration mode, the levitation distance of the flat object h can be expressed as

$$h = cu\sqrt{\frac{1+\gamma}{4w}\rho}, \qquad (34.2)$$

where c is the speed of sound in air, u is the displacement amplitude of the vibrating plate, w is the weight of the levitated object, γ is the ratio of the isobaric specific heat to the isovolumetric specific heat in air, and ρ is the density of air. In Eq. 2, the levitation distance h is proportional to u. Figure 34.5 shows the relationship between the levitation distance of the slider and the vibration displacement amplitude at 68 and 69 kHz. The vibration displacement amplitude indicates the maximum displacement amplitude of the slider, and the levitation distance of the slider indicates that in the steady state. The measurement points of the levitation distance of the slider were the ends of the slider in the x-direction, i.e., x = 0 and 30 mm.

Greater levitation distances h could be obtained with the greater displacement amplitude u at both frequencies, and h is proportional to $u^{1.1}$ and $u^{1.5}$ at 68 and 69 kHz. The levitation distance at x = 30 mm is approximately ten times greater than that at x = 0 mm because the vibration displacement amplitude is larger. The slider is tilted approximately 0.1° from the x-y plane when the slider is levitated. Around the driving frequencies of 68 and 69 kHz at which the slider could be moved, the thrust of the slider was measured. Figure 34.6 shows the frequency characteristics of the thrust force of the slider. The plots and error bars express the average values and the standard deviations as calculated from the experimental results on 10 samples at each frequency. The slider could be moved in both the negative and positive x-direction around the driving frequencies of from 67.7 to 68.0 kHz and 68.2 to 68.9 kHz, respectively. The slider could not be levitated at 68.1 kHz because the vibration displacement amplitude of the slider was small. The moving directions of the slider corresponded to the direction predicted by FEA, as shown in the previous section, at both frequencies. Therefore, the moving direction of the slider could be changed by controlling the driving frequency, even though the frequency difference was only approximately 1 kHz. The maximum thrust forces of the slider excited with 80 V_{pp} were 2.8 and 1.8 mN in the negative and positive directions, respectively.

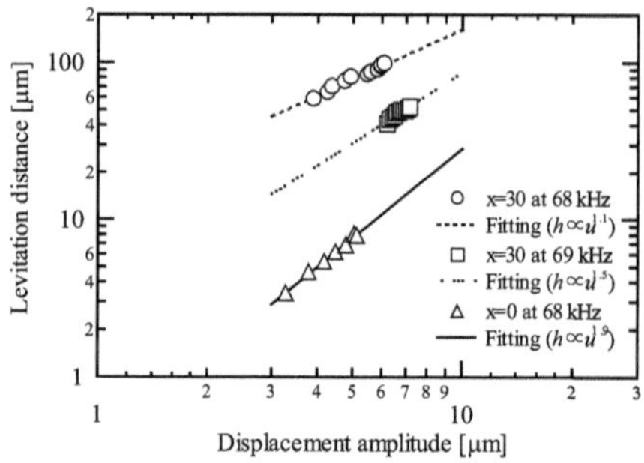

Fig. 34.5 Relationship between the levitation distance of the slider and the vibration displacement amplitude at 68 kHz and 69 kHz

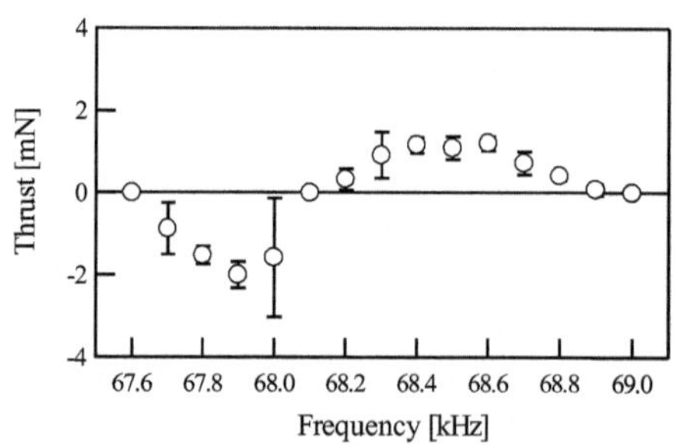

Fig. 34.6 Frequency characteristics of the thrust force of the slider

34.5 Configuration of the 2D Stage

We propose the ultrasonically levitated 2D stage, in which the four self-running linear sliders are integrated to a base plate, and the vibrations of these four sliders will control the thrust of the 2D stage in the four directions, i.e., positive x-, y-, and negative x-, y-directions. To isolate the performances of each embedded slider, the four sliders with different lengths were designed to have different resonance frequencies. The lengths of these sliders were determined via the FEA. The FEA was performed with a single slider shown in Fig.34.1 (a) having the different length. No fixing condition was applied in the slider model while four sliders will be connected via a 1-mm-thickness rectangular base plate. Figure 34.7 shows the relationship between the slider length and the resonance frequency of the vibration mode A calculated by the FEA. With the slider length of 25 to 30 mm, the lower resonance frequencies of the mode A can be obtained with the larger slider length. From the experimental result shown in Fig.34.4, the resonance frequency of the vibration mode B was approximately 1 kHz higher than that of the mode A. Now we focus on the vibration mode A on the slider since the larger thrust will be obtained than in the vibration mode B, as mentioned above. From the computed results by the FEA, the lengths of the four sliders were set at 27, 28, 29 and 30 mm with resonance frequencies of 79, 77, 75 and 72 kHz, respectively. Figure 34.8 shows the configuration of the 2D stage. The proposed 2D stage was machined from a rectangular aluminum plate ($41\times41\times1$ mm^3), and the four vibrating plates, which are replaced by the previous sections, were integrated around the base plate. The vibrating plates are cantilever-type and PZT elements ($10\times10\times1$ mm^3) are attached on their fixed-end of the vibrating plates and electric wires are connected to these four PZT elements. The mass of the stage was 7.4 g. The four vibrating plates, two pairs both in x- and y-directions, on the channels 1, 2, 3 and 4 have different lengths of 30, 29, 28 and 27 mm, respectively, to isolate the performances of each vibrating plate.

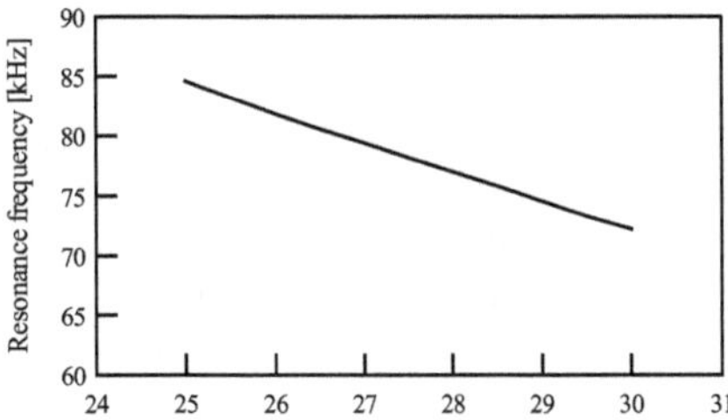

Fig. 34.7 Computed (FEA) relationship between the beam length of the slider and the resonance frequencies of the vibration mode A

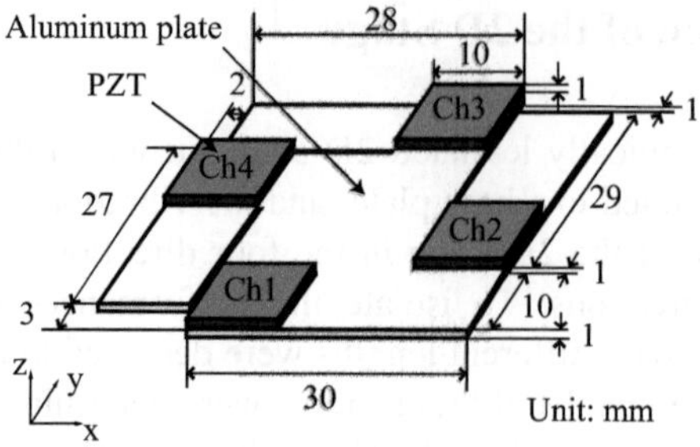

Fig. 34.8 2D self-running ultrasonically levitated stage

34.6 Levitation and Propulsion Performance of the 2D Stage

The four PZTs were excited with the same in-phase driving signal. By sweeping the driving frequency from 68 to 80 kHz, the vibration mode of the 2D stage was investigated. The vibration displacement amplitude of the four vibrating plates was measured by using the LDV. Figure 34.9 shows the frequency characteristics of the vibration displacement amplitude of the four vibrating plates. The measured position on the vibrating plate is the center in the width direction and the opposite side from the PZT, as shown in same figure by the appearance of the loop of flexural vibration in mode A. There were several peaks on the four vibrating plates from 68 to 80 kHz, and the vibration mode A is excited on the vibrating plates 1, 2, 3 and 4 at 71, 72, 74 and 77 kHz, respectively. These experimental resonance frequencies are close to the expected values compared with the predicted values by FEA in the previous section. The vibration displacement amplitude distributions on the entire 2D stage were measured using the LDV at 71, 72, 74 and 77 kHz, and the experimental results are shown in Fig.34.10. At the frequencies of 71, 72, 74 and 77 kHz, the displacement amplitude of the vibrating plates on the channels 1, 2, 3 and 4 were prominent compared with those of the others, and vibration mode A could be observed. However, the coupling vibration was seen in some cases: for example, the displacement amplitude on the channel 2 was also large at 74 kHz when vibration mode A was excited in the channel 3. In this case, the vibration mode B on the channel 2 was excited since vibration mode B is excited at approximately 2 kHz higher than the resonance frequency of mode A at 72 kHz.

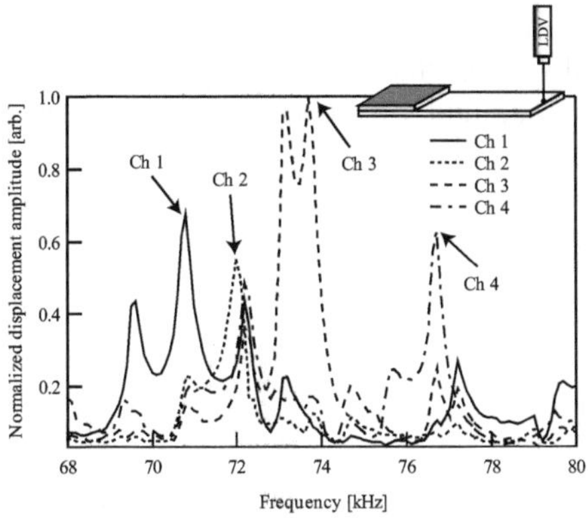

Fig. 34.9 Frequency characteristics of the vibration displacement amplitude of the four vibrating plates on the channels 1 to 4. The vibration was measured at the center in the width direction of the vibrating plate (upper right)

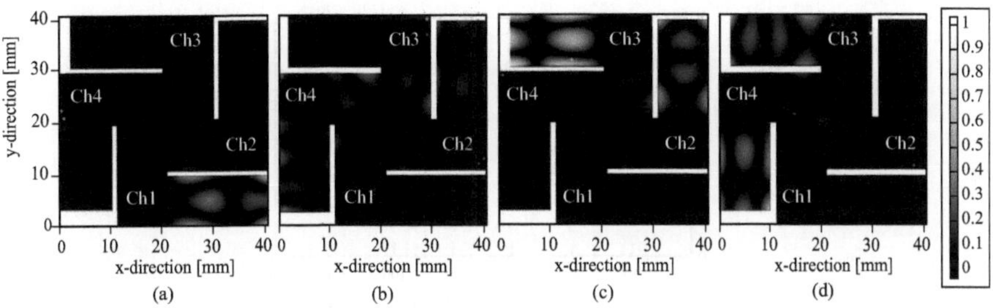

Fig. 34.10 Vibration amplitude distribution of the stage at (a) 71 kHz, (b) 72 kHz, (c) 74 kHz, and (d) 77 kHz

The levitation characteristics of the stage were investigated. The stage on the flat substrate could be levitated with the driving voltage of 80 V_{pp} at several frequencies from 60 to 100 kHz including above-mentioned four frequencies, 71, 72, 74 and 77 kHz. At most of these resonance frequencies, a stable thrust force could not be obtained although the stage could be just levitated. The representative levitation property, which is the relationship between the vibration displacement amplitude and the levitation distance of the stage at 74 kHz, was shown in Fig.34.11. The driving voltages of the four PZTs were varied from 60 to 130 V_{pp} and the vibration displacement amplitude indicates the maximum value on the stage. The levitation distance indicates that at (x, y)=(2 mm, 41 mm) in Fig.34.10. A greater levitation distance h can be obtained with a larger displacement amplitude u, and h is proportional to $u^{1.1}$. Under the same driving conditions, the levitation distance differed with the measuring position: this means the stage was statically inclined owing the vibration distribution shown in Fig.34.10. From the result at 74 kHz shown in Fig.34.10 (c), the vibration displacement amplitude on the other three channels 1, 2 and 4 were smaller than that on channel 3. The levitation distances on the channels 1, 2 and 4, which were measured at (x, y)=(41 mm, 0 mm), (41 mm, 40 mm) and (0 mm, 3 mm), respectively, were also smaller than that shown in Fig.34.11. These results mean that the stage was inclined around the x- and y-axes, and the inclination around y-axis will assist the stage motion in x-direction [12,13].

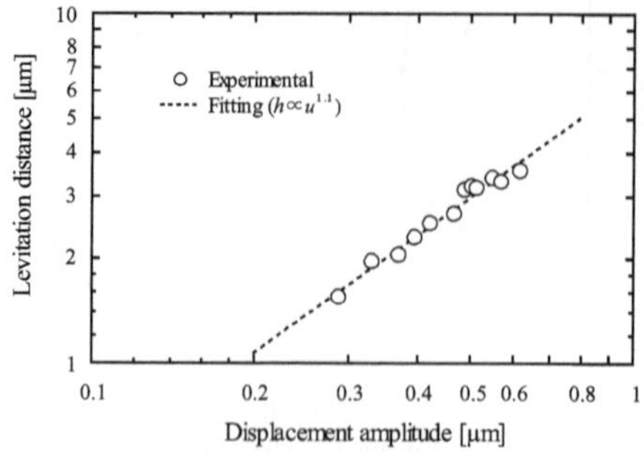

Fig. 34.11 Levitation distance versus the vibration displacement amplitude of the stage at 74 kHz

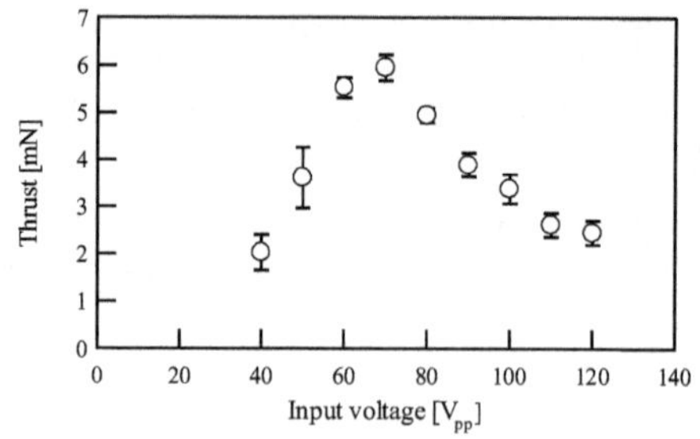

Fig. 34.12 Thrust force of the stage versus the input voltage to four PZTs

The propulsion characteristics of the stage were investigated. The stage was moved with the diving voltage of 120 V_{pp} at several frequencies at which the stage could be levitated. The four PZT elements were excited with same in-phase driving signal. The thrust characteristics of the stage are summarized in Table 34.1. When the stage movement was restricted to the straight motion in the x-direction, the stage rose and moved in the negative and positive x-directions at 71 and 74 kHz with the thrust of 2.0 and 2.7 mN, respectively. Likewise, when the stage movement was restricted in the y-direction, the stage moved in the negative and positive y-directions at 72 and 77 kHz with the thrust of 2.0 and 1.7 mN, respectively. With the single-phase driving of the four PZTs, the stage motion curved, and it is difficult to realize straight motion without the guide rails since the vibration distribution of the other three vibrating plates, except that with vibration mode A affects the stage motion. Straight motion may be realized by independently controlling the driving voltages and the frequencies of each PZT and using vibration mode B, which induces the thrust in the opposite direction from that of the mode A. By changing the driving frequency, the moving direction of the stage could be controlled in two dimensions with single-phase drive. It is obvious that the moving direction is attributed to flexural mode A on the vibrating plate with the prominent vibration amplitude, and the moving directions of the stage correspond to those of our linear slider. The thrust of the stage was measured with changing the driving voltage. Figure 34.12 shows the relationship between the driving voltage and the thrust in the positive x-direction at 74 kHz. The experimental values and the error bars indicate the averaged values of 5 times and the standard deviations. The stage could not be levitated by a driving voltage under 40 V_{pp} owing to the small vibration amplitude. When the stage was excited with the driving voltage of 40 to 70 V_{pp}, the thrust could be increased with the input voltages. With the driving voltage over 80 V_{pp}, the thrust was decreased with an increase in the input voltage, and a maximum thrust of 6.3 mN could be obtained with 70 V_{pp}, an electric power consumption of 1.9 W, and a levitation distance of 2 μm. Since the levitation distance increases with the input voltage, i.e. the vibration amplitude, as shown in Fig.34.11, acoustic streaming is decreased and the thrust is also decreased by increasing the input voltage. When the vibration amplitude is decreased, the acoustic streaming is also decreased owing the effect of the viscosity boundary layer. Therefore, the thrust of the stage has an optimum value against input voltage.

Table 34.1 Propulsion characteristics of the stage

Vibration mode A	Frequency (kHz)	Moving direction	Thrust (mN)
Ch 1	71	neg. x	2.0
Ch 2	72	neg. y	2.0
Ch 3	74	pos. x	2.7
Ch 4	77	pos. y	1.7

34.7 Conclusions

A novel non-contact self-running ultrasonically levitated 2D stage was proposed, and the levitation and propulsion performances were investigated. The linear slider can be levitated and moved by acoustic radiation force and acoustic streaming. The configuration of the slider is remarkably simple and thin, consisting of a vibrating aluminum plate and a PZT plate. The moving directions of the slider were could be predicted via the FEA. The asymmetric vibration modes of the slider were excited at the expected driving frequencies by the FEA results, and the slider could be levitated. By switching the driving frequency, the direction of the slider could also be changed. The 2D stage consists of the base plate and the four vibrating plates, which have similar configuration as the linear slider with the different lengths. The four cantilever-type vibrating plates were designed via the FEA to isolate the vibration characteristics of each vibrating plate. The 2D stage could be levitated and moved at expected frequencies around 70 to 80 kHz. The flexural lattice vibration mode assists the stage movement, and the noncontact movement of the stage in two dimensions could be achieved by controlling the driving frequency. The maximum thrust of 6.3 mN could be obtained at 74 kHz with 70 V_{pp}.

Acknowledgments This work was partially supported by Ono Acoustics Research Fund, Fluid Power Technology Promotion Foundation, Electro-Mechanic Technology Advancing Foundation and the Ministry of Education, Science, Sports and Culture, Grant-in-Aid for Scientific Research on Priority Areas, No. 438, 2007 and 2008.

References

1. Ide T, Friend R J, Nakamura K, Ueha S (2005) A low-profile designe for the noncontact ultrasonically levitated stage. Jpn J Appl Phys 44 No. 6B:4662-4665
2. Koyama D, Ide T, Friend R J, Nakamura K, Ueha S (2007) An ultrasonically levitated noncontact stage using traveling vibrations on precision ceramic guide rails. IEEE Trans Ultrason Ferroelect Freq Contr 54:597-604
3. Koyama D, Nakamura K, Ueha S (2007) A stator for a self-running, ultrasonically-levitated sliding stage. IEEE Trans Ultrason Ferroelect Freq Contr 54:2337-2343
4. Yamazaki T, Hu J, Nakamura K, Ueha S (1996) Trial construction of a noncontact ultrasonic motor with an ultrasonically levitated rotor. Jpn J Appl Phys 35:3286-3288
5. Minikes A, Bucher I (2004) Levitation force induced by pressure radiation in gas squeeze films. J Acoust Soc Am 116:217-226
6. Minikes A, Bucher I (2003) Noncontacting lateral transportation using gas squeeze film generated by flexural traveling waves-Numerical analysis. J Acoust Soc Am 113:2464-2473
7. Nyborg L W (1958) Acoustic streaming near a boundary. J Acoust Soc Am 30:329-339
8. Lee C, Wang T (1989) Near-boundary streaming around a small sphere due to two orthogonal standing waves. J Acoust Soc Am 85:1081-1088
9. Rudenko V O, Soluyan I S (1977) Acoustic Streaming. In: Theoretical Foundations of Nonlinear Acoustics, Consultants Bureau, New York
10. Chu B, Apfel E (1982) Acoustic radiation pressure produced by a beam of sound. J Acoust Soc Am 72:1673-1687
11. Lee C, Wang T (1993) Acoustic radiation pressure. J Acoust Soc Am 94:1099-1109
12. Koyama D, Takei H, Nakamura K, Ueha S (2008) A self-running standing wave-type bidirectional slider for the ultrasonically levitated thin linear stage. IEEE Trans Ultrason Ferroelect Freq Contr. 55:1823-1830
13. Koyama D, Nakamura K (2009) Noncontact self-Running ultrasonically levitated two-dimensional stage using flexural standing waves. Jpn J Appl Phys 48 No. 7: in press

Chapter 35
Development of Shape Memory Actuator for Cryogenic Application

Koichi TSUCHIYA[1], Tamotsu KOYANO[2], Seiichiro II[1], Yoshikazu TODAKA[3] and Minoru UMEMOTO[3]

Abstract Shape memory alloy (SMA) is one of actuator materials which converts thermal energy into mechanical energy. Since it has much larger output strain and stress compared to piezoelectric materials or magnetostrictive materials. Thus SMA is suitable as actuator materials for extreme environment and in the areas which require a large motion and high energy density. The purpose of present study is to develop the shape memory alloys which can operate at temperatures below liquid nitrogen temperature. Effect of grain size and chemical composition on martensitic transformation were assessed in TiNi and Cu-Al-Mn alloys.

35.1 Introduction

Shape memory alloy (SMA) is one class of actuator materials which convert thermal energy into mechanical energy. Characteristics of various actuator materials are summarized in Table 35.1. SMA has much larger output strain and stress compared to piezoelectric materials or magnetostrictive materials. Thus SMA is suitable as actuator materials for the areas which require a large motion and high energy density. Magnification system of displacement is not necessarily. Another advantage of SMA is that it can be driven simply by heating the actuator elements. Because of these two characteristics the simplification and miniaturization of an actuation system can be easily achieved. Therefore the SMA actuators are suitable for those used under various extreme conditions. The focus of the present study is to develop a SMA actuator for cryogenic conditions.

[1] Koichi TSUCHIYA and Seiichiro II
Hybrid Materials Center, National Institute for Materials Science

[2] Tamotsu KOYANO
Graduate School of Pure and Applied Sciences, University of Tsukuba

[3] TODAKA and Minoru UMEMOTO
Dept of Production Systems Engineering, Toyohashi University of Technology

Table 35.1 VARIOUS ACTUATOR MATERIALS

	Strain	Stress (MPa)	Power Density (Jm^{-3})
shape memory alloy (TiNi)	$8x10^{-2}$	~400MPa	$3x10^{7}$
piezoelectric ceramics (PZT)	$\sim 3x10^{-4}$	~20MPa	$6x10^{3}$
giant magnetostrictive materials (Terfenol-D)	$1.7x10^{-3}$	~70MPa	$1.8x10^{4}$

Fig.35.1 is the liquefaction temperatures of various gases and possible area of application for such actuators. Cryogenic shape memory actuators can be useful for gas liquefier (H, He, N, O etc.), gas separation device, compressor, cryostat, temperature controller, liquid pumps, and so on. Hence they are useful for coming hydrogen society. They may be also various space system applications.

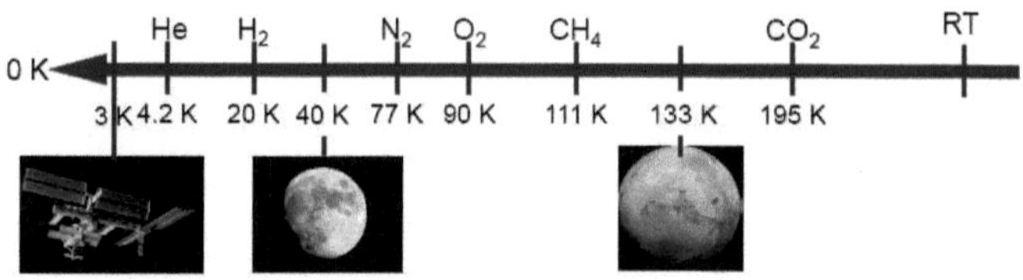

Fig. 35.1 Liquefaction temperatures for various gases

Astonishingly, there are very few studies regarding cryogenic shape memory alloy. Since the underlying mechanism of shape memory effect is shape change by martensitic transformation (MT), the information on MT temperature is very important. However, only limited data is available for MT temperatures below LN_2 temperature (~77 K). It has been reported that for TiNi martensitic transformation is suppressed if the Ni content exceed 52mol% [1]. Prado *et al.* reported the Cu-Al-Mn exhibit MT down to 17K [2]. Very few investigations were done on the shape recovery at cryogenic condition.

In general MT temperature is controlled by changing chemical composition of the alloys. Another method to change the MT temperature is by thermomechanical treatment. The present report describes the effect of grain refinement to nanoscale by severe plastic deformation on MT temperature in TiNi SMA.

35.2 Effect of Severe Plastic Deformation on Microstructure of TiNi Shape Memory Alloy

35.2.1 Experimental Procedures

Samples (25 x 10 x 1 mm^3) were cut from hot-rolled sheets of Ti-50.2at%Ni and Ti-50.9at%Ni (KIOKALLOY-R, Daido Steel, Co., Ltd.). They were homogenized at 1173 K for 3.6 ks, and quenched into room temperature water.

Severe plastic deformation by high pressure torsion (HPT) was applied on these samples. A disc sample of 10mm diameter and 0.85 mm thickness was used. The sample was placed between a pair of anvils with a depression of 0.25 mm depth. One of the anvils was rotated at 1 rpm under an applied quasi-hydrostatic pressure of 5 GPa at room temperature. The sample was deformed up to 10 turns of rotation. They were then subjected to post-deformation aging at temperatures ranging from 573 K to 773 K for 3.6 ks. Martensitic transformation temperatures were measured by RIGAKU DSC-8230L differential scanning calorimeter (DSC). A sample was first heated to 473 K from room temperature and then a cooling/heating cycle was performed in a temperature range of 150 ~ 473 K with a heating/cooling rate of 0.17 Ks^{-1}. X-ray diffractometry was done using a Cu-K_α radiation on RIGAKU RINT-2500 (40 kV-250 mA). Microstructural observations were made on HITACHI H-800 and JEOL JEM-2010 transmission electron microscope (TEM) both operated at 200 kV. TEM samples were perforated using STRUER Tenupol-3 at 253 K with electrolyte of H_2SO_4/methanol solution.

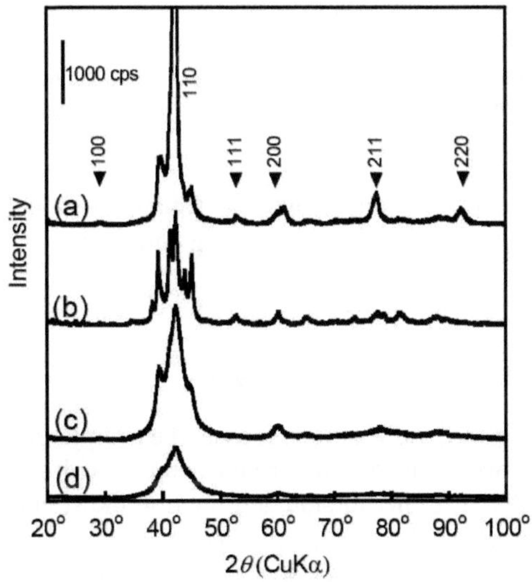

Fig. 35.2 X-ray diffraction profiles of sample after HPT deformation (a) before deformation (b) after compression(c) N=0.125 (d) N=10

35.2.2 Results and Discussion

Evolution in XRD profiles by HPT deformation of Ti-50.2mol%Ni is shown in Fig.35.2. The sample was in the B2 parent phase state with a small amount of B19' martensite. The sample transformed to B19' martensite on the application of compressive stress. Significant peak broadening and decrease in intensity with the deformation can be seen after HPT deformation of $N = 0.125$. Crystallite size after $N = 10$ was estimated to be 4.6 nm by the Scherrer equation.

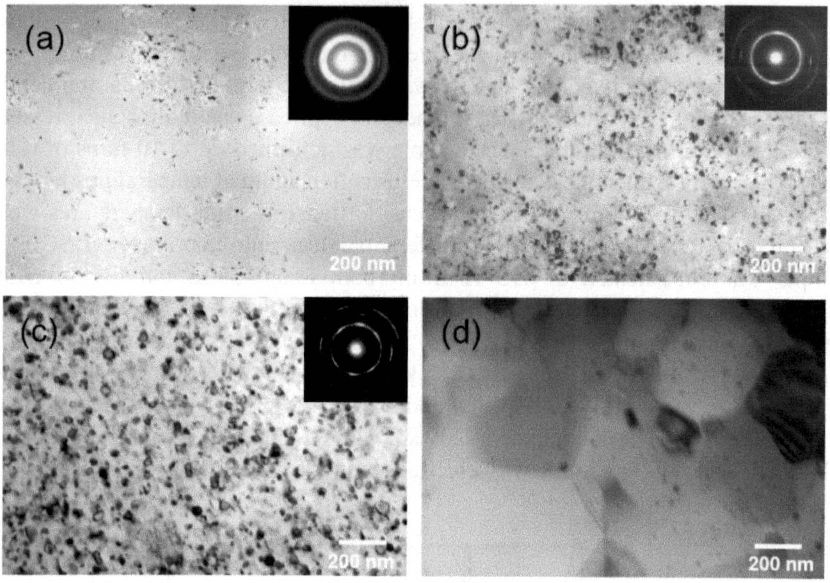

Fig. 35.3 TEM micrographs of Ti-50.2N after HPT deformation and post deformation aging for 3.6 ks (a) after HPT(N=10) (b)aged at 573 K (c) aged at 673 K. (d) aged at 773 K

Amount of deformation induced by HPT depends on radial position, r. Equivalent shear strain, γ_{eq}, is given by the following equation:

$$\gamma_{eq} = \frac{2\pi rN}{\sqrt{3}t}, \qquad (35.1)$$

where N is the number of rotation, and t is the sample thickness [3]. The TEM and DSC samples were cut from deformed samples with their centers at around $r = 3$ mm. By applying further torsion deformation the structure refinement occurs and after $N = 10(\gamma_{eq} = 128)$, the sample becomes amorphous almost completely. The volume fraction of amorphous was estimated to be about 96 % from the recovery/recrystallization DSC peak.

TEM micrographs of HPT deformed sample (N = 10) after aging at different temperatures are shown in Fig.35.3(a)~(d). Fig.35.3(a) is the bright field TEM micrograph of sample after deformation. The sample is composed of amorphous/nanocrystalline area as the SAD indicates. The volume fraction of amorphous was estimated to be about 96 %. Amorphous region remains after aging at 573 K for 3.6 ks. The SAD pattern exhibits a hallo ring from the amorphous region and the Debye rings from the nanocrystals. Aging at 673 K eliminated the amorphous area and the whole sample became nanocrystals of the B2 phase as seen in Fig.35.3(c). As shown in Fig.35.3(d) grains grew into submicron size when aged at 773 K. X-ray diffraction measurements were done on these HPT deformed and aged samples. Crystallite sizes were evaluated by the Scherer equation using the full widths of half maxima of 110(B2) peaks and plotted in Fig.35.4. The obtained data are in a reasonable agreement with the TEM observations. These observations clarify that a combination of severe plastic deformation and heat treatment is an effective tool to realize nanostructured TiNi.

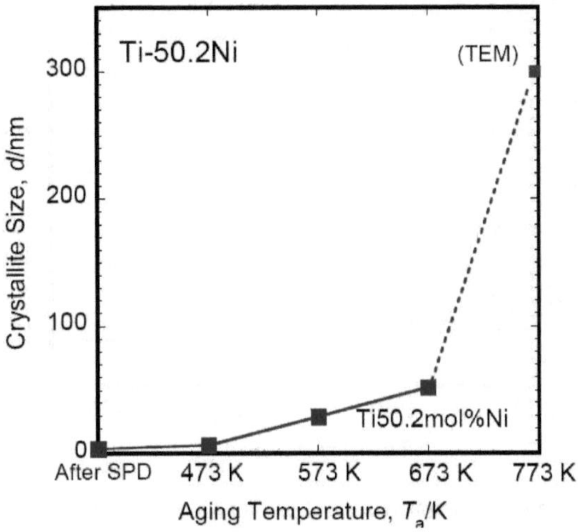

Fig. 35.4 Effect of post-deformation aging temperature on grain size in Ti-50.2Ni

Phase transformation in HPT deformed and aged TiNi was studied by DSC measurements and the results obtained were shown in Fig.35.5. Transformation temperatures determined from the DSC curves are plotted as a function of aging temperature in Fig.35.6. Here M_s, M^* and M_f denote martensitic start temperature, peak temperature and finish temperature, respectively. A_S, A_f, A^* denote transformation temperatures for reverse martensitic transformation, and R^* stands for R phase transformation peak temperature on cooling.

The DSC profile of as-deformed sample (not shown here) and the one aged at 573 K indicates no clear sign of martensitic transformation. These samples composed of amorphous and strained B2 nanocrystals. After aging at 673 K, a single peak on a cooling run and double peaks on a heating run are seen. The small hysteresis suggests that the peak on cooling and the one at higher temperature correspond to R phase transformation. The small and broad peak on heating should correspond to transformation from B19' to R or to B2. Then the absence of the forward B19' transformation peak may be due to an increase in the transformation interval (= M_s-M_f). In order to overcome the stress field due to lattice distortion and dislocations, and elastic constrains by grain boundaries, a large under cooling is required for the transformation to occur. Also during the transformation, interaction of the existing stress field with the one created by the formation of B19' phase increases the required under cooling. This will lead to a large transformation interval and the forward transformation peak may not be detectable by DSC. It should be also noted that the R phase peak temperature decreases with aging temperatures, which suggests that the strain field tends to stabilize the R-phase. These behaviors are quite similar to TiNi deformed by cold rolling [4, 5].

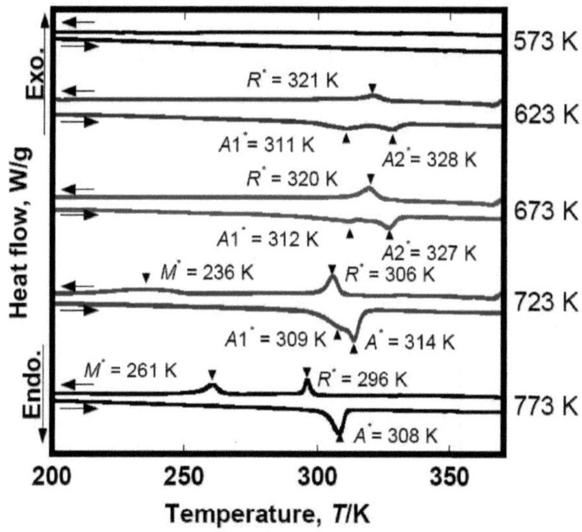

Fig. 35.5 Effect of post-deformation aging on DSC curves of Ti-50.2Ni

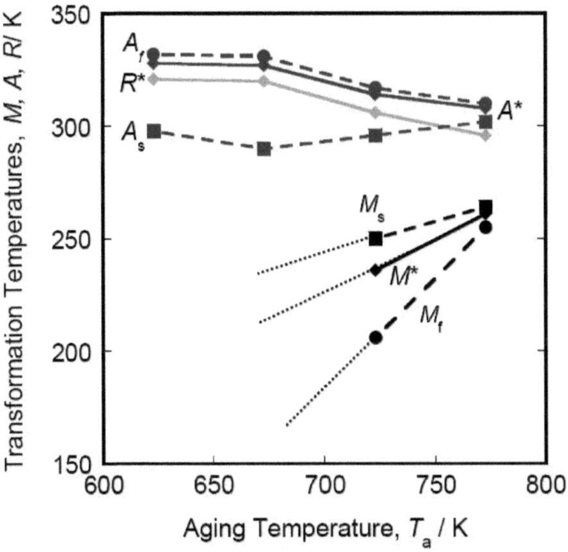

Fig. 35.6 Effect of aging temperature on transformation temperatures

Effect of grain refinements by HPT deformation and subsequent aging were investigated on martensitic transformation temperatures. It was revealed that reduction of grain size below 50 nm leads to pronounced decrease in forward martensitic transformation temperature, M_s and M_f, and to an increase in the transformation temperature interval , M_s-M_f . Reverse transformation temperature, which correspond to shape recovery temperature is less affected by nanoscale grain size. R phase is stabilized by the structural refinements. These results suggest that the structural refinement has limitation to control the shape recovery temperature.

35.3 Martensitic Transformation in Cu-Al-Mn Alloys at Cryogenic Temperatures

35.3.1 Experimental Procedures

Mother ingots of Cu-Al-Mn alloys with several chemical compositions shown in Table 35.2 are produced by high-frequency melting furnace, and homogenized for 3.6 ks at 1173 K. They were hot and cold rolled to the sheets with 0.5 mm thick. Samples for various measurements were cut from the sheets and final heat treatment of 300 s at 1173 K and water quench was applied. Selected samples were further aged at 373 K for 600 s followed by water quench. Martensitic transformation temperatures were measured by differential scanning calorimetry with heating/cooling speed of 0.17 Ks^{-1} in the temperature range between 150 K and 450 K. In order to assess the martensitic transformation below liquid nitrogen temperature, electrical resistivity measurements were done by 4 probe DC method in the temperature range of 4 ~ 300 K with cooling/heating rate of 1 Ks^{-1}.

Table 35.2 Chemical compositions of alloys

	mol%		
	Cu	Al	Mn
10Mn	73	17	10
12Mn	71	17	12
14Mn	69	17	14
16Mn	67	17	16
18Mn	65	17	18

35.3.2 Results and Discussion

Fig.35.7(a) shows DSC profiles of as-homogenized Cu-Al-Mn samples. 12Mn~18Mn alloys did not exhibit any sign of martensitic transformation but 10Mn alloy exhibit exothermic transformation peak at around 358 K on the cooling run after heating to 450 K.

Fig.35.7(b) shows DSC profiles of aged Cu-Al-Mn. DSC profiles of 14~18Mn alloy did not exhibit transformation peak, while 12Mn exhibited marked transformation peaks on heating and cooling runs. Martensitic transformation peaks were determined to be M_s = 192 K, M^* = 190 K, M_f = 183 K, A_s = 200 K, A^* = 204 K, A_f = 208 K.

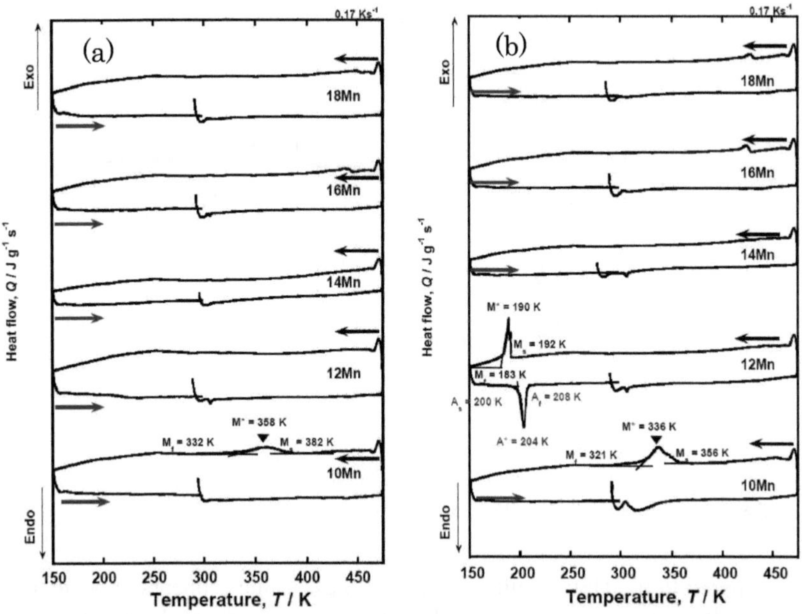

Fig. 35.7 (a) DSC profiles of as-homogenized Cu-Al-Mn (b) DSC profiles of aged Cu-Al-Mn

Meanwhile, 10Mn alloy exhibited forward transformation peak although the peak temperature was about 20 K lower as compared to the one in as-homogenized sample.

Fig.35.8 shows the temperature dependence of electrical resistivity for aged samples. Resistivity at each temperature was normalized by the value at room temperature for each alloy. 14~16Mn alloys exhibited a typical metallic behavior, in which the resistivity decrease monotonically with temperature. However 12Mn alloy exhibited marked increase at around 100 K on cooling and corresponding decrease on heating. This is an indication that the martensite phase with high specific resistance formed at the temperature. The transformation temperatures are determined to be $M_s = 117$K, $M_f = 100$ K, $A_s = 119$ K, $A_f = 131$ K.

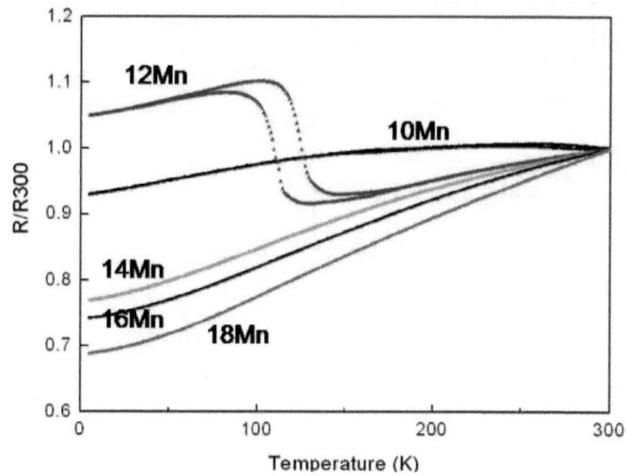

Fig. 35.8 Temperature *vs.* electrical resistivity

Compared to the DSC results in Fig.35.7 these values appears to be about 80 K lower. Although the samples were heat treated and aged at the same conditions, the resistivity measurements were done about 1 month after the aging while the DSC was measured immediately after the aging. Therefore the observed discrepancy of the transformation temperature can be attributed to the room temperature aging effect.

Similar aging behavior has been observed in other Cu-base shape memory alloys, such as Cu-Zn-Al [6, 7]. High density of quenched-in vacancies can migrate before annihilation results in the slight change in long or short range ordered structures. This will change the free energy balance between the parent phase and martensite phase, leading to the change in martensitic transformation temperatures. In the case of Cu-Al-Mn alloy thermo-dynamically stable structure is the $L2_1$ ordered structure[8, 9]. After the heat treatment and quench the degree of long range order is expected to be very low. The present results suggest that the degree of order increases by aging at 373 K or at room temperature and transformation temperature decreases. Therefore it can be expected that the further adjustment of aging time at 373 K will lead to lower martensitic transformation temperatures which is suitable for cryogenic applications.

35.4 Summary

(1) Suppression of martensitic transformation temperatures was investigated in nanocrystalline TiNi shape memory alloys produced by severe plastic deformation and heat treatment. Amorphous Ti-50.2mol%Ni was made nanocrystalline of about 10 nm by aging at 673 K for 3.6 ks. It was revealed that the nano-grain size

causes a marked decrease in M_s and M_f for B19' martensite and also an increase in transformation interval ($\Delta M = M_s - M_f$); however, it leads to a slight increase in reverse transformation temperatures.
(2) Effect of chemical composition and heat treatment was investigated for Cu-Al-Mn alloys. Martensitic transformation temperatures can be lowered by increasing the degree of long range order in the parent $L2_1$ phase. It was found that the Cu-14Al-12Mn alloys can be a good candidate for the cryogenic shape memory alloy.

Acknowledgments A part of the work was supported by a Grant-in-Aid for Scientific Research on Priority Area, No. 438 "Next-Generation Actuator Leading Breakthroughs". The authors are grateful for the experimental assistance by Dr. A. Sandu.

References

1. Kakeshita T, Fukuda T, Tetsukawa H, Saburi T, Kindo K, Takeuchi T, Honda M, Endo S, Taniguchi T, Miyako Y (1998)Negative Temperature Coeffcient of Electrical Resistivity in B2-Type Ti-Ni Alloys. Jpn. J. Appl. Phys.37 :2535-2539
2. Prado OM, Decorte MP, Lovey F (1995)Martensitic Transformation in Cu-Mn-Al Alloys. Scrip. Metall. Mater.33 :877-883
3. Wetscher F, Vorhauer A, Stock R, Pippan R (2004)Structural Refinement of Low Alloyed Steels during Severe Plastic Deformation. Materials Science and Engneering A :387-389, 809-816
4. Nakayama H, Tsuchiya K, Z -G Liu, Umemoto M, Morii K, Shimizu K (2001)Process of Nanocrystallization and Partial Amorphization by Cold Rolling. Mater. Trans.42: 1987-1993
5. Nakayama H, Tsuchiya K, Umemoto M (2001)Crystal Refinement and Amorphisation by Cold Rolling in TiNi Shape Memory Alloys. Scripta Mater.44 :1781-1785
6. Tsuchiya K, Miyoshi D, Tateyama K, Takezawa K, Marukawa K (1994)Changes in the Ordered Structure in Cu-Zn-Al Martensites by Isothermal Aging. Scrip. Metall. Mater.31 :455-460
7. Tsuchiya K, Takahashi K, Marukawa K (1996)Study of Aging Kinetics in Copper Based Alloy Martensites. Mater. Trans. JIM 37 :304-308
8. Kainuma R, Satoh N, Liu JX, Ohnuma I, Ishida K (1998)Phase Equilibria and Heusler Phase Stability in the Cu-rich Portion of the Cu-Al-Mn system. J. Alloys. Comp. 266 :191-200
9. Kainuma R, Takahashi A, Ishida K (1995)Thermoelastic Martensite and Shape Memory Effect in Ductile Cu-Al-Mn Alloys. Metall. Mater. Trans,27 :2187-2195

Chapter 36
Development of Environmental-Friendly Lead-Free Piezoelectric Materials for Actuator Uses

Hiroaki TAKEDA[1], Enzhu LI[1], Takashi NISHIDA[2], Takuya HOSHINA[1], Takaaki TSURUMI[1]

Abstract We have currently investigated the piezoelectric properties of bismuth tungstate, Bi_2WO_6 (BWO), mono-domain crystals and baium titanate, $BaTiO_3$ (BT), based ceramics for use at high temperature. For BWO, the crystals grown by a slow cooling technique were characterized. The resonance response revealed that the BWO crystals maintained their piezoelectricity up to 400°C. In the solid solution ceramics of the BT-$(Bi_{1/2}Na_{1/2})TiO_3$ (BNT) system, we disclosed that the Curie temperature T_c can be controlled from 130 to approximately 220°C. BNT-substituted BT ceramics showed a low temperature dependence of the piezoelectric constant up to 160°C. We also synthesized $(K,Na)NbO_3$–$LiNbO_3$–CuO lead-free piezoelectric ceramics that show a high mechanical quality factor Q_m of 1400, a high d_{15} of 207 pC/N, and a high k_{15} of 0.72. Using the shear vibration mode of the ceramics, an ultrasonic motor driven with four piezoelectric ceramic plates was developed. The highest revolution speed of 486 rpm was achieved at 34.5 kHz with the input voltage of approximately 180 $V_{p\text{–}p}$ (peak to peak).

36.1 General Introduction

Recently, piezoelectric actuator materials for high temperature use are definitely required for gas injectors or combustion sensors directly placed in the cylinders of engines. The lead zirconate titanate, $Pb(Zr,Ti)O_3$ (PZT), ceramic is a well-known piezoelectric material. However, recent progress in environmental safeguards has required environmentally-friendly, that is, lead-free materials whose electric properties are comparable to those of the lead-containing ones.

[1] Hiroaki TAKEDA, Enzhu LI, Takuya HOSHINA and Takaaki TSURUMI

Graduate School of Science and Technology, Tokyo Institute of Technology

[2] Takashi NISHIDA

Graduate School of Materials Science, Nara Institute of Science and Technology

Under such conditions, we have attempted to develop lead-free piezoelectric materials usable at high temperatures. For this purpose, we have evaluated the $BiWO_6$ mono-domain crystals and $BaTiO_3$-based ceramics. Moreover, we fabricated ultrasonic motors using $(K,Na)NbO_3$-based ceramics. In this report, we present our recent research results.

36.2 Bismuth Tungstate Mono-Domain Crystals

36.2.1 Backgraound

The titled compound, bismuth tungstate, Bi_2WO_6 (BWO), is one of the BLSF compounds having the high Curie temperature of T_c= 940°C [1]. Recent progress in environmental safeguards requires environmentally-friendly, that is, lead-free materials whose electric properties are comparable to those of the lead-containing ones. Therefore, BWO is expected to be a lead-free material for use in piezoelectric devices suitable for high temperatures as long as its piezoelectric properties are retained.

We have attempted to measure the piezoelectric properties using crack- and strain-free BWO mono-domain crystals. Plate-like crystals with sizes up to 1.0 mm thick were obtained by the slow cooling technique, considering the influence of the starting material volumes on the thickness of the BWO crystals grown using the BWO-$Li_2B_4O_7$(LBO) solution [2]. In ref.[2], the relationship between the morphology of the crystal and the crystallographic *abc* - rectangular *XYZ* axes was clarified. This information provides very useful information to determine the orientation of the sample, which requires the cutting of the crystals during the sample preparation for the electric measurements. On the basis of this relationship, in this study, we characterized the dielectric and piezoelectric properties of the BWO mono-domain crystals.

36.2.2 Crystal Growth of BWO

The detailed growth process of BWO mono-domain crystals were reported in refs.[2, 3]. We obtained plate-like crystals with sizes up to 1.0 mm thick by the slow cooling technique. Using polarizing microscopic and etch pit observations, it was confirmed that these BWO crystals consists of a mono-domain fabric.

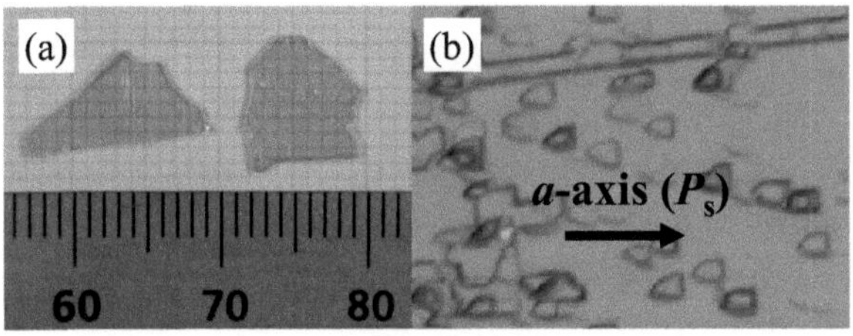

Fig. 36.1 (a) As-grown BWO mono-domain crystals and (b) surface morphology of <001> surfaces after immersion in HNO_3 solution at 80°C for 0.5 h

Figure 36.1(a) shows the typical BWO plate-like crystals obtained by the slow cooling technique. The crystals are transparent, light yellow in color, and have a well-developed smooth {001} surface. We confirmed that no domain structure and few inclusions were found inside the grown crystals. Therefore, it was found that all the grown BWO crystals consisted of a mono-domain fabric. Fig.36.1(b) shows the etch patterns of the {001} surface of the BWO mono-domain crystal after immersion in a 5 N HNO_3 solution at 80°C for 0.5 h. We adopted the HNO_3 solution as the etchant, because we knew that BWO crystals were easily dissolved in other acids such as the hydrochloric and sulfuric acids. The shape of the pits is like the pentagon piece used in a Japanese chess game. The shapes also confirmed that there is a *2*-fold rotation axis and mirror plane along the *Z* axis. According to *the IEEE standard on piezoelectricity* [4], in point group *mm2*, *+Z* is chosen so that the corresponding piezoelectric constant, d_{33}, is positive. Although the other axes are trivial, in this study, *+X* and *+Y* were set to form a right-handed system (see Table 36.1). By using the static piezoelectric measurement, the geometric relationship between the etch pit shape on the {001} surface and the crystallographic *abc* - rectangular *XYZ* axes were determined. This information is very useful for the piezoelectric characterization and actually used for determination of the material constants.

36.2.3 Physical Properties of BWO

The BWO belongs to the orthorhombic point group *mm*2. Independent material constants consist of three dielectric, five piezoelectric and nine elastic compliance constants (ε_{ij}, d_{ij} and s_{ij}, respectively). In this study, we determined the piezoelectric constants, d_{31} and d_{33}, because these constants are of significant interest. The material constants were determined using an impedance/gain phase analyzer (HP 4194A: Agilent) as reported in ref.[3]. The electromechanical coupling factor, k_{ij}, and the piezoelectric modulus were evaluated by measuring the mechanical series resonance frequency, f_s, and parallel resonance frequency, f_p, of the equivalent resonators. The equivalent resonators were fabricated in the form of plates according to various vibration modes. The dielectric constants, ε_{ij}, were determined by measuring the capacitances of the resonators by taking the parasitic capacitance into account. The changes in d_{31} and d_{33} in the temperature range from room temperature to 500°C were investigated.

Table 36.1 Relathionship between crystallographic and rectangular axes of BWO crystal. The point group symmetry is set at *2mm*

Axis identification		Comments
Crystallographic	Rectangular	
a	*Z*	polar (d_{33}) axis
b	*X*	–
c	*Y*	Longest axis

Figure 36.2 shows the frequency dependence of the impedance |Z| and phase θ measured in the k_{31} mode. The size of the BWO mono-domain crystal was about $l4.39 \times w0.63 \times t0.47$ mm^3. If the ideal poling state is obtained, the impedance phase angle θ approaches 90° in the frequency range between the resonance and anti-resonance frequencies. The phase angle of the impedance was limited to approximately 89°. The k_{31} and mechanical quality factor values of the BWO crystals were 24% and 4000, respectively. In this study, all the material constants at room temperature of the BWO mono-domain crystals were successfully determined. The dielectric constants, ε_{ij}, of the BWO crystals were ε_{11}=100, ε_{22}=70, ε_{33}=70. The piezoelectric constants were d_{31}=-17.0, d_{32}=-4.4 and d_{33}=27 pC/N. The remainder of these constants will be reported in a related publication.

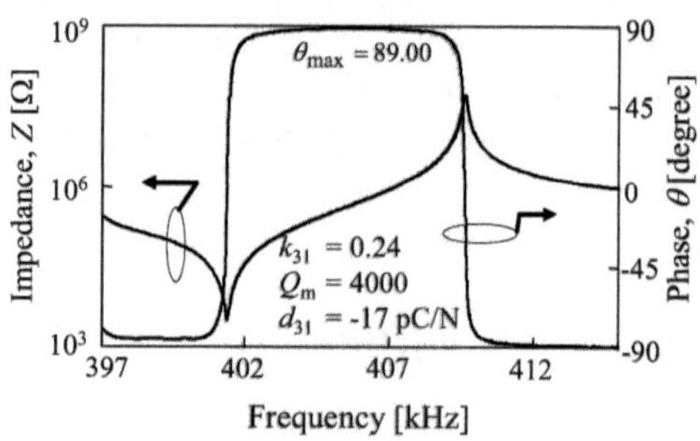

Fig. 36.2 Impedance response of length-extensional vibration mode

Figure 36.3 shows the temperature dependence of the piezoelectric constants, d_{31} and d_{33}, of the BWO crystals. In this figure, the relative piezoelectric constant, d_{ij}^*, represents the ratio of the *ij* value at the measurement temperature and that at room temperature (25°C). Both d_{31} and d_{33} increased with temperature. The resonance response confirmed the piezoelectric behavior of the BWO crystals up to 400°C. Hence, we found that the BWO crystals can be used as sensors and transducers in the middle temperature range.

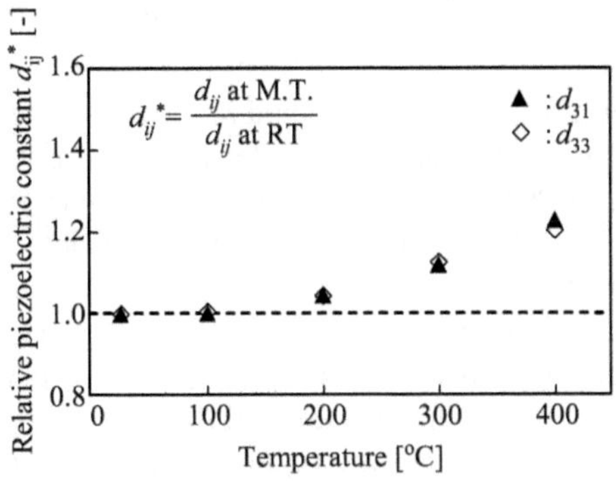

Fig. 36.3 Temperature dependences of a piezoelectric constant d_{ij}^* of the BWO crystal

36.3 $BaTiO_3$-Based Ceramics

36.3.1 Backgraound

Barium titanate ($BaTiO_3$; BT)-based ceramics are widely utilized as capacitor and PTCR (Positive Temperature Coefficient of Resistivity) materials. The aim of this study is to develop lead-free piezoelectric materials. For this goal, we selected a ferroelectric lead-free perovskite-type compound $(Bi_{1/2}Na_{1/2})TiO_3$ (BNT, T_c= 320°C[5]) as another end member of the BT-based solid solutions. The BNT ceramics are very attractive lead-free piezoelectric materials. The ceramics with the BT:BNT=6-7:94-93 molar ratio compositions, corresponding to the morphotropic phase boundary (MPB) one, show superior piezoelectric properties [6]. Since many studies have focused their attention on BNT-based solid solutions [7, 8], there is little information about the BT-BNT solid solution ceramics with a BT-rich composition.

In this section, we describe the synthesis and electric properties of the BT-BNT ceramics with a BT-rich composition compared to the pure BT one. We showed an increase of a T_c by increasing the $(Bi_{1/2}Na_{1/2})$ content as well as its operation limit at high temperature. This operation limit relates the depolarization temperature of the BT-BNT ceramics by the di-, pyro-, and ferro-electric measurements. Moreover, the temperature dependence of the piezoelectric property is also disclosed.

36.3.2 Preparation and Basic Properties of Ceramics

A conventional ceramics fabrication technique was used to prepare the (1-x)BT-xBNT solid-solutions. As the starting powders, we used $BaCO_3$, Na_2CO_3, TiO_2, of 99.99% and Bi_2O_3 of 99.9% purity. The composition (x) was set at x=0.00-1.00. The powders were mixed in acetone, dried, and then calcined at 800-1000°C for 2 h. The calcined powders with a binder (2wt% poly vinyl alcohol) were uniaxially pressed into a disk with a 15-mm diameter and 1-mm thickness at 190 MPa. The disc samples directly placed on a platinum sheet were placed in a furnace and heated at 1300-1350°C for 2 h. The sintered disk ceramics with an 11.0 mm diameter were polished for characterization of their electric properties. For the phase identification using the powder X-ray diffraction (XRD) technique, the ceramic samples were ground and pulverized. The densities of the ceramics were measured by the Archimedes method using distilled water.

All diffraction peaks of the BT-BNT powders were successfully indexed on the basis of the published data for the host $BaTiO_3$ (ICCD #05-0626) and then demonstrated that all the BT-BNT ceramics consisted of a single phase with a tetragonal symmetry. In the BT-BNT systems, the tetragonal phase with the BT structure is maintained up to x=0.9. These data were supported by the phase diagram reported in ref.[6]. The lattice parameters, the a- and c- axis lengths, of the BT-BNT ceramics decreased with the increasing BNT content. Both of them monotonously decreased from 3.997(2) to 3.895(4) Å for the a-axis and 4.024(2) to 3.953(5) Å for the c-axis, in which the parentheses represent the standard deviation. These results are reasonable from the viewpoint of the ionic size (Ba^{2+} (r^{VIII} = 1.42 Å), Na^+ (1.18 Å) and Bi^{3+} (1.17 Å) [10]) and suggested that the Na^+ and Bi^{3+} cations were introduced into the Ba site. We found the optimal sintering temperatures of the BT-BNT ceramics with the relative density of D_r>90%, where D_r is defined as the ratio of the specimen density to the theoretical density. The density of the ceramics significantly affects the electric properties. For fair comparison of the electric properties, therefore, we tried to obtain a density >90% for all the solid-solution ceramic specimens by changing the sintering temperature. The sintering temperature decreased with the increasing BNT content. This was because the Bi_2O_3 content with a low melting temperature in the starting material powders increased with the BNT content.

The electrodes were formed on the face of the disks by fire-on Ag paste or sputtered Au films. The frequency dependence of the dielectric constant, ε_s, and dielectric loss factor, tanδ, of the ceramic samples were measured in the frequency range of 10 kHz to 1MHz using an impedance analyzer (HP4194A). We performed the poling treatment on the sample before the pyroelectric measurement. The samples were poled at 100°C in a silicone oil bath with an electric field of over 20 kV/cm for 30 min. Referring to the Japan Industrial Standard R1651[9], a deporalization current was measured using a digital resistance meter (ADVANTEST R8340) during heating from room temperature (RT) to 500°C, and then the pyroelectric coefficient, γ, was calculated. The piezoelectric measurement was performed using the same technique described in the previous chapter.

36.3.3 Curie Temperature and Piezoelectric Property of BT-BNT Ceramics

We investigated the T_c change in the BT-BNT ceramics with the BNT content. Fig.36.4 shows the temperature dependence of the dielectric constant, ε_S, and dielectric loss tangent, tanδ, measured at the frequency of 1MHz for the BNT 50 mol% ceramics. In this figure, the pyroelectric coefficient, γ, change during the heating process is also listed. The values of ε_S and tanδ for the ceramics at room temperature and at 1MHz were 390 and 0.04, respectively. The plots of ε_s and tanδ show a peak at 240 and 220°C, respectively. The plot of γ shows a peak at 230°C (T_p). In this report, we defined the peak temperatures of ε_S, tanδ, and γ as T_m, T_d and T_p, respectively. Up to a 40 mol% BNT content, T_m agreed with T_d and T_p, and did not depend on the measurement frequency. However, as shown in Fig.36.4, in the 50 mol% BNT ceramics, T_m was different from T_d and T_p.

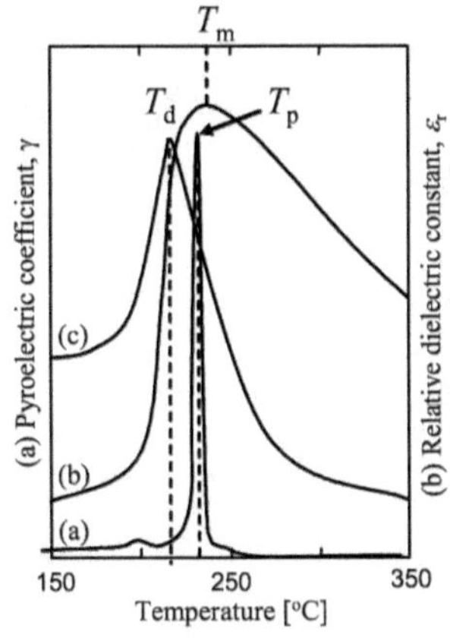

Fig. 36.4 Temperature dependence of (a) pyroelectric current, (b) dielectricity, and (c) dielectric loss of the BNT 50 mol% ceramics. This figure was used in this study to determine Tm, Td, and Tp, which indicate the temperatures with the maximum values

We obtained the T_m, T_d and T_p values in the composition range from 5 to 50 mol% BNT content, and constructed a BT-BNT binary phase diagram. We investigated the relationship between the phase transition temperature and chemical composition. Fig.36.5 shows the constructed phase diagram for the BT-BNT binary system. Up to a 10 mol% BNT content, T_c drastically increases from 130 to 200 °C. At a 20 mol% BNT content, the T_c reaches 220 °C. From 20 to 40 mol%, the slope is gentle, and T_c gradually increases up to 230°C.

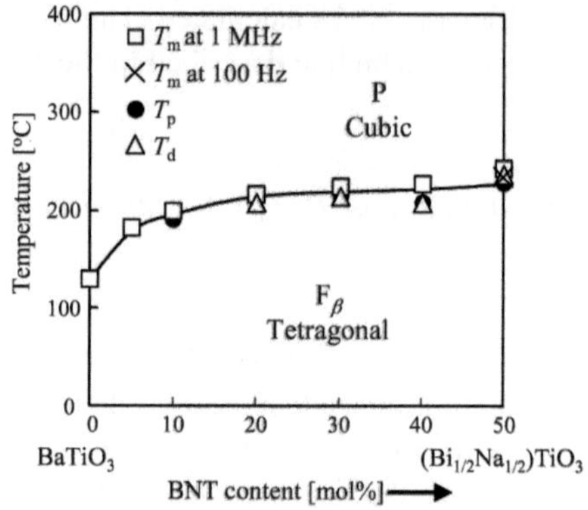

Fig. 36.5 Phase diagram of a BT-BNT binary system constructed from the results of di-, pyro-, and ferro-electric measurements of BT-BNT ceramics

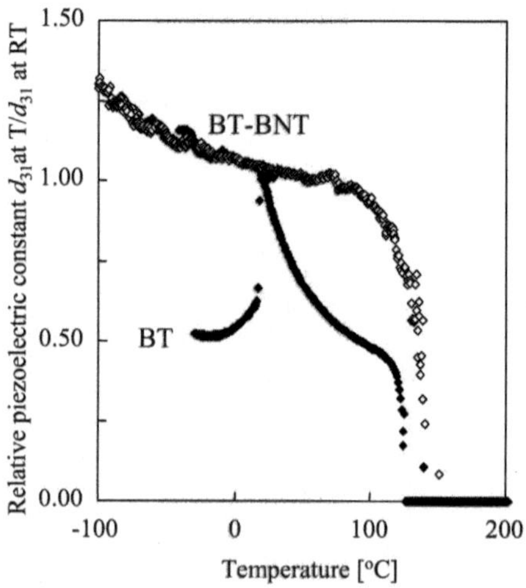

Fig. 36.6 Temperature dependences of the piezoelectric constant d_{31} of BT-BNT and 6BT ceramics

Figure 36.6 shows the temperature dependence of the piezoelectric constants, d_{31}, of the BT-BNT ceramics, compared with the BT one. In this figure, the relative piezoelectric constant is adopted for the emphasizing the changes in the value at the measurement temperature and that at room temperature (25°C). From -20 to 100°C, the d_{31} of the BT ceramics drastically changed by about 50%. On the other hand, the change in the d_{31} value of BT-BNT is no more than 25% from -100 to 100°C. Moreover, the resonance response confirmed the piezoelectric behavior of the BT-BNT ceramics up to 160°C. Therefore, we expect that the BT-BNT piezoelectric ceramics should be promising materials usable over a broad temperature range.

36.4 Ultrasonic Motors Using Share-Mode Vibration

36.4.1 Background

In mobile equipment such as digital still cameras and cellular phones demand the miniaturization of motors. Ultrasonic motors (USM) driven with piezoelectric ceramics are better than electromagnetic motors in the mini-motor area because the efficiency of USMs is almost unrelated to their sizes [11]. Moreover, other merits such as high resolution of displacement control, absence of parasitic magnetic fields, frictional locking at the power-off stage, and high thrust-to-weight ratio make them a good candidate for use in precision micromechanical systems. However, almost all the motors are now utilizing PZT-based materials, containing a large amount of Pb, the use of which should be restricted in the future because of the demand of environment protection. It had been considered that it is difficult to use lead-free piezoelectric ceramics as the actuators because their piezoelectric properties are not comparable with PZT. To date, USMs driven with either d_{33} longitudinal mode or d_{31} transverse mode of lead-free materials were reported by Tamura *et al.* [12] and Doshida *et al.* [13, 14]. The material employed in the former study was $LiNbO_3$ single crystal; the latter one employed multilayer-type piezoelectric elements as well as a displacement-enhancing mechanism to obtain sufficient displacement for the motor operation.

It is interesting to note that although the d_{15} which couples to the shear mode is the highest coefficient in soft PZT ceramics and was known for a long time, it was seldom utilized in actuator and transducer applications. USMs based on the shear mode of PZT-based piezoelectric ceramics were only reported by Aoyagi *et al.* [15], Uchino *et al.* [16] and Imabayashi and Funakubo [17]. To increase the application field of USMs, USMs made of plate of lead-free $(K,Na)NbO_3$-based ceramics are desired at present.

36.4.2 Synthesis of Lead-Free Piezoelectric Ceramics[18]

A high Q_m of piezoelectric ceramics is essential for the application of high-power ultrasonic motors because the displacement of piezoelectric elements is proportional to Q_m at a resonance frequency, and a large mechanical loss ($1/Q_m$) results in a heat generation during the high-power operation. Therefore, we tried to enhance the Q_m of $(K,Na)NbO_3$–$LiNbO_3$–CuO ceramics by controlling the process conditions to obtain ceramics with fine grains, and also tried to develop an ultrasonic motor using high *d* and *k* values obtained in the shear mode of piezoelectric elements. The Q_m of KNN-based ceramics was enhanced by controlling the process conditions, such as calcination temperature and sintering temperature, and co-doping of Li and Cu. The Q_m increased with decreasing grain size, which was interpreted using a model based on the internal bias field owing to the movements of space charges to suppress the domain wall motion. The detailed process of the enhancement is described in ref.[18]. The ceramics with the composition of $[(Na_{0.5}K_{0.5})_{0.98}Li_{0.02}]NbO_3$+0.45 mol% CuO showed a Q_m of 1396 and d_{15} of 207 pC/N, which were employed in the shear mode ultrasonic motor.

36.4.3 Fabrication of Disc-Type Traveling-Wave USMs

Figure 36.7 showed the structure of the disc-type motor driven with the shear mode of the lead-free ceramics. It consists of a rotor, a spring and a stator. The rotor contacting with the stator is driven to rotate through a frictional force between the rotor and the stator. The spring gives a pre-pressure between the stator and the rotor. The stator has a mid-cylinder on a metal disc to changes the vertical movement to the horizontal movement. Four pieces of piezoelectric ceramic which is a quarter of a piezoelectric ceramic disc were attached at the bottom of the stator. The ceramic pieces were polarized along the radial direction: two of them are from inner to outer direction and the other two are in the opposite direction as shown in Fig.36.8. To drive the motor, a pair of alternating voltages with a phase shift of 90 degrees, $V_o \sin\omega t$ and $V_o \cos\omega t$, was applied to the bottom electrode segments from a power source. The shear strain can induce a bending mode (shear-shear mode) with a large driving force in a piezoelectric beam under resonance drive with a free-free boundary condition. Since the applied electric field is out of phase, the applied voltages $V_o \sin\omega t$ and $V_o \cos\omega t$ will induce the piezoelectric disc to produce two bending modes leading to the rotation. The shear-shear mode flexure traveling wave was produced in the stator causing a portion of the rotor in contact with the stator to rotate through a frictional force between the rotor and the stator due to the traveling wave deformation of the stator. And in actual work, the motor was fixed by two crocodile clamps as shown in Fig.36.9. Fig.36.10 exhibits the photo of the disc-type motor using shear mode of the lead-free piezoelectric ceramics.

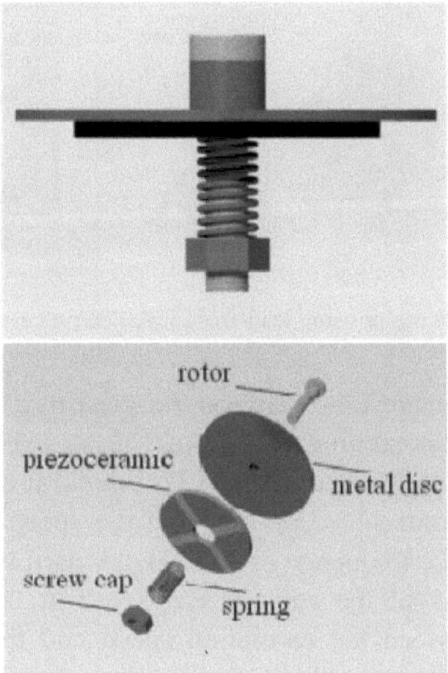

Fig. 36.7 Sketch of rotary shear mode motor driven with lead-free piezoelectric ceramics

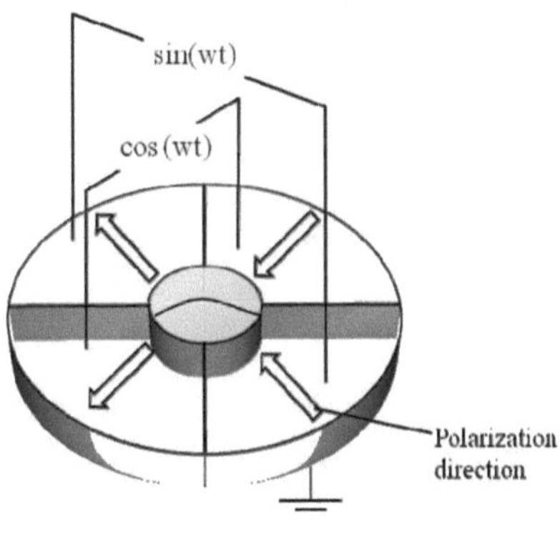

Fig. 36.8 Sketch of piezoelectric ceramic driven the shear mode motor

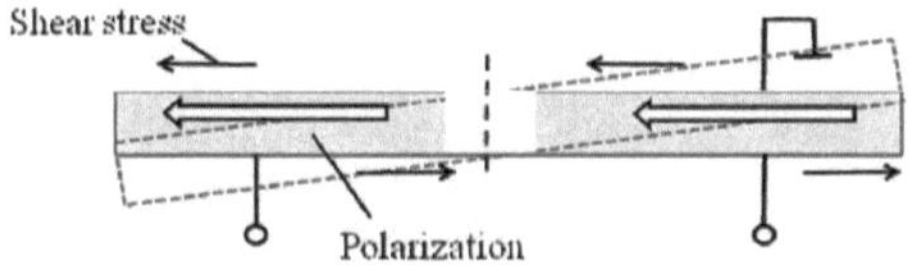

Fig. 36.9 Sketch of bending mode excited by the shear-shear strain

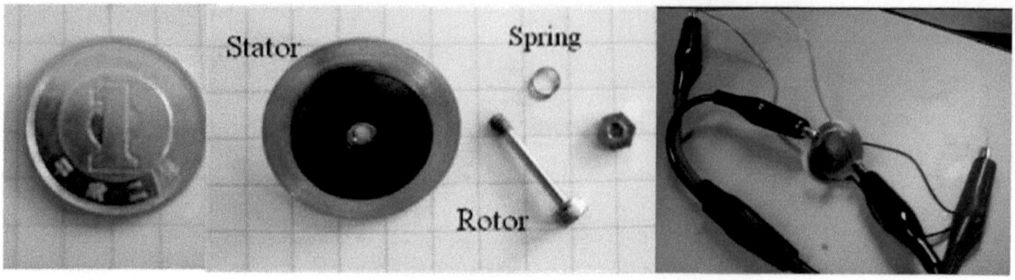

Fig. 36.10 The shear mode motor using lead-free piezoelectric ceramics

Figure 36.11 (a) shows relation between the shear motor's revolution speed and the exciting frequency at exciting voltage of 180 $V_{p\text{-}p}$. The revolution speed was measured by an optical tachometer (AD-5172, Edenki Inc). The motor could work at frequency ranging from 34～35.3 kHz and the speed reached a maximum of 486 rpm at the resonance frequency of 34.5kHz. A narrow frequency range is due to the 'hard' characters of the ceramic employed in this motor. Fig.36.11 (b) shows the relation between the revolution speed and the exciting voltage. The revolution speed was 56 rpm at 80$V_{p\text{-}p}$ and increase sharply with the exciting voltage, and the saturated trend was not observed when the exciting voltage increased to the maximum voltage of the amplifier (180 $V_{p\text{-}p}$).

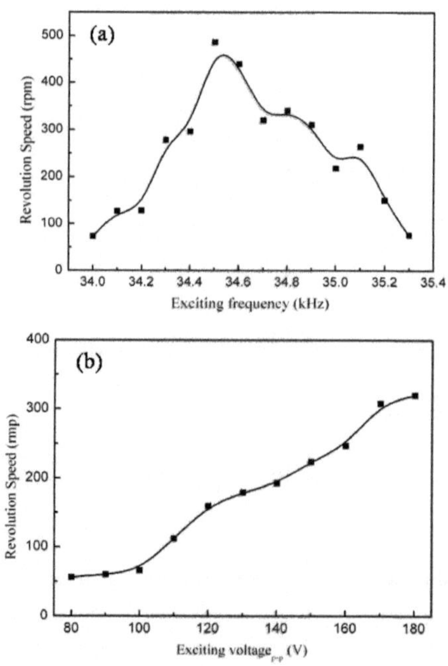

Fig. 36.11 Variation of the revolution speed against the exciting frequency (a) and voltage (b)

36.5 Summary

Using the BWO mono-domain crystals with a 1.0 mm thickness, etch pit observations and static piezoelectric measurements were performed. The relationship between the etch pit shape and the crystallographic-rectangular axes provided useful information for the electrical measurements. By using this relationship, all the dielectric and some piezoelectric constants of the BWO crystal were evaluated. The BWO crystal maintained its piezoelectricity up to 400°C. The BWO mono-domain crystal is an attractive candidate for piezoelectric materials usable at high temperature.

In lead-free piezoelectric ceramics using the BT-BNT system, we disclosed that the T_c can be increased up to approximately 230°C. The BT-BNT piezoelectric ceramics have a high temperature stability from -100 to 100°C, and apparently maintains a higher piezoelectricity than that (130°C) of BT.

The shear mode ultrasonic motor using a single plate lead-free (K,Na)NbO_3-based piezoelectric ceramic was developed successfully.

Acknowledgments This work was financially supported by a Grant-in-Aid for Scientific Research on Priority Areas, No.438 "Next-Generation Actuators Leading Breakthroughs".

References

1. Wolfe RW, Newnahm NE, Kay MI (1969) Crystal structure of Bi_2WO_6. Solid State Comm 7:1797-1801
2. Takeda H, Nishida T, Okamura S, Shiosaki T (2005) Crystal growth of bismuth tungstate Bi_2WO_6 by slow cooling method using borate fluxes. J Euro Ceram Soc 25:2731-2734
3. Takeda H, Nishida M, Nshida T, Shiosaki T (2007) Growth of bismuth tungstate Bi_2WO_6 mono-domain crystals and chemical etching for measuring their electrical properties. Trans Mater Res Soc Jpn 32 :11-14
4. (1987)IEEE Standard on Piezoelectricity 176-1987
5. Smolensky GA, Isupov VA, Agranovskaya AI, Krainik NN (1961) New ferroelectrics of complex composition IV. Sov Phys Solid State 2:2651-2654
6. Takenaka T, Maruyama K, Sakata K (1991) $(Bi_{1/2}Na_{1/2})TiO_3$-$BaTiO_3$ system for lead-free piezoelectric ceramics. Jpn J Appl Phys 30:2236-2239
7. Sheets SA, Soukhojak AN, Ohashi N, Chiang YM (2001) Relaxor single crystals in the $(Bi_{1/2}Na_{1/2})_{1-x}Ba_xZr_yTi_{1-y}O_3$ system exhibiting high electrostrictive strain. J Appl Phys 90:5287-5295.
8. Nagata H, Takenaka T (2001) Additive effects on electrical properties of $(Bi_{1/2}Na_{1/2})TiO_3$ ferroelectric ceramics. J Euro Ceram Soc 21:1299-1302
9. The Japan Industrial Standard (2002) Method for measurement of pyroelectric coefficient of fine ceramics. JIS R1651 [in Japanese]
10. Sanson A, Whatmore RW (2002) Properties of $Bi_4Ti_3O_{12}$-$(Na_{1/2}Bi_{1/2})TiO_3$ piezoelectric ceramics. Jpn J Appl Phys 41:7127-7130
11. Uchino K (1997) Piezoelectric Actuators and Ultrasonic Motors. Kluwer Academic, Boston
12. Tamura H, Iwase M, Hirose S, Aoyagi M, Takano T, Tomikawa Y (2008) Measurement of $LiNbO_3$ rectangular plate under large vibration velocity of the first longitudinal and second flexural modes. Jpn J Appl Phys 47:4034-4040

13. Doshida Y, Kishimoto S, Ishii K, Kishi H, Tamura H, Tomikawa Y, Hirose S (2007) Miniature cantilever-type ultrasonic motor using Pb-free multilayer piezoelectric ceramics. Jpn J Appl Phys 46:4921-4925
14. Doshida Y, Kishimoto S, Irieda T, Tamura H, Tomikawa Y, Hirose S (2008) Double-mode miniature cantilever-type ultrasonic motor using lead-free array-type multilayer piezoelectric ceramics. Jpn J Appl Phys 47:4242-4247
15. Aoyagi M, Murasawa Y, Ogasawara T, Tomikawa Y (1997) Experimental characteristics of a bolt-clamped short-cylindrical torsional vibrator using shear mode piezoceramics inserted in the axial direction. Jpn J Appl Phys 36:3126-3129
16. Uchino K, Dong S, Strauss MT (2006) U.S. Patent 7095160B2
17. Imabayashi H, Funakubo T (1994) Jpn Kokai Tokkyo Koho JP 1994113565. [in Japanese].
18. Li E, Kakemoto H, Hoshina T, Tsurumi T (2008) A shear-mode ultrasonic motor using potassium sodium niobate-based ceramics with high mechanical quality factor. Jpn J Appl Phys 47:7702-7706

MIX
Papier aus verantwortungsvollen Quellen
Paper from responsible sources
FSC® C105338

If you have any concerns about our products,
you can contact us on
ProductSafety@springernature.com

In case Publisher is established outside the EU,
the EU authorized representative is:
Springer Nature Customer Service Center GmbH
Europaplatz 3, 69115 Heidelberg, Germany

Printed by Libri Plureos GmbH
in Hamburg, Germany